Cases on Managing Dairy Productive Chains

Eduardo Guilherme Satolo
São Paulo State University, Brazil

Priscilla Ayllen Bustos Mac-Lean
São Paulo State University, Brazil

A volume in the Advances in Logistics, Operations, and Management Science (ALOMS) Book Series

IGI Global
PUBLISHER of TIMELY KNOWLEDGE

Published in the United States of America by
IGI Global
Engineering Science Reference (an imprint of IGI Global)
701 E. Chocolate Avenue
Hershey PA, USA 17033
Tel: 717-533-8845
Fax: 717-533-8661
E-mail: cust@igi-global.com
Web site: http://www.igi-global.com

Library of Congress Cataloging-in-Publication Data

Names: Satolo, Eduardo, editor. | Mac-Lean, Priscilla, editor.
Title: Cases on managing dairy productive chains / Eduardo Guilherme Satolo, Priscilla Mac-Lean.
Description: Hershey, PA : Engineering Science Reference, [2023] | Includes bibliographical references and index. | Summary: "The publication of a book involving Management, Ambience and Good Practices in the Dairy Productive Chain is relevant because: Allows readers to broadly broaden their view on a topic. For academics and researchers, the relevance and topicality of the themes, and their applications in one of the main food production chains, such as milk, allows us to visualize new insights for research, in the various aspects of applicable management. For entrepreneurs (from the various links in the chain, from rural producers to processors), the presentation of cases in dairy segments instigates their application, an entrepreneurial spirit, and, as a result, innovation in management areas. For teachers and students, the cases presented serve as a subsidy for use in a teaching environment, illustrating the theoretical concepts"-- Provided by publisher.
Identifiers: LCCN 2022049522 (print) | LCCN 2022049523 (ebook) | ISBN 9781668454725 (hardcover) | ISBN 9781668454732 (paperback) | ISBN 9781668454749 (ebook)
Subjects: LCSH: Dairy processing. | Dairy products industry. | Dairy farming.
Classification: LCC SF250.5 .C377 2023 (print) | LCC SF250.5 (ebook) | DDC 636.2/142--dc23/eng/20221118
LC record available at https://lccn.loc.gov/2022049522
LC ebook record available at https://lccn.loc.gov/2022049523

This book is published in the IGI Global book series Advances in Logistics, Operations, and Management Science (ALOMS) (ISSN: 2327-350X; eISSN: 2327-3518)

British Cataloguing in Publication Data
A Cataloguing in Publication record for this book is available from the British Library.

Advances in Logistics, Operations, and Management Science (ALOMS) Book Series

John Wang
Montclair State University, USA

ISSN:2327-350X
EISSN:2327-3518

MISSION

Operations research and management science continue to influence business processes, administration, and management information systems, particularly in covering the application methods for decision-making processes. New case studies and applications on management science, operations management, social sciences, and other behavioral sciences have been incorporated into business and organizations real-world objectives.

The **Advances in Logistics, Operations, and Management Science** (ALOMS) Book Series provides a collection of reference publications on the current trends, applications, theories, and practices in the management science field. Providing relevant and current research, this series and its individual publications would be useful for academics, researchers, scholars, and practitioners interested in improving decision making models and business functions.

COVERAGE

- Production Management
- Networks
- Computing and information technologies
- Risk Management
- Services management
- Political Science
- Decision analysis and decision support
- Marketing engineering
- Organizational Behavior
- Information Management

IGI Global is currently accepting manuscripts for publication within this series. To submit a proposal for a volume in this series, please contact our Acquisition Editors at Acquisitions@igi-global.com or visit: http://www.igi-global.com/publish/.

The Advances in Logistics, Operations, and Management Science (ALOMS) Book Series (ISSN 2327-350X) is published by IGI Global, 701 E. Chocolate Avenue, Hershey, PA 17033-1240, USA, www.igi-global.com. This series is composed of titles available for purchase individually; each title is edited to be contextually exclusive from any other title within the series. For pricing and ordering information please visit http://www.igi-global.com/book-series/advances-logistics-operations-management-science/37170. Postmaster: Send all address changes to above address.

Titles in this Series

For a list of additional titles in this series, please visit: http://www.igi-global.com/book-series/

Moving From R&D Development to Production and Sales in Start-Up Companies
Amiram Porath (APC, Israel)
Business Science Reference • © 2023 • 181pp • H/C (ISBN: 9781799856856) • US $195.00

Advanced Research and Real-World Applications of Industry 5.0
Mahmoud Numan Bakkar (Abu Dhabi Vocational Education and Training Institute (ADVETI), Institute of Applied Technology (IAT), UAE) and Elspeth McKay (Cogniware.com.au, Australia)
Business Science Reference • © 2023 • 315pp • H/C (ISBN: 9781799888055) • US $250.00

Cases on International Business Logistics in the Middle East
Islam Abdelbary (Arab Academy for Science, Technology, and Maritime Transport, Egypt & University of Plymouth, UK) Sandra Haddad (Arab Academy for Science, Technology, and Maritime Transport, Egypt) and Ghada Elkady (Arab Academy for Science, Technology, and Maritime Transport, Egypt)
Business Science Reference • © 2023 • 342pp • H/C (ISBN: 9781668446867) • US $240.00

Examining the Vital Financial Role of SMEs in Achieving the Sustainable Development Goals
Dalila Taleb (University of Abou Bakr Belkaïd, Algeria) Mohammed El Amine Abdelli (University of Brest, France) Afef Khalil (University of Tunis, Carthage, Tunisia) and Asma Sghaier (University of Sousse, Tunisia)
Business Science Reference • © 2023 • 275pp • H/C (ISBN: 9781668448342) • US $250.00

Lean Thinking in Industry 4.0 and Services for Society
Katarzyna Antosz (Rzeszow University of Technology, Poland) Jose Carlos Sa (School of Engineering, ISEP, Polytechnic of Porto, Portugal) Małgorzata Jasiulewicz-Kaczmarek (Poznan University of Technology, Poland) and Jose Machado (University of Minho, Portugal)
Business Science Reference • © 2023 • 312pp • H/C (ISBN: 9781668456064) • US $250.00

IGI Global
PUBLISHER of TIMELY KNOWLEDGE

701 East Chocolate Avenue, Hershey, PA 17033, USA
Tel: 717-533-8845 x100 • Fax: 717-533-8661
E-Mail: cust@igi-global.com • www.igi-global.com

Editorial Advisory Board

Table of Contents

Detailed Table of Contents

In the first stage of the study a literature review was performed, along with a study of the normative instructions (NI) related to the quality of milk, as well as maximum limits of quality indicators, such as somatic cell count (SCC) or standard plate count (SPC) and total bacterial count (TBC). These all represent factors that consequently generate milk quality, as any variation in milk production will affect production costs. The case studies were performed through on site observations at the properties using interview techniques between the researcher and producer, as through the dissemination of knowledge and development of the producers in the application of tools for Quality Improvement for Process Control with milk quality parameters as SPC, SCC and TBC in line with Normative Instruction 62, in force at the time of implementation of the projects in the regions surrounding Goiânia-GO and Pontal-MG, as well as good milking practices.

Milk is a complete feed, composed of carbohydrates, proteins, vitamins, and minerals of great importance for human nutrition, in addition to playing a representative role in aspect economy. However, milk production is very affected by market variations, such as production costs, influences of exportation activities of the product and

surrounding conditions. The act was implemented in the industry with ten principles which can promote the quality product production which will include the premises & personnel's hygiene, the cleanliness of the raw material section along with processing and storage facilities. As dairy industries plays major role among world population, it is necessary for the industries and the technologist to follow the GMP principles. Hence, at this chapter, the detailed discussion over GMP along with its principles in the fore coming sections.

One of the greatest challenges of dairying in tropical regions of the world is heat stress. Upon the uncertain effects of climate change might have in dairy cattle productivity in the tropics, intrinsic advantages exhibited by heat tolerant breeds, in combination with adaptive nutritional management for heat stress, and modifications to the tropical environment, could be valuable in the quest to assure a robust and profitable dairy operation in our region. In the American continent, farmers have selected bovines better adapted to the region, like the hot and humid environment of Puerto Rico, where we can find "purebred" Holstein cattle that phenotypically have a very short and slick hair. This chapter intends to present the main managements practices carried out in dairy herds in the tropics upon the climate changes.

The milk composition and quality are related to the animal's health (cow, buffalo, goat, and sheep). The mammary gland disease negatively impacts agricultural economies by reducing milk production and quality, treatment costs, discarded milk, and reduced longevity of animals. Therefore, the chapter reports how and which anatomical, physiological, and behavioral differences inherent to each species can influence the occurrence, treatment, and prevention of mastitis. Thus, early detection and antibiotic resistance are essential factors to exploit new detection methods and more natural remedies associated with reducing antimicrobial therapies such as antioxidants, nanoparticles, or essential oils base-plant. Furthermore, healthy animal selection and being genetically resistant to mastitis associated with reasonable hygienic procedures, training programs, and periodic equipment maintenance are crucial steps to ensure the supply of good quality milk and dairy products without risks to the population. In addition, these measures must respect the specificities of each milk-producing species.

The dairy chain is remarkably important in mammals' nutrition, highlighting the human food through the production of milk and its derivatives. Thus, antibiotics have become essential for life and health maintenance of animals, being widely used for treatment of bacterial infectious diseases. However, antimicrobial resistance resulted from uncontrolled usage of antibiotics for prophylactic purposes and animal growth promoter. The challenge of antimicrobial resistance and persistence of antibiotic residues in milk still occurs, which presents a potential risk to human and animal health. Therefore, it is crucial to develop alternative strategies to overcome the inappropriate use of antibiotics, especially for dairy farming. Thus, this chapter intends to raise awareness and clarify issues related to the use of antibiotics in dairy cattle, the major problems of antimicrobial resistance, residues in milk, and to report about the main alternative solutions to the use of conventional antibiotics in dairy production with scientific studies.

In a tropical climate, air temperature can be stressful for Holstein cows throughout the year. This chapter's objective was to determine how the use of ventilation and fogging on a freestall barn can positively change the behavior, milk production, and milk quality of Holstein cows in all seasons. Twenty-eight lactating cows were divided into two groups: cooling group (CG) and non-cooled group (NCG). For different weather patterns throughout the year, behaviors (position, posture, and activity) were observed every 30 min during the day. Milk production was measured and milk samples were collected for analysis of protein, fat, lactose, total solids, and somatic cell count. All animals spent most of the day standing in the shade eating, ruminating, and idling regardless of the season (P<0.05). Also, animals under cooling system showed greater milk production, fat content, and higher somatic cell count (P<0.01). Provision of cooling system during the warmer months of the year proved to be efficiently increasing the feeding time, milk production, and milk quality of lactating cows.

Milk and its derivatives are among the foods most consumed by the population, capable of meeting important nutritional needs of humans. As the consumer market is increasingly demanding in terms of quality of the milk produced, the difficulty in maintaining and controlling these points is an issue present in several studies. Quality can be influenced from the type of production system in which the milk is produced, passing through nutritional and sanitary management, to the environment in which the animals remain during their productive life. This chapter presents a brief characterization of the dairy production chain in Brazil and the federal legislation that governs the production and quality of the milk produced. Some results of research on milk quality in shared tanks carried out by the Kamby Research Group with small rural producers in the Tupã/SP region are also presented. Finally, factors that affect the quality of milk and future actions to improve the dairy production chain are pointed out.

Preface

The dairy production chain is an important part of the global food industry and plays a significant role in providing food, livelihoods, and economic benefits to millions of people worldwide. Dairy production involves a complex supply chain that includes animal breeding, feeding, milking, transportation, processing, and distribution, and it has social, environmental, economic, and political impacts that affect both consumers and producers.

World milk production has increased over the last few decades, mainly driven by increased demand for dairy products in developing countries and increased productivity of dairy cows.

According to data from the Food and Agriculture Organization of the United Nations (FAO), world milk production has increased from around 650 million tonnes in 2000 to around 810 million tons in 2020. The main milk producers are India, the United States, China and Brazil.

Milk production can be divided into two main categories: cow milk production and milk production from other animals such as buffaloes, goats, sheep and camels. Cow's milk production accounts for the vast majority of world milk production, representing about 85% of the total.

The dairy sector is important to the economy of many countries, especially for small-scale producers who depend on dairy production for their livelihood. However, milk production also faces challenges, such as concern for animal welfare, the environmental impacts of production, and concerns about consumer health and food safety.

Small-scale producers play a crucial role in the dairy chain, in addition to contributing to the local economy by providing employment and income opportunities. They often sell their milk and dairy products directly to consumers, which helps to keep money circulating in the local economy. For sustainable agriculture, small-scale producers are often more environmentally sustainable than large-scale producers because they use fewer resources and have a smaller carbon footprint. They may use more traditional and natural farming practices, which can help to preserve biodiversity and soil health. The products generated by small-scale producers are often high-quality dairy products because they can pay close attention to the care of their animals and the processing of their products.

They may use traditional methods and recipes that have been passed down through generations, resulting in unique and flavorful products. These producers are responsible for the preservation of cultural heritage because they have a strong connection to their cultural heritage and the traditions of their region. By preserving these traditions and techniques, they help to maintain cultural diversity and promote a sense of community pride. Finally, small-scale producers are often more resilient to market fluctuations and disruptions than larger producers because they have diversified income streams and are less reliant on a single market or buyer. This can help to ensure the stability of the local food system and reduce the risk of food shortages.

The milk production chain, in addition to its economic importance, also has an impact in social, environmental, and political terms on the various actors who act directly or indirectly.

From a social point of view, the dairy production chain provides employment and income opportunities for many people, particularly in rural areas where there may be few other sources of income. It also contributes to food security and helps meet the nutritional needs of millions of people worldwide. However, the working conditions in some parts of the dairy industry can be poor, and there are concerns about the welfare of animals used in the production process.

The political impact of the dairy chain is influenced by a range of political factors, including government policies and trade agreements. Governments often provide subsidies and other forms of support to the dairy industry, while trade agreements can affect the competitiveness of dairy products in global markets. Political decisions can also have an impact on environmental regulations, which can affect the sustainability of dairy production.

From an environmental point of view dairy production can have significant environmental impacts, including greenhouse gas emissions from livestock, deforestation for feed production, and water pollution from manure runoff. There are also concerns about the impact of dairy farming on biodiversity, as well as the potential for dairy production to contribute to soil degradation and erosion.

The dairy chain faces a range of challenges, including increasing demand for dairy products, changing dietary patterns, environmental concerns, and the need to improve the welfare of animals used in the production process. Other challenges include the need to improve the efficiency of the production process, reduce waste and emissions, and address the impact of climate change on the industry. The industry also needs to address concerns about the use of antibiotics and other veterinary drugs in dairy farming and address the issue of antimicrobial resistance. Finally, there is a need to improve the livelihoods of small-scale dairy farmers, particularly in developing countries, by improving access to finance, technology, and markets.

Understanding the importance and complexity of this production chain, *Cases on Managing Dairy Productive Chains* discusses the relevance and timeliness of themes and their applications in this chain, and allows researchers to visualize new insights for research in the various aspects of applicable management. It also presents cases in dairy segments and investigates their applications, entrepreneurial strategies, and innovation in management areas. Covering topics such as cooling systems, raw materials and sustainability indicators, this casebook is an excellent resource for farmers, managers, engineers, extension workers, higher education students and faculty, librarians, researchers, and academics.

The present book can be used as a research book for final undergraduate in animal science, food engineering, biosystems engineering, agricultural technology, and technical courses (e.g., agribusiness, quality management, chemistry, industrial processes, industrial engineering, etc.) or as a subject on postgraduate level. Also, this book can serve as a useful reference for academics, researchers, managers, engineers, and other professionals in matters related to engineering and management of industrial processes in the dairy chain. In terms of the market, we can highlight: Primary Market: Farmers, Engineers, Managers, Extensionists; Secondary Market: Researchers and Postgraduate Students.

The first chapter, conducted by Cynara Mendonça Tinoco and Lays Freitas aimed to improve milk quality by disseminating best practices in milking. To achieve this, they conducted a literature review and studied normative instructions related to milk quality, including maximum limits of quality indicators such as Somatic Cell Count (SCC) and Total Bacterial Count (TBC). These factors are crucial as they impact milk quality and production costs. The researchers also performed case studies by visiting properties and conducting interviews with producers. Through the dissemination of knowledge and the application of tools for quality improvement, producers were able to improve their processes and meet the milk quality parameters outlined in Normative Instruction 62. The study was conducted in the regions surrounding Goiânia-GO and Pontal-MG and also emphasized the importance of Good Milking Practices.

The second chapter, conducted by Luiz Roma Júnior, Lívia Castelani, and Thatiane Mitsunaga aimed to explore the role of milk quality in improving dairy production. Milk is a crucial source of carbohydrates, proteins, vitamins, and minerals that play a vital role in human nutrition and the economy. However, milk production is heavily influenced by market variations, production costs, and international market activities. To address these challenges, all components in the production chain must focus on sustainability and profitability, especially the producer. The study presents the relationship between indexes and different areas of the milk production chain to aid in the sustainability and profitability of dairy farms. It also discusses each component and its relation to obtaining maximum information possible to contribute to the dairy industry.

In the third chapter, Eduardo Guilherme Satolo, Bruna Ferreira, Priscilla MacLean, Juliani Gotardi, and Isabela Santos aimed to identify the technical quality requirements evaluated in the supply of raw milk by dairy companies in the state of São Paulo, Brazil. The quality requirements are established through regulations 76 and 77, and the chapter aims to use the voice of the customer (VOC) to identify compatible internal requirements to drive the improvement process. The results of the study highlight the need for actions to raise awareness of the agents involved in the dairy production chain, including the role of the dairy in measuring and controlling legal parameters, the importance of superior milk quality in producing better quality products, and the need for action plans to be carried out in partnership with rural producers for collective gain along the production chain.

Chapter 4, authored by Mohana Priya Rajendran and Muthuminal R, provides an overview of Good Manufacturing Practices (GMP) in the dairy industry. GMP refers to the practices followed by industries to produce safe and quality products and was established in 1995 by the Food and Drug Administration (FDA). The chapter discusses the rules and regulations introduced under the FDA to regulate cleanliness and product quality in the industrial environment, which is essential to maintain the quality of edible products. The implementation of GMP in the industry involves ten principles to promote the production of quality products, including premises and personal hygiene, cleanliness of raw material section, and processing and storage facilities. As the dairy industry plays a significant role in the world population, it is essential for industries and technologists to follow GMP principles. The chapter provides a detailed discussion of GMP principles in the following sections.

In chapter 5, Guillermo Ortiz-Colón discusses management practices for dairy herds in the tropics under climate change. Dairying in tropical regions faces the challenge of heat stress in cattle, and climate change can exacerbate this problem. To maintain a robust and profitable dairy operation, heat-tolerant breeds, adaptive nutritional management, and modifications to the tropical environment can be valuable. In the American continent, farmers have selected bovines better adapted to the region, such as "purebred" Holstein cattle with short, slick hair. This chapter presents the main management practices carried out in dairy herds in the tropics to address the effects of climate change.

The chapter 6 focuses on the impact of mammary gland diseases on milk production and quality in dairy species such as cows, buffalos, goats, and sheep. The authors, Bruna Rekowsky, Matheus Oliveira, Uiara Paim, José Silva and Marion Costa, discuss how anatomical, physiological, and behavioral differences inherent to each species can affect the occurrence, treatment, and prevention of mastitis, a common mammary gland disease that negatively impacts agricultural economies. The chapter emphasizes the importance of early detection and antibiotic resistance, as well as natural remedies such as antioxidants, nanoparticles, and essential oils. The selection of genetically resistant and healthy animals, as well as hygienic procedures, training programs, and equipment maintenance, are also discussed as crucial steps to ensure the supply of good quality milk and dairy products without risks to the population. Overall, the chapter underscores the need to respect the specificities of each milk-producing species in controlling and preventing mastitis.

Chapter 7, presented by Wendell Leite, Mariane Severino and Eduardo Vicente, discusses the use of antibiotics in dairy farming and the challenges it presents, such as antimicrobial resistance and the presence of antibiotic residues in milk, which can be harmful to human and animal health. The chapter emphasizes the need for alternative strategies to overcome the inappropriate use of antibiotics in dairy farming. It aims to raise awareness and clarify issues related to the use of antibiotics in dairy cattle, highlighting the major problems of antimicrobial resistance and residues in milk. The chapter also reports on the main alternative solutions to the use of conventional antibiotics in dairy production, supported by scientific studies.

In the chapter 8, the authors Cristiane Titto, Evaldo Titto, Rafael Titto, Alfredo Pereira, Messy Pantoja and João Negrão, discuss the effects of using evaporative cooling systems on Holstein cows in a tropical climate. The study observed the behavior, milk production, and milk quality of 28 lactating cows, which were divided into two groups: a cooling group and a non-cooled group. The behavior of the cows was observed every 30 minutes during the day, and milk samples were collected for analysis of protein, fat, lactose, total solids, and somatic cell count. The results showed that regardless of the season, all cows spent most of the day standing in the shade eating, ruminating, and idling. However, the cows in the cooling group showed greater milk production, fat content, and higher somatic cell count. Therefore, the study concluded that providing cooling systems during the warmer months of the year can efficiently increase feeding time, milk production, and milk quality of lactating cows.

The chapter 9, written by Priscilla Mac-Lean, Eduardo Satolo, Julia Barbosa, Isabela Santos, Leonardo Campos and Ligia dos Santos discusses the importance of milk and its derivatives as a highly consumed food that provides important nutritional benefits to the human body. The composition of milk includes various solids such as lactose, fat, protein, minerals, and vitamins. However, maintaining and controlling the quality of milk has become increasingly challenging due to the market's demand for high-quality products. Quality can be influenced by various factors, such as the production system, nutritional and sanitary management, and the environment in which the animals remain during their productive life. The chapter's objective is to present the main criteria established by the legislation for the production, transport, and processing of milk in Brazil. It also aims to describe the laboratory and field analyses used to evaluate the quality and microbiological parameters of milk. The chapter further intends to detect possible adulterations carried out during the collection or delivery process that modify the composition of the product. Overall, the chapter provides valuable insights into the importance of maintaining the quality of milk and its derivatives in Brazil.

Chapter 10, written by Mayara Murata, Josemeyre Bonifacio da Silva Marques, Luiz Ito Morioka and Hélio Suguimoto, explores the potential applications of cheese whey, the main by-product of the dairy industry, for the production of high-value-added products. Currently, only half of the cheese whey produced is used for established processes, such as animal feed, fresh product production, and the extraction of high-value compounds. The rest is often discarded, leading to environmental concerns and waste. However, researchers have been investigating strategies to make use of this by-product and transform it into new biotechnological products, such as biofuels, biopolymers, and functional food ingredients. The chapter highlights the challenges involved in implementing these processing systems and the importance of collaboration between different players in the dairy industry to transform cheese whey waste into high-value products, reduce waste, and comply with environmental legislation and public policies.

In Chapter 11, the authors, Mirian Fabiana da Silva and Ferenc Bánkuti, aimed to evaluate the perception of Agricultural Science undergraduate students from Brazilian institutions on dairy farm sustainability indicators. A total of 351 survey respondents participated in the study. The results showed that technical and environmental indicators were rated as the most important by undergraduate students from public institutions. Students who had previous contact with rural production gave lower importance scores to environmental indicators and higher scores to economic and technical indicators. Agricultural Engineering students rated environmental indicators as less important compared to Agronomic Engineering and Veterinary Medicine students. The authors concluded that the students' perceptions on dairy farm sustainability indicators were influenced by their on-farm experiences, differences in syllabus between Agricultural Science courses, and the public or private nature of the educational system.

The publication of the book *Cases on Managing Dairy Productive Chains*, involves this form, involves important contents such as Management, Ambience and Good Practices in the Milk Production Chain is relevant because:

- Allows readers to broaden their view on a topic.
- For academics and researchers, the relevance and topicality of the themes, and their applications in one of the main food production chains, such as

milk, allows viewing new perspectives for research, in the various aspects of applicable management

- For entrepreneurs (from the various links in the chain, from rural producers to processors), the presentation of cases in the dairy segments encourages their application, the entrepreneurial spirit and, consequently, innovation in management areas.
- For teachers and students, the cases presented serve as a subsidy for use in a teaching environment, illustrating the theoretical concepts.

Eduardo Guilherme Satolo
São Paulo State University, Brazil

Priscilla Ayllen Bustos Mac-Lean
São Paulo State University, Brazil

Acknowledgment

The construction of a literary work is never carried out without partners who believe in its proposal. In this way, the editor's express gratitude to the publisher IGI Global and the chapter authors and other partners involved in the cases reported in this book.

Eduardo Guilherme Satolo
São Paulo State University, Brazil

Priscilla Ayleen Bustos Mac-Lean
São Paulo State University, Brazil

Chapter 1
Dissemination of Best Practices in Milking Associated With Better Milk Quality

Cynara Mendonça Moreira Tinoco
https://orcid.org/0000-0002-6493-2451
Federal University of Goias, Brazil

Lays Carvalho Freitas
São Paulo University, Brazil

EXECUTIVE SUMMARY

In the first stage of the study a literature review was performed, along with a study of the normative instructions (NI) related to the quality of milk, as well as maximum limits of quality indicators, such as somatic cell count (SCC) or standard plate count (SPC) and total bacterial count (TBC). These all represent factors that consequently generate milk quality, as any variation in milk production will affect production costs. The case studies were performed through on site observations at the properties using interview techniques between the researcher and producer, as through the dissemination of knowledge and development of the producers in the application of tools for Quality Improvement for Process Control with milk quality parameters as SPC, SCC and TBC in line with Normative Instruction 62, in force at the time of implementation of the projects in the regions surrounding Goiânia-GO and Pontal-MG, as well as good milking practices.

DOI: 10.4018/978-1-6684-5472-5.ch001

INTRODUCTION

The worldwide production of milk in 2021, even after considering the problems faced with the COVID-19 pandemic, presented a slight increase of 1.1%, reaching 887 million tons. This increase is attributed to the expansion in production in India and Pakistan as a result of the continual increase in the number of dairy herds, as well as availability of forage aided by favorable monsoon rains (OECD/FAO,2022).

However, the importation of dairy products does not follow the same pattern of growth, especially in the question of powdered milk, owing to problems linked to markets closure due to COVID, physical distance between countries, economic depression, along with falls in the price of petroleum. The worldwide production of milk breaks down as follows 81% derived from cows, 15% from buffalos and 4% from a combination of goats, sheep and camels. The United States, India, Brazil and Germany are the highest milk producers worldwide, while New Zealand, the European Union and the United States are the main dairy exporters. The increases in world dairy trade were pushed mainly by the strong demand in the People's Republic of China (henceforth "China"), the highest worldwide importer of dairy products (OECD/FAO, 2020, 2022).

China, even under pressure from COVID, has been working since 2018 in the consolidation of its countrywide dairy production, and through such has improved reliability and efficiency in its scale of production, via the implemented standardization. These factors have produced an increase in the internal consumption of milk from 2020, along with generating consumer confidence regarding the product (OECD/FAO,2021).

In the case of Brazil, according to the Brazilian Institute of Geography and Statistics, the industry and activity of dairy production produces an important economic and social contribution in Brazil, where it is considered as the fifth largest milk producer worldwide with an estimated 25.5 billion liters being industrialized in Brazil, in 2020 alone. This activity presents itself as one of the principle economic sources of the country, as it plays a significant role in generating employment and income, with the highest percentage of small and medium producers in the dairy production chain. In consonance with the aforementioned, there are an estimated one million dairy producers in the field, who collectively generate millions of jobs along the various segments of the expanded chain. These factors constitute greatly toward the structuring of the rural economy, as well as providing support to the livelihoods of farmers (OECD/FAO, 2021).

Dairy farming plays a significant role toward social economic development. Across the whole world, around 150 million families work in the production of milk. The majority of these consist of smallholder farmers in developing countries, where this is the main activity for their livelihoods (OECD/FAO, 2021).

Dairy products are, by nature, highly perishable, with an inherent variation in milk collection throughout the year, along with a large number of small producers, which makes the production chain a highly complex one. However, a closer look at the industry reveals a lack of standardization in dairy production, with problems attached to primary processes, such as the cleaning of equipment and production sites, along with vaccination control and antibiotic application. These factors lead to a low quality in Brazilian produce, and as such investments are necessary for improvements to the milk management chain, aimed at the development and planning of production (sales and operation), which are seen as fundamental for meeting the demand for dairy products (Tinoco & Freitas, 2014; Tinoco et al., 2015).

Milk quality requirements in Brazil established by Ministry of Agriculture, Livestock and Supply (MAPA) at NI 76 (MAPA, 2018a) and 77 (MAPA, 2018b), limit parameters in farm raw milk for individual or community tank quarterly geometric averages should, from 01 June of 2019, to Standard Plate Count (SPC) (the old Total Bacteria Count - TBC) limit to 300 thousand CFU/mL and SCC lower than 500 cells/mL. Since 2011 up to 2019, accordance with the Normative Instruction 62 (MAPA, 2011), the Somatic Cell Count (SCC) maximum limits was 600 thousand CFU/mL and 600 thousand cells/mL, in the South, Southeast and Center East of Brazil. In European Union (EU), since 1992, permits a TBC of maximum 100 thousand CFU/mL at 30 °C and an SCC maximum of 400 thousand cells/mL (Berry et al., 2006).

The cleaning and sanitization of milking and herd environments are factors that most influence bacterial contamination (TBC) according to (M. A. V. P. Brito, 2013). In addition, the temperature and the storage period have a direct influence on the multiplication of microorganisms present in milk, which has an impact on the total bacterial count present in milk. The somatic cells present in milk are defense cells produced by the organism to combat infections in the mammary gland (Guerreiro et al., 2005). The total bacterial count (TBC) of the milk sample from a tank is indicative, among other elements, of the health condition of the herd, sanitation, cleanliness of equipment and storage temperature of the milk (Berry et al., 2006). As such, the SCC is commonly used as an indicator to aid in the monitoring of milk quality and dairy cow health due to the incidence of clinical and sub-clinical mastitis in the herd, seen as an inflammatory reaction of the mammary gland. This can produce physical and chemical changes in the milk, increasing the number of somatic cells due to the presence of pathogenic microorganisms, along with changes such as loss of functionality (Machado et al., 2000; MAPA, 2018a). For this reason, the somatic cell count (SCC) is one of the most influential parameters in determining the health of the udder and milk quality.

The SCC in milk increases in direct proportion to the gravity of the infectious disease. In milk that does not contain subclinical mastitis, the SCC is low (<100,000 cells/mL) (NI 76). The factors that may directly contribute to SCC are the time of year (rainy periods), milking stage and cow age (cellular response). However, the age and milking stage do not impact SCC in cows not infected, that is, the increase of SCC, at the end of milking, is related to the probability of the animal having become infected during milking and as it ages (Santos & Botaro, 2008). Milk with high SCC possesses lower yield in its industrialization due to the reduction in the concentration of fat, lactose, protein, potassium and calcium at the expense of the increase in chlorine, sodium and whey protein. Additionally, this can cause a number of quality defects such as rancidity, changes in aroma, flavor, texture and reduction in milk storage time (Santos, 2000).

Fat and protein are highly relevant elements in the composition of milk, as it is from these components that dairy products are derived, and likewise quality is directly proportional. In this sense, one understands that it is necessary that these be measured and monitored (Gigante & Costa, 2008). Through such, high quality milk can be defined as possessing an excellent composition (fat, protein, lactose, vitamins and minerals), free from pathogens, without physical-chemical contaminants and a low microbial count. As an essential requirement for good quality products, the quality of milk initiates with the herd as the primary condition for acquiring good products (Benbrook et al., 2013).

Normally, the udder of a healthy animal is inhabited by the genera Streptococcus, Staphylococcus and Micrococcus, representing more than 50% of the total microbiota of raw milk. The microbial count of milk obtained aseptically from healthy animals is less than 100 thousand CFU/mL (Boor et al., 2017). However, after milking, the initial microbial count of raw milk can increase due to the introduction of microorganisms from the environment, poorly sanitized milking equipment and contact with handlers. Consequently, the consumption of contaminated milk and its derivatives is one of the main causes of foodborne illness. Therefore, proper milking procedures are essential for preventing intoxications and infections, as well as reducing microbial contamination (Pinto et al., 2019).

Mesophilic microorganisms in raw milk are unavoidable as they are present on the cow udder, equipment, hands of the handler, water and air. However, viable cell counts greater than 500 thousand CFU/mL of raw milk indicate poor hygienic quality when obtaining milk (Perin et al., 2019). In Brazil, legislation allows for mesophilic aerobes counts in raw milk of 300 thousand CFU/mL (Marioto et al., 2020). One way of minimizing contamination by mesophilic microorganisms is to implement good manufacturing practices, rapid cooling of milk after milking and maintenance of the entire cold chain until milk industrialization (Perin et al., 2019).

In order to reduce the microbial spoilage by mesophilic microorganisms and the consequent increase in the shelf life of milk and dairy products, at the end of the 1950's bulk tanks were implanted to rapidly decrease the temperature of the extracted milk and store it under refrigeration on the farms before collection for industry (Moatsou & Moschopoulou, 2014). In addition, the use of refrigerated tank trucks and cold storage in the industry were implemented, which reduced the mesophilic microbial load. However, refrigeration led to the development of a psychotropic microbiota that went on to dominate the microbiota of refrigerated milk in the United States and in other large milk producers (Holzapfel et al., 2022). When milk with psychotropic microorganisms is placed under refrigeration the bacteria continue to develop, and as such the storage time of the product is reduced for both refrigerated and non-refrigerated storage (Lorenzetti, 2006).

The development of improvements in milk production aimed at higher quality brings measurable results both for producers and for the dairy products, thus guaranteeing consumer confidence concerning food safety and competitiveness on the market. With a view to expanding milk consumption and, consequently, greater profit generation, it is necessary to increase production and guarantee its quality, factors that can be worked on by improving the standards incumbent processes (J. R. F. Brito & Dias, 1998). According to Tinoco et al (2015) and Deepa & Krishnan (2021) the Methodology Six Sigma can be used in the systematic improvement of the process in favor of reducing the variability of defects. The method to cycle of improving, optimizing and stabilizing business processes and projects must be directed by DMAIC. In order to present a best practice about milk production in Brazilian, an examination is made of the case study performed at five milk producers, distributed in the dairy basin in the region around Goiânia-GO and Pontal-MG.

METHODS

Study Design and Technological Concerns

The study was based on the action research methodology (Thiollent & Silva, 2007) used for the development of practices in extension projects in search of technological development in rural areas with implementation of the Six Sigma methodology. This methodology is a tool that improves effectiveness and efficiency in organizations in order to reduce variability in indicators (Lopes & Tinoco, 2016). First, a bibliographical and document investigation was performed, concerning Good Milking Practices that have already been developed by research groups in the area of milk and of dairy companies, which are known through the quality of the acquired product. In order to develop the project, a partnership was sought with dairy producers of the region, the Finance Department of the Municipality of Project Implementation in 2015 and the Livestock Union. The project focused on small and medium-sized family farmers, as a result of keeping the rural producer in the field and for allowing a change in the culture of the agricultural producer, proposed by the Six Sigma methodology (Tinoco, 2015; Tinoco et al., 2015). This cultural change in producers seeks out food security, through improved product quality at the start of the Milk Chain, as well as improvements to indicators in a way that allows for better use of the raw materials in dairy industries, which are factors of importance for sustainability of the Dairy Chain (Freitas et al., 2021). As a support method for implementing the methodology, the DMAIC method was employed in 6 properties improving daring best practices.

Case Description

The states of Minas Gerais and Goiás are the highest producers of Brazilian milk, where these represent respectively 27.11 per cent and 9.13 per cent of the total production of milk in 2021 (EMBRAPA, 2021). In this sense, extension projects were carried out linked to the Federal University of Uberlândia (UFU) with funding via the UFU/Community Extension and Integration Program with finance (PEIC/UFU) and the Federal University of Goiás (UFG) via the UFG/ Culture and Extension Volunteer Program with financing (PROVEC /UFG). The project counted on the participation of fourteen (14) students from the Production Engineering course, as well as support from two (2) veterinarians for the improvement of the milking process from 2015 to 2019.

With more than six (6) producers attended to, in addition to an agreement with two (2) Dairy farmers in the regions of Ituiutaba/MG and Goiânia/GO, the 6 producers (Table 1) who interacted the most with the project participated in all stages of the DMAIC – Define, Measure, Analyse, Improve, Control (Deepa & Krishnan, 2021; Tinoco et al., 2015, 2017). In addition to the producers who were monitored via the project, in partnership with one Dairy farmer and SENAR MG, a workshop was held with Dairy producers to disseminate the improvement of milking processes and financial management of properties.

The six properties, cases covered in this study, are made up of only small and medium-sized producers, with mechanical milking, with almost all being family producers, as only in Cases 4 and 5, the producers do not work directly in milking, but manage the property.

Table 1. Case studies

Properties	Case 1	Case 2	Case 3	Case 4	Case 5	Case 6
Number of people who depend on income	Four people, but just the owner takes care of milking	Three people, but just two take care of milking	Five people, but only the owner and his wife take care of milking	Four people, but three employees take care of the milking	Four people, but just two employees take care of the milking	Three people, the owner and an employee take care of the milking
Property size (m^2)	121,000	72,100		2,904,000		6,098,400
Land lease	169,400 m^2	Does not pay rent	Does not pay rent	Does not pay rent	Does not pay rent	Does not pay rent
Land use	Corn plantation and pasture on both lands	Pasture and sugarcane cultivation	Pasture and manufactured feed made up of corn, soybean meal, salt. Cillage is corn bran. Dry six monghs of year adicional than 2kg/cow of cottonseed and 4kg/cow of soybean hulls	Corn plantation and pasture, but the cows are in confinement.	Pasture and manufactured feed made up of corn, soybean meal, salt.	Pasture and manufactured feed made up of corn, soybean meal, salt.
Number of animals (lactating cows)	202	35	50	130	32	Cillage is corn bran.
Animal breed	Girolando, Holstein and Jersey cows	Holstein cows	Holstein cows	Holstein cows	Holstein cows	Girolando and Holstein cows
Daily lactation (milk producting cows)	50	28	50	130	32	27
Liters of milk/day	1200	500	1200	1700	448	273
Quantity of milkings/day	2	2	2	2	2	1
Milking interval	12 hours	10 hours	12 hours	12 hours	12 hours	24 hours
Average productivity/cow/day - liters/day	24.48	17.86	24	13	13	10.11
Chiller capacity (liters)	3200	700	3200	3200	700	700
Milk collection interval (hours)	48	24	48	48	48	48

continues on following page

Table 1. Continued

Properties	Case 1	Case 2	Case 3	Case 4	Case 5	Case 6
Milking period	9 to 10 months, with 2 months of rest for the next lactation.	9 to 10 months, with 2 months of rest for the next lactation.	7 months, with 2 months of rest for the next lactation.	7 months, with 2 months of rest for the next lactation.	9 months	9 months
Feeding animals during the milking period	No feed, but Feed and silage after milking, wet season, pasture addition	Feed during milking period and pasture after feeding, soybean husk, cottonseed, dry period: urea and sugarcane	In the dry season the cows come out of milking and go to the trough. In the rainy season they are released.	No feed, but after milking feed and silage made by the farmer.	Feed during milking period and pasture after feeding, soybean husk, cottonseed, dry season: urea and sugarcane	Feed during milking period and soybean husk, cottonseed after milking during dry period: urea and sugarcane
Corral desingn	Cattle handling on concrete floor, without the presence of other animals.	Cattle handling on concrete floor, but the rating area room is pasture, and without the presence of other animals.	Cattle handling on concrete floor	Cattle handling on concrete floor and the waiting room in concrete floor.	Cattle handling on concrete floor, but the waiting room is on land, and without the presence of other animals.	Cattle handling on concrete floor, but the waiting room is on land, and without the presence of other animals.
Drinking fountain in the corral or resting place	Non-existent	Has, but just for one cow at a time	Non-existent	Non-existent	Non-existent	Non-existent
Control tank temperature	Yes, between 3°C and 4°C	Yes, between 3°C and 4°C	Yes, between 3°C and 4°C.	Ready feed is made up of corn, soybean meal, salt, monenzinho and bicarbonate. Cillage is corn bran. 2kg/cow of cottonseed and 4kg/cow of soybean hulls.	Yes, between 3°C and 4°C	Yes, between 3°C and 4°C
Sizing of corridors	Four animals in each aisle, but no disease control	Three animals in each corridor	Four animals in each aisle, but no disease control	Four animals in each aisle, but no disease control	Three animals in each aisle	It has one corridor
Milking type	Semi-automatic mechanics	Semi-automatic mechanics, with reservoir in the middle of process without control temperature	Semi-automatic mechanics	Semi-automatic mechanics	Semi-automatic mechanics	Semi-automatic mechanics

continues on following page

Table 1. Continued

Properties	Case 1	Case 2	Case 3	Case 4	Case 5	Case 6
Number of people at milking	2	2	2	2	2	1
Vaccination	All mandatory, controlled by the owner.	All mandatory, controlled by the owner accompanied by a veterinarian.	All mandatory, controlled by the owner.	All mandatory, controlled by the owner.	All mandatory, controlled by the owner.	All mandatory, controlled by the owner.
Problems with acidity and antibiotics	No problem	No problem	No problem	No problem	No problem	No problem
Mastitis problems	Few occurrences, and when it occurs, the infected animal is directed to the last batch until it is cured.	Few occurrences, and when it occurs, it is directed to the last batch until it is cured.	Control pattern for cattle, cow discarded if it has chronic mastitis, first cow cut from herd.	Mastitis. Problem with hoof: Laminitis, caused by a fungus, a good control would be the hoof is treated with copper sulphide. Clean, put the sulphide and bandage for about 1 week	Few occurrences, and when it occurs, it is directed to the last batch until it is cured.	Few occurrences, and when it occurs, it is directed to the last batch until it is cured.
TBC (thousand CFU/mL)	197	14	11	10	32	372
SCC (thousand cells/mL)	1927	581	1147, but there occurred this month	433	637	255
Other animals breeds	Chicken, pigs and horses.	Dog	Not applicable	Dog	Dog	Dog
The industry for whom it provides benefits good SCC and TBC results	No	Yes	No	Yes	No	No

Technology Components

In order to define the project support manuals, the process manuals already known as Good Practices on Farms - BPF (created by DPA – Dairy Partners Americas in partnership with SENAI and EMBRAPA) were used (Morini & Braga, 2009) to improve productivity and quality of milk, while maintaining the goal of reaching quality indicators of Normative Instruction 62 (NI 62), Somatic Cell Count (SCC), Total Bacteria Count (TBC) and the prevention of Trypanosomiasis (Tinoco et al., 2015).

In addition to the manuals, used as support for training the project team were veterinarians, farm owners, along with the development of the tools (Flowchart, Fish diagram, 5W1H-who, where, why, what, when, how) addressed at each stage of the project that used the DMAIC (Deepa & Krishnan, 2021; Tinoco et al., 2017) (Figure 1 – Stage 1).

Technological Concerns

In stage D - Define (Figure 1 – Stage 2), the agreement is inspected between the property and industry that does the milking on the rural property. In this part verifications are made into whether there exists a bonus regarding the quality of milk (TBC and SCC), an interval between milk collections on the part of the collecting industry, investment limits made available by the producer, if there exists paid labor on the farm besides the owner, as well as tank volume for milk cooling.

In stage M – Measure, the quality indicators TBC, SCC, fat, protein, total solids (TS), lactose, and productivity in light of the quality norms for milk defined by the Ministry of agriculture were all verified. Identification of the level of suitability concerning the property and milking location in terms of the rules defined in the Good Practices manual (Tinoco et al., 2017), the parts verified through the checklist – verification list (plan for monitoring the activities of the milking process - milking management; mastitis control; equipment maintenance; cleaning of equipment, tank and utensils; milk cooling). The dairy farms did not have dates (TBC and SCC) from the past, and because of this did not know how was the effect of actual Roadmap.

In stage A – Analyze the problems identified from the indicators compared to NI 62 (MAPA, 2011), with obligation levels of SCC< 600,000 cell/mL, and TBC < 400,000 CFU/mL, where the fundamental causes that led to the effect should be verified, i.e., indicators outside the values agreed with the dairy producers and current legislation. In this stage, Cause and Effect diagrams or Fish diagram (Figure 2), Pareto chart, Fault diagram were used (Tinoco et al., 2017). To obtain the SCC and TBC indicators for the properties, an agreement was made with one of the dairy producers, in order to allow tests to be performed in a laboratory capable of carrying out milk tests, both in Goiás and Minas Gerais.

Figure 1. Project diagram

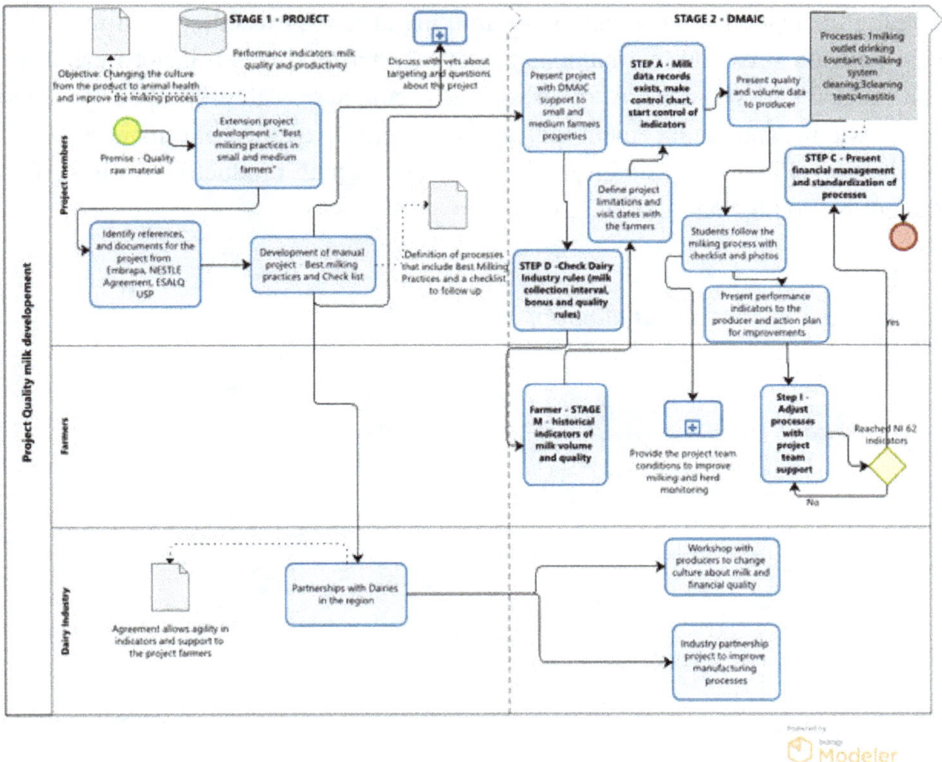

The dairy producer that pays by quality only pays with a bonus over the value paid per liter of milk, if the TBC and SCC indicators are lower than the norm of European dairy management, which is a TBC maximum of 100,000 CFU/mL, at 30 °C and SCC maximum of 400,000 cells/mL (Berry et al., 2006). This value may reach 1% higher than the value paid per liter of milk to the producer; however, in cases where values are higher than European limits, the same percentage is taken from the value paid per liter of milk. In this sense, this dairy producer is provided support from a veterinarian and zootechnician in the search for establishing feeding and management rules for the animals, and as a result, a higher value in the amount received for the milk. However, in case of non-compliance with the established management format, the producer is penalized, receiving a lower amount than the amounts paid for other dairy products.

Among the procedures adopted by rural producers, some are common across all properties. The route of the herd to the corral and the waiting areas for milking/ resting are open muddy ground, with little attention being given to this stage, which can cause both mastitis and hoof problems. Operators have the practice of holding the teat cups at the end of the milking process, exceeding the time, which can generate stress on the mammary quarters, mastitis, which if not treated well can cause the loss of teats. Another verified factor, across all properties, is the lack of knowledge and management indicators concerning quality and finances of milk production (Figure 2), as well as problems concerning productivity arising from the production format (Table 2).

Figure 2. Cause and effect diagram

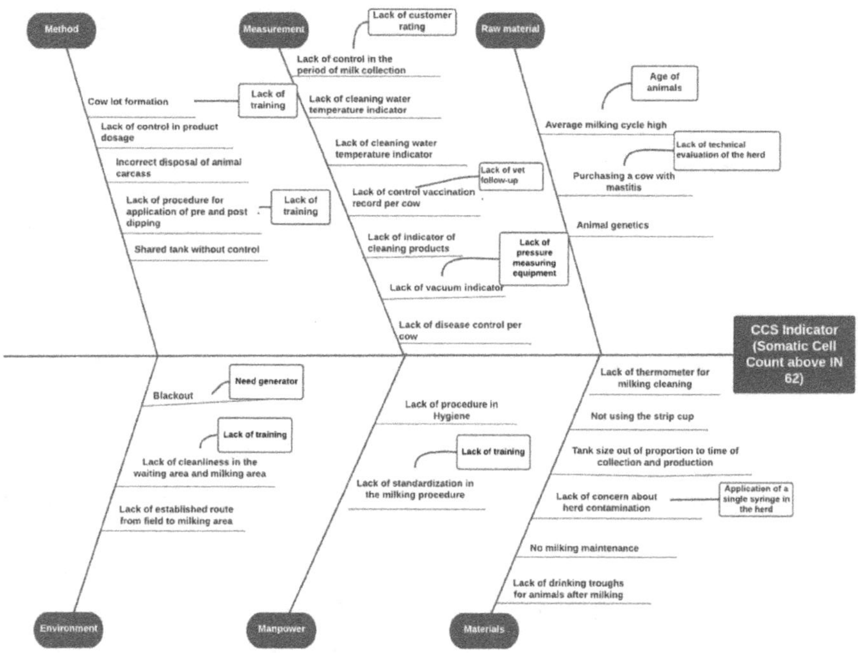

The most problematic points on the checklist were verified through the Pareto chart (check items at each visit to the project), which consists of the parts: milking management, mastitis control, equipment maintenance, cleaning of equipment, tank and utensils, and cooling system. From the problems identified as a priority in improving milk quality indicators, one notes Somatic Cell Count (SCC) and Total Bacteria Count (TBC). As for SCC, this is a primordial factor in animal health and herd productivity, where SCC of the tank with 100 thousand cells/mL can generate a reduction of 90 kg/365 days in the volume of milk produced by each cow (Table 2). In regards to TBC, its alteration results directly from the contamination of the milking system associated with problems in cleaning and maintenance. The factors of SCC and TBC are defined and adjusted through the Manual of Good Practices that is directly responsible. As a conclusion of the "ANALYSIS" of the processes and performance indicators of each property, short, medium and long-term action plans were established. From the observations made on the property, concerning the milking procedures, equipment and utensils, the short, medium and long-term Action Plan (5W1H) was presented to the producers (Table 3).

In stage I – Implement or Improve, in this stage the plans of action, defined in stage A, were implemented. This implementation was negotiated with the producer, with training concerning the change in processes, change in products for cleaning of the milking system, when deemed necessary. The performing of follow-ups and filming of the implementation of each stage of the action plan, while providing an understanding of the difficulties to the producer over the changes in essential processes. Through films and photos, the problems and good points were verified, while improving productivity and allowing the producer to go over all actions, and in the case of employees perform recycling and training of new employees in case of turnover. This Roadmap and indicator are known and can be control, its average and standard deviation.

Table 2. The volume of loss in milk production SCC per year

SCC from the tank (thousand cell/ml)	Reduction in milk production (Kg/305 dias)
50	-
100	90
200	180
400	270
800	360
1,600	450

In stage C – Check or Control, verification was made into the implemented actions from the action plan, and its results concerning the performance indicators SCC and TBC, in a way as to identify those actions that allow for the meeting of IN 62 requirements, as well as the goal defined by the dairy producers. As one of the dairy producers pays for milk quality, where one has as a limiting factor the European norm, with values lower than IN 62, while both pay for milk components (Fat, Protein, Total Solids, Lactose).

These factors, with bonuses resulting from quality and milk components, are factored in very well, since non-compliance burdens the producer, such bonuses can be used as an extra gain in monthly production, which can be transformed into property improvements. Those actions that had the desired effect on the indicators need to be standardized and defined as tasks for milking management, while those that did not need to be identified changed and a newly designed application processes created.

Management and Organizational Concerns

Properties 1 and 2 supply milk to different companies, the dairy industry that collects the milk from property 1 (Table 4) works with the Brazilian limits for TBC and SCC (TBC- 600,000 CFU/mL and SCC – 600,000 cells/mL). However, in accordance with the values presented for SCC, the milk could not be collected, but property 1 continues to supply milk without any prior warning from the company.

In addition, the need for improvements in milking processes (Table 3), together with the importance of managing the SCC and TBC indicators of NI 62 was discussed with the producers. As such, it was established that the cooling of raw milk (in natura) should occur within two hours after milking, at a temperature of 4°C, and should reach the industry at up to 7°C.

Noted here is that the procedures implemented in Case 1 presented problems between the months of January and March of 2016, the return of milk in the milking system due to a problem in the pressure level of the teat cups. As this problem was not identified, it generated a colony of bacteria in the milk pipe, leading to an increase in TBC. Due to the examination being monthly, and as the result is delivered 15 days after the examination, there was a delay in adjusting the milking system. One of the most efficient changes made was in the cleaning process, where the product brand used for cleaning the milking system was changed to the brand defined by the veterinarian. According to the producer, the product in question was priced much higher than the one used prior, and in the analysis, Table 5, product used on the property, and Table 6, product suggested by the veterinarian, one notes that in terms of cost and profitability, the product suggested by the veterinarian ended up being cheaper per volume spent for each cleaning of the milking system used by the producer.

Table 3. Plan – 5W1H

What	Who	When	Where	Why	How	How Much
Improve after milking	Everyone	Short term Depends on producer	In the corral / after milking	Cows consume 50% of the water post-milking and milk is 80% water	Install water troughs for animals after milking	Creation of troughs and workforce
Cow teat contagion	Everyone	Short term Depends on producer	In the corral / On the teats of the cows	In order not to contaminate the teat cups, passing on diseases to the herd	Pre-dipping before putting on the liners	Product and applicator cost
				Do not contaminate other teats, either of the animal itself, or of others and the milk employee	Avoid passing your hands on the cow teats, wipe with disposable paper towels	Paper cost
Use of the strip test cup	Everyone	Short term Depends on producer	In the corral / On the teats of the cows	For the identification of clinical mastitis	Remove the first three streams of milk from each teat to identify mastitis lumps	Manpower
5S	Everyone	Short term Depends on producer	In the Corral	For better workplace organization and efficiency in the milking process	Sort materials and keep milking environment clean	Manpower
Improve milking system cleanliness	Everyone	Short term Depends on producer	Milking	For cleaning water with high temperature not to damage the pipe or water with low temperature so as not to accumulate grease	Control the temperature of the pipe cleaning water at the inlet (between 75 and 80 °C) and at the outlet (above 45 °C)	Labor and thermometer purchase
Train on how long the vacuum should remain on the teats	Everyone	Short term Depends on producer	Milking	It wastes time in the process and causes injuries to cow teat	The liners cannot exceed 7.5 min for each 20 liters milked and 6 min for cows with less	Operator
Using the CMT Test Racket and CMT Solution	Everyone	Medium term: Depends on producer	In the corral / On the teats of the cows	For checking subclinical mastitis	Remove 2.5 ml of milk from the plate and add 2.5 ml of CMT reagent and shake for one minute in circles. Determine the infection status by the degree of clotting obtained, compare to the standard	Manpower
Database creation	Everyone	Medium term: Depends on producer	On the Farm	For infection control and compare with previous data for improvement	From the test results (Racket, strip cup), make a note of the cows that tested positive for mastitis	Manpower

continues on following page

Table 3. Continued

What	Who	When	Where	Why	How	How Much
Machine and environment cleaning	Everyone	Medium term: Depends on producer	Corral and objects	To reduce outbreaks that can generate infection and prevent mastitis and bacteria from increasing in the herd/environment	For the environment: Scrape and wash the environment after each milking, with plenty of water and the appropriate cleaning products. And for equipment cleaning, use specific products in hot water for some and cold water for others.	Manpower
5S	Everyone	Medium term: Depends on producer	Corral and objects	Decrease indicator (TBC) and increase volume	Train and standardize on the use of milking equipment.	Manpower
Pre- dipping	Everyone	Long term Depends on producer	In the corral / On the teats of the cows	With pre-dipping, the teat is disinfected and the teat does not spread the contagion	Applying pre-dipping to the base of the teat before placing the liners.	Manpower
Personal hygiene	Everyone	Long term Depends on producer	In the corral and housing	In order not to pass infection from the operator to the herd.	Perform hand hygiene with each milked cow, thus preventing the spread of infection.	Manpower
5S	Everyone	Long term Depends on producer	On the farm	5S Induces Discipline, Identifying Problems, and Driving Improvements How to Decrease SCC Rates	1) Sense of Use (Separate what is useful and what is not within the milking process and use it efficiently.	Manpower
					2) Sense of Organization (Save utensils in the best way for the next milking)	Manpower
Correct handling during milking	Everyone	Long term Depends on producer	In the Corral	To have a healthy herd with good quality milk with a low SCC index.	1)Use strip cup test to identify clinical mastitis	Manpower
					2) Apply pre-dipping	Manpower
					3) Need to apply post-dipping to close the teat sphincter	Manpower
					4) Check that the cows are drinking water after milking	Manpower

Table 4. Milk indicators Case 1

Milk Collection	TBC (thousand CFU/mL)	SCC (thousand cells/mL)	Fat Content (% m/m)	Protein Content (% m/m)	Lactose Content (% m/m)	Total Solids Content (% m/m)
Jan/15	197	1927	3.65	3.14	4.31	12.17
Mar/15	32	858	3.69	3.19	4.43	12.29
May/15	9	574	3.73	3.35	4.43	12.48
Jul/15	19	305	3.86	3.29	4.39	12.47
Sep/15	10	496	3.67	3.26	4.51	12.44
Nov/15	11	514	3.85	3.29	4.41	12.53
Jan/16	316	841	3.67	3.15	4.52	12.34
Mar/16	356	643	3.73	3.35	4.26	12.32
May/16	8	1925	3.76	3.21	4.24	12.24
Average (mar - nov)	16	549.4				
St dev (mar - nov)	9.68	199.80				
Evolution: Mar/15 - Jan/15 (%)	515.62	124.59	-1.08	-1.56	-2.71	-0.98
Evolution: Mar/15 - Jul/15 (%)	-40.62	-64.45	4.60	3.13	-0.90	1.46
Evolution: Mar/15 - Nov/15 (%)	-65.62	-40.09	4.33	3.13	-0.45	1.95
Evolution: Mar/15 - Jan/16 (%)	887.5	-1.98	-0.54	-1.25	2.03	0.41
Evolution: Mar/15 - Mar/16 (%)	1012.5	-25.05	1.08	5.01	-3.84	0.24
Evolution: Mar/15 - May/16 (%)	-75	124.35	1.89	0.62	-4.29	-0.41

Table 5. Cleaners and Sanitizer for milk equipment

Brand A	Price	Quantity used at each milking	Total value
Foaming Detergent (20L)	99.65	50 ml	0.25
Alkaline Detergent (20L)	89.95	180 ml	0.8
Acid (20L)	160	180 ml	1.22
		Total:	2.27

Table 6. Cleaners and Sanitizers for milk equipment, another brand

Brand B	Price	Quantity used at each milking	Total value
Foaming Detergent (20L)	160	50 ml	0.3
Alkaline Detergent (20L)	140	50 ml	0.35
Acid (20L)	120	50 ml	0.4
		Total:	1.05

Source: Prepared by the authors

In the case of SCC, this increase, from November up to February, normally occurs in all producers as a result of the rainy season and the formation of mud in the passageway from the pasture to the corral, at the point where the animals leave. The veterinarian identified this as an important point for improvement, as the cows leave the milking procedure with the sphincter open and lie down in the mud, leading to clinical mastitis. Additionally, the delay in the exchange of teat cups and control of the vacuum of the milking system generates teat stress and sub-clinical mastitis.

The cleaning milking system with non-standard temperature and process, problems in post and pre-dipping, the lack of cleanliness in the equipment were some of the factors listed for the imbalance of the mean, and variability of TBC. Furthermore, the acquisition of animals with mastitis contributes to SCC. Analyzing the SCC and TBC during the project, in the Case 1, the milk quality indicators were 549.4 thousand cells/ml (standard deviation in 199.80) and 16 thousand CFU/ml (standard deviation in 9.68), respectively, which become capable to control after the implementation of the necessary actions. Besides taking into account that the milk produced, is still below NI 62, considered at the time of the project, 2015, as well as NI 76, the current standard.

In the case of property 2, the dairy industry that collects the milk, pays for reductions in TBC and SCC values, and works with the limits set by Europe (TBC- 100,000 CFU/mL and SCC – 400,000 cells/mL). At the beginning of the study, property 2 presented indicators for Brazilian standards (Table 7), but as the company to which it supplies subsidizes indicators based on European Union standards (Berry et al., 2006), there is a need for process improvements.

The TBC, an indicator of problems regarding the disinfection of the milking system, can generate increases in the indicator, but small variations at Roadmap can result from contaminants and dirt on the teat that were not cleaned before milking. Temperature is one of the control factors of bacteria in raw milk, which after milking has around 9 thousand bacteria, and after 24 hours at a temperature of 4°C, the milk can have up to 10 thousand bacteria due to psychotropic bacteria.

Through the production of enzymes, lipases and proteases, milk has negative changes to its properties, and these factors are not easily reversed, even after the thermal processes, to which milk is subjected, in the dairy industry during pasteurization. However, it can present a bitter and rancid taste, both in the milk as well as in stored by-products (Lorenzetti, 2006).

Table 7. Milk indicators Case 2

Milk Collection	TBC (thousand CFU/mL)	SCC (thousand cells/mL)	Fat Content (% m/m)	Protein Content (% m/m)	Lactose Content (% m/m)	Total Solids Content (% m/m)
Jan/15	16	731	4.29	3.35	4.65	13.62
Mar/15	14	581	4.51	3.42	4.68	13.65
May/15	23	1442	4.17	3.27	4.41	12.8
Jul/15	31	719	4.14	3.24	4.42	12.75
Sep/15	21	387	3.43	3.39	4.58	12.4
Nov/15	13	569	4.78	3.46	4.68	13.85
Jan/16	92	801	4.62	3.29	4.36	13.34
Mar/16	24	564	4.54	3.4	4.28	13.23
May/16	38	655	4.29	3.44	4.41	13.2
Average (mar - nov)	32	714.8				
St dev (mar - nov)	25.60	318.30				
Evolution: Mar/15 - Jan/15 (%)	12.5	25.82	-4.88	-2.05	-0.64	-0.22
Evolution: Mar/15 - Jul/15 (%)	121.43	23.75	-8.20	-5.26	-5.56	-6.59
Evolution: Mar/15 - Nov/15 (%)	-7.14	-2.07	5.99	1.17	0	1.47
Evolution: Mar/15 - Jan/16 (%)	557.14	37.87	2.44	-3.80	-6.84	-2.27
Evolution: Mar/15 - Mar/16 (%)	71.43	-2.93	0.67	-0.59	-8.55	-3.08
Evolution: Mar/15 - May/16 (%)	171.43	12.74	-4.88	0.59	-5.77	-3.30

Another factor considered in the changes made to the processes of the properties is the importance of the water trough at the exit of the milking area. As this is noted as the period of highest water consumption, which brings about an improvement in the body temperature of the animal, while also increasing the volume of water in the milk, as water accounts for around 85% of milk composition (Perissinotto et al., 2004).

Through the monitoring of property 2 via control charts, it was noted that a majority of the problems that led to increases in SCC stem from herd acquisition (March 2015), without any attention being given to animal livestock/breeding records and its genetics concerning mastitis. This is another identified factor, in regards to the procedures and milking area in the wet season, which lasts from November to January and alters both the SCC and the TBC.

The cleaning milking system with non-standard temperature and process, problems in post and pre-dipping, the lack of cleanliness in the equipment were some of the factors listed for the imbalance of the mean, and variability of TBC. Furthermore, the acquisition of animals with mastitis contributes to SCC. The SCC and TBC during the project, in the Case 2, was 714.8 thousand cells/ml (standard deviation in 318.30) and 32 thousand CFU/mL (standard deviation in 25.60), respectively, which become capable for TBC but It did not capable for SCC in most of months.

Producer 3, Case 3, does not follow up on production records history and, as a result of the change in the dairy, does not have data on milk properties (Table 8) until November, are from another dairy industry. Thus, results from tests (Table 8) carried out through the project established with the current dairy industry to whom the milk producer delivers the milk produced on the property. As in the case of other producers, this one also has the bad practice of holding the teat mechanical milking system cups at the end of the milking process pressing them down, in an attempt to increase the milk obtained from the cow, which can cause cracks in the cow teats, leading to mastitis. In addition, the milking time, which should be around 6 min, is increased to around 8 min (Freitas, 2015).

In addition to these changes in the milking process, another significant factor for the reduction of SCC is the strip cup test, where the first three jets of milk from each teat, directed in the cup, are discarded to verify clinical mastitis, which represents 30% of the incidence of mastitis disease. However, the test for subclinical mastitis, 70% of mastitis, is not performed. This disease generates economic impacts for milk producers, related to the quality and components of milk, in addition to demanding a lot of time from the producer in terms of its treatment. In all types of mastitis, there may be a reduction in production from the affected teat, as well as there may be a total deterioration of the tissue of the mammary gland, leading to the loss of the teat, where it is no longer able to produce milk. If the disease reaches an advanced stage and is not treated properly, the cow may be lost. Through the appropriate consideration of these points, the producer began to put into action the necessary activities, and as such achieved superior management results in the milking process (Tinoco, 2015).

Table 8. Milk indicators Case 3

Milk Collection	TBC (thousand CFU/mL)	SCC (thousand cells/mL)	Fat Content (% m/m)	Protein Content (% m/m)	Lactose Content (% m/m)	Total Solids Content (% m/m)
Sep/15	1	349				
Oct/15	1	220				
Nov/15	3	242				
Dec/15	11	1147	3.22	4.55	12.38	3.22
Jan/16	4	136	3.05	4.49	11.99	3.05
Feb/16	6	164	3.31	4.54	12.87	3.31
Mar/16	24	190	3.21	4.4	12.92	3.21
Average (sep 15 - may 16)	7.14	349.71				
St dev (sep 15 - may 16)	8.19	358.18				
Evolution: Dec/15 - Oct/15(%)	-90.9	-69.57				
Evolution: Dec/15 - May/16 (%)	118.18	-83.44	-0.31	-3.30	4.36	-0.31

The cleaning milking system with non-standard temperature and process, problems in post and pre-dipping, the lack of cleanliness in the equipment were some of the factors listed for the imbalance of the mean, and variability of TBC. Furthermore, the acquisition of animals with mastitis contributes to SCC. The SCC and TBC, in Case 3, during the project was 349.71 thousand cells/ml (standard deviation in 358.18) and 7.14 thousand CFU/ml (standard deviation in 8.19), respectively, which become capable to control after the implementation of the necessary actions. Besides taking into account that the milk produced, is still below NI 62 (MAPA, 2011), considered at the time of the project, 2015, as well as NI 76 (MAPA, 2018a), the current standard.

The evolution presented in the Cases 1, 2 and 3 (Tables 4, 7 and 8) both in TBC and in SCC, shows that through adequate training of producers, problems can be identified, and the correct course of action taken, which brings about improvements and reductions in the variability of the indicators.

In the analysis of property 4 (Table 9), in the case of the TBC indicator, the major cause of alterations in the process, which was responsible for 80% of the problems and sensitive to abrupt changes, is the sealing ring of the milk cooling tank. The sealing ring of the tank should be cleaned in the correct manner, as this is the perfect place for bacteria to proliferate and produced colonies. A rubber seal change was performed, along with the cleaning of the seal placement area, the other factor that does not allow for reductions in the TBC is the place where the cooling tank is situated, i.e., closed area, which does not allow for correct ventilation, resulting in improper operation of the cooling compressor. In addition, the organization of the storeroom and milking area, as in the removal of unnecessary items, as presented through the 5S methodology, set in order, shine, standardize, sustain the cycle, with the goal being to always maintain the work environment clean and organized.

Table 9. Case 4

Milk Collection	TBC (thousand CFU/mL)	SCC (thousand cells/mL)	Milk (L)
Nov/15	18	667	29.97
Dec/15	12	407	30.58
Jan/16	21	111	30.80
Feb/16	10	433	29.60
Mar/16	14	313	24.76
Apr/16	9	361	24.81
May/16	24	415	29.34
Jun/16	23	399	31.56
Jul/16	14	293	39.21
Aug/16	226	353	52.99
Sep/16	6	201	57.98
Oct/16	339	210	61.61
Nov/16	8	220	57.08
Average (nov 15 - nov 16)	55.69	337.15	
St dev (nov 15 - nov 16)	5.60	146.72	
Evolution Feb/16 - Nov/16 (%)	80	54.04	1.26
Evolution Feb/16 - May/16 (%)	140	-4.16	-0.88
Evolution Feb/16 - Aug/16 (%)	2.16	-18.48	79.05
Evolution Feb/16 - Nov/16 (%)	-20	-49.19	92.86

Some factors were presented for improving the management of the SCC, as in control of the animal genetics, as these are generated via insemination. Another factor was the training of those responsible for milking to improve the process of pre- and post-dipping, as these products are purchased, but with staff turnover, trained staff leave, and purchases are no longer made.

The cleaning milking system with non-standard temperature and process, problems in post and pre-dipping, the lack of cleanliness in the equipment were some of the factors listed for the imbalance of the mean, and variability of TBC. Furthermore, the acquisition of animals with mastitis contributes to SCC. The SCC and TBC, in Case 4, during the project was 337.15 thousand cells/ml (standard deviation in 146.72) and 55.69 thousand CFU/mL (standard deviation in 5.60), respectively, which become capable to control after the implementation of the necessary actions. Besides taking into account that the milk produced, is still below NI 62 (MAPA, 2011), considered at the time of the project, 2015, as well as NI 76 (MAPA, 2018a), the current standard.

Producer 5, Case 5, implemented the mechanical milking system with support from the project, while taking into consideration the part of controlling the teat cup pressure, thus not producing stress on the teats of the animal. However, in the acquisition of the herd, some animals were later found as having mastitis problems, which presented SCC problems (Table 10). In the area of TBC, as the milking operative was still undergoing training, temperature control for the inlet temperature to clean the milking system, around 75 °C, and the water temperature at the outlet that needs to be around 45 °C was not optimized.

Table 10. Case 5

Milk Collection	TBC (thousand CFU/mL)	SCC (thousand cells/mL)
Jul/16	58	1037
Aug/16	32	637
Sep/16	23	320
Oct/16	83	330
Nov/16	-	551
Average (jul 16 - nov 16)	49	459.5
St dev (jul 16 - nov 16)	27.09	159.28
Evolution Aug/16 - Jul/16 (%)	81.25	-99.83
Evolution Aug/16 - Nov/16 (%)	-	1.621.87

The cleaning milking system with non-standard temperature and process, problems in post and pre-dipping, the lack of cleanliness in the equipment were some of the factors listed for the imbalance of the mean, and variability of TBC. Furthermore, the acquisition of animals with mastitis contributes to SCC. The SCC and TBC, in Case 5, during the project was 459 thousand cells/mL (standard deviation in 159.28) and 49 thousand CFU/mL (standard deviation in 27.09), respectively, which become capable to control after the implementation of the necessary actions. Besides taking into account that the milk produced, is still below NI 62 (MAPA, 2011), considered at the time of the project, 2015, as well as NI 76 (MAPA, 2018a), the current standard.

Table 11. Case 6

Milk Collection	TBC (thousand CFU/mL)	SCC (thousand cells/mL)	Fat Content (% m/m)	Protein Content (% m/m)	Total Solids Content (% m/m)	Milk (L)
Aug/18	14	247	4.65	3.53	9.09	3.86
Sep/18	6	345	4.44	3.54	9.01	
Oct/18	24	296	4.3	3.51	9.04	3.21
Nov/18	4	398	4.41	3.57	8.93	4.50
Dec/18	55	354	3.79	3.29	8.67	4.29
Jan/19	372	255	3.86	3.39	8.81	4.29
Feb/19	786	229	3.76	3.19	8.59	5.02
Mar/19	165	418	3.98	3.26	8.75	4.29
Apr/19	94	187	4.3	3.52	9	3.89
May/19	167	154	4.3	3.52	9	4.69
Jun/19	188	181	4.6	3.45	8.89	5.58
Jul/19	277	126	4.58	3.48	8.99	5.51
Aug/19	195	157	4.81	3.52	8.75	6.41
Sep/19	56	175	4.49	3.31	8.93	5.37
Average (aug 18 - sep 19)	171.64	251.57				
St dev (aug 18 - sep 19)	208.43	96.13				
Evolution Jan/19 - Mar/19 (%)	-55.64	63.92	3.11	-3.83	-0.68	0
Evolution Jan/19 - Abr/19 (%)	-74.73	-26.66	11.39	3.83	2.15	-9.32
Evolution Jan/19 - Jul/19 (%)	-25.53	-50.58	18.65	2.65	2.04	28.38
Evolution Jan/19 - Sep/19 (%)	-84.94	-31.37	16.32	-2.35	1.36	25.19

As for the TBC on the part of producer 6 (Table 11), Case 6, problems were listed in the milking system, which was misaligned, forming a colony of bacteria in the milk conduction pipe, problems were also found in regard to the inlet and outlet temperature concerning the hygiene of the milking system. Additionally noted were problems in post and pre-dipping, as found in the absence of the strip cup procedure, equipment that allows for the rapid identification of clinical mastitis. Conversely, the producer at times calls upon an assistant to take on tasks in moments of absence; during which times, the cleaning and hygiene procedures are not carried out for the milking system nor for the animals (Oliveira, 2020). This type of process contributes for the high TBC, 171,64 (thousand CFU/mL) (standard deviation 208,43)

This producer works with organic products and sees in the project the support needed to change the treatments used in the field from allopathy to homeopathy, thus being able to sell milk and its by-products (cheese, butter) as organic products. But because the unbalance treatments.

In the same format as presented for producer 1, 2, 3, 4 and 6, the producers invested in the insemination segment to improve the herd, but these do not manage the genetics of the animals that will be selected for milking, thus raising the issue of animal records, namely the introduction of mastitis. The genetic evaluation for resistance to mastitis (SCC) has a high genetic correlation, due to factors linked to greater resistance in animals such as udder health (Urioste et al., 2011).

As a result, in Case 1, the owner was willing to accept the improvement proposals in order to improve the milking process on the property. Here the owner understood that process control could be used with the industry to raise the price of the milk, as this constituted milk as possessing an advantage when subjected to quality control. However, when these practices were actually put in place following the manual mounted with their participation, difficulties began to appear, mainly due to the changing of their current milking process, associated with the increase in the process time. However, through visits and presentations showing the advantages and importance of some techniques concerning the herd and for milk quality, the owner accepted a portion of the changes.

The full acceptance of changes came after the owner saw the results for the first milk test, where a 50% reduction in the indexes for TBC and SCC was noted, besides the partnership made with the owner's wife, who started to help with the control indicators.

Property 2 is the one that presented greatest ease in the implementation of practices, as the owner is very concerned with the quality of the milk, due to its influence on its profitability. However, difficulty was noted in the part of the process that affects the quality of the herd, which can be verified through the property SCC index, which remains above the value desired by the collection company.

CURRENT CHALLENGES FACING THE ORGANIZATION

Producers 3 and 5 were trained and present good knowledge concerning the implemented improvements, although they state that only prior proposals will be implemented, without implementing any scheme of continued improvement. On property 4, in terms of SCC, one of the requests was the implementation of the California Mastitis Test (CMT) at least once every two weeks, however, the pilot test was executed and employees trained, but without continuity of the process, and as the property is within the indicators of NI 62, the owner declined to continue. Additionally, the rotation of employees was checked, which creates the continual need for training and the eventual hampering in the advancement of processes.

Another difficulty is presented in the lack of financing on the part of the University and Dairy Industry, which do not make financing available for projects that do not work directly with disadvantaged people that need a budget for transportation, food and insurance for the students and teachers involved.

In the TBC indicator, only producer 6 could not maintain the average below 100 thousand CFU/mL, a value compatible with the level of control in Europe, which is lower than Brazil NI 62 (MAPA, 2011), more than this, maintains standard deviation in low variability. This shows better process control, but require improvements. These shows the efficiency of training with the producers and effectiveness of the project. In the SCC indicator, producer 2 was unable to meet the indicator Brazil NI 62 (MAPA, 2011) for most of the year, due to herd acquisition with a history of chronic mastitis, which shows the relevance about the evaluation of mastitis in the purchase. Other fact about SCC, the rainy season, December to March, producers 1, 2 and 3, presented mud problems on the way to milking and in the exit of the corral, causing dirt in the fourth breast, as well as affecting the hoof of cows.

SOLUTIONS AND RECOMMENDATIONS

World agricultural production determines high quality and safe products for human consumption, and is an ongoing process related to all members of the dairy chain. This process begins at milking and is intrinsically applied in order to guarantee the best conditions for achieving a product of excellent quality.

The improvement in the hygienic quality of milk is accomplished by means of a simple process, which presents prompt results that start with the improvement of milking practices in order to avoid milk contamination, while maintaining the storage tanks under conditions of perfect hygiene.

The application of the Six Sigma methodology, with the aid of the DMAIC method for improving processes in dairy production, led to a reduction in SCC indicators at properties 3, 4 and 6, in addition to an increase in protein and fat of around 5%. These indicators generate improvements in quality and yield from the raw material used by the industries in the production of dairy products. On properties 1, 2 and 3, even with not reaching the IN 62 requirements, better knowledge on the part of the producer concerning factors that lead to increase in SCC was demonstrated, along with those measures that should be put into practice for both prevention and correction.

Through such a setup, there was an increase in food safety due to greater control of bacteria and diseases in cattle, such as the prevention of Trypanosomiasis, where some producers stopped using products that stimulate oxytocin and allocated individual syringes for each animal when applying medication. Verified also was that SCC can show significant increases, even with follow-ups over the long term, in the process and with good results, as occurred on property 3. Hence, greater control of the processes is needed, as in the organization of milking batches and the maintenance of rest areas in a manner that avoids disease and its spread. Through this analysis, the apparent need to integrate the components of the chain was understood, in order to improve the management of those processes associated with it. Thus, to take advantage of the moment, Brazil needs to expand its production and acquirement (milk collection), through increasing cow productivity as well as improving milk quality, while targeting both the domestic and foreign markets.

One of the paths presented in this study becomes the initiation of the process for improving the milk chain, through its application of best practices in dairy production with a reduction in TBC and SCC, generating through such an increase in the sales value of the product and a reduction in loss. The implementation of such measures leads to improvements in earnings of the families involved in the process, along with improvements in milk traceability through the increase of indicators for the collection of the product, improvements in the quality of dairy products and greater reliability and trustworthiness of the raw material along the milk chain. However, it is still difficult to control procedures at rainy season about cleaning of the milking system and mastitis, largely due to the care on the herd's path to milking and the exit of the corral.

REFERENCES

Benbrook, C. M., Butler, G., Latif, M. A., Leifert, C., & Davis, D. R. (2013). Organic Production Enhances Milk Nutritional Quality by Shifting Fatty Acid Composition: A United States–Wide, 18-Month Study. *PLoS One*, *8*(12), e82429. doi:10.1371/journal.pone.0082429 PMID:24349282

Berry, D. P., O'Brien, B., O'Callaghan, E. J., Sullivan, K. O., & Meaney, W. J. (2006). Temporal Trends in Bulk Tank Somatic Cell Count and Total Bacterial Count in Irish Dairy Herds During the Past Decade. *Journal of Dairy Science*, *89*(10), 4083–4093. doi:10.3168/jds.S0022-0302(06)72453-5 PMID:16960086

Boor, K. J., Wiedmann, M., Murphy, S., & Alcaine, S. (2017). A 100-Year Review: Microbiology and safety of milk handling. *Journal of Dairy Science*, *100*(12), 9933–9951. doi:10.3168/jds.2017-12969 PMID:29153181

Brito, J. R. F., & Dias, J. C. (1998). A qualidade do leite (EMBRAPA; TORTUGA (ed.)). EMBRAPA; TORTUGA.

Brito, M. A. V. P. (2013). *Identificando fontes e causas de alta contagem bacteriana total do leite do tanque*. Embrapa Gado de Leite.

Deepa, O. S., & Krishnan, S. M. (2021). Impact of Six Sigma in Dairy Production for Enhancing the Quality. In K. Tiwari & M. S. Khan (Eds.), *Lecture Notes in Mechanical Engineering* (pp. 339–350)., doi:10.1007/978-981-33-4320-7_31

EMBRAPA. (2021). Empresa Brasileira de Pesquisa Agropecuária. *Anuário leite* 2021.

Freitas, L. F. (2015). *Aplicação do método dmaic no acompanhamento da CCS para qualidade do leite*. UFU.

Freitas, L. F., Pião, R. C. S., & Lourenzani, A. E. B. S. (2021). Objetivo de desenvolvimento sustentável 12 no contexto de perdas e desperdicios na cadeia do leite. XIII ENGEMA - *Encontro Internacional Sobre Gestão Empresarial e Meio Ambiente*, 2359–1048.

Gigante, M. L., & Costa, M. R. (2008). Influência das células somáticas nas propriedades tecnológicas do leite e derivados. II Congresso Brasileiro de Qualidade Do Leite. *CCS Gráfica e Editora*, 1, 161–174.

Guerreiro, P. K., Machado, M. R. F., Braga, G. C., Gasparino, E., & Franzener, A. S. M. (2005). Qualidade microbiológica de leite em função de técnicas profiláticas no manejo de produção. *Ciênc. Agrotec. Lavras*, *29*(1), 216–222.

Holzapfel, W. H., Todorov, S. D., & Cogan, T. M. (2022). History of Dairy Bacteriology. In Encyclopedia of Dairy Sciences (3 rd, Volu, pp. 123–132). Elsevier. doi:10.1016/B978-0-08-100596-5.23031-X

Lopes, J. D., & Tinoco, C. M. M. (2016). Aplicação da metodologia Seis Sigma em uma indústria de cosméticos localizada no Brasil. *XXXVI Encontro Nacional de Engenharia de Producao* - ENEGEP.

Lorenzetti, D. K. (2006). *Influência do tempo e da temperatura no desenvolvimento de microrganismos psicrotróficos no leite cru de dois estados da região sul.* Universidade Federal do Paraná.

Machado, P. F., Pereira, A. R., & Sarríes, G. A. (2000). Composição do Leite de Tanques de Rebanhos Brasileiros Distribuídos Segundo sua Contagem de Células Somáticas. *Revista Brasileira de Zootecnia*, *29*(6), 1883–1886. doi:10.1590/S1516-35982000000600038

MAPA. (2011). Instrução Normativa No 62..

MAPA. (2018a). INSTRUÇÃO .

MAPA. (2018b). INSTRUÇÃO NORMATIVA No 77.

Marioto, L. R. M., Daniel, G. C., Gonzaga, N., Mareze, J., Tamanini, R., & Beloti, V. (2020). Potencial deteriorante da microbiota mesófila, psicrotrófica, termodúrica e esporulada do leite cru. *Ciência Animal Brasileira*, *21*. Advance online publication. doi:10.1590/1809-6891v21e-44034

Moatsou, G., & Moschopoulou, E. (2014). Dairy Microbiology and Biochemistry. In B. Ozer & G. Akdemir-Evrendilek (Eds.), Dairy Microbiology and Biochemistry (1 st Editi). CRC Press. doi:10.1201/b17297

Morini, R. M., & Braga, C. A. S. B. (2009). Manejo consultoria agropecuária: Qualidade do leite e manejo de ordenha. Universidade Federal de Goiás, OECD/FAO. (2020). Dairy and dairy products. In OECD-FAO agricultural outlook 2020-2029.

OECD/FAO. (2021). Dairy and dairy products. In OECD-FAO agricultural outlook 2021-2030.

OECD/FAO. (2022). Dairy and dairy products. In OECD-FAO agricultural outlook 2022-2031.

Oliveira, M. R. (2020). *Estudo e aplicação de ferramentas da qualidade no processo de ordenha em uma pequena propriedade rural.* UFG.

Perin, L. M., Pereira, J. G., Bersot, L. S., & Nero, L. A. (2019). The Microbiology of Raw Milk. In *Raw Milk* (pp. 45–64). Elsevier., doi:10.1016/B978-0-12-810530-6.00003-1

Perissinotto, M., Moura, D. J., Silva, I. J. O., & Matarazzo, S. V. (2004). Influência do ambiente Influência do ambiente na ingestão ingestão de água de água por vacas leiteir acas leiteiras. *Revista Brasileira de Engenharia Agrícola e Ambiental*, *9*(2).

Pinto, C. L. O., Martins, M. L., Campos, N. L. F., Martins, A. D. O., Benevenuto, W. C. A. N., Martins, E. M. F., Silva, H. L. A., & Cruz, A. G. (2019). Microbiologia do leite cru. In Elsevier (Ed.), Microbiologia, higiene e controle de qualidade no processamento de leites e derivados (Coleção Lá, pp. 5–42).

Santos, M. V. (2000). *Contagem de células somáticas e qualidade de queijos*. Milkpoint.

Santos, M. V., & Botaro, B. (2008). *A mastite e os outros fatores que afetam a CCS*. Milkpoint.

Silva, J. C. (2015). *Efeito do tipo de ordenha e ambiente sobre a qualidade do leite cru com base na contagem de células somáticas na mesorregião do sul goiano*. Universidade Federal de Uberlândia. doi:10.14393/ufu.di.2015.47

Thiollent, M., & Silva, G. D. O. (2007). The Use of Action Research in the Management of Environmental Problems. *Reciis*, *1*(1), 91–98. doi:10.3395/reciis.v1i1.37en

Tinoco, C. M., & Freitas, L. C. (2014). Comportamento da cadeia de lácteos brasileira Behavior of the Brazilian dairy chain. *IV Congresso Brasileiro De Engenharia De Produção*.

Tinoco, C. M. M. (2015). *Oportunidades de melhorias na produção leiteira: trabalho em propriedades familiares de Ituiutaba e região*. Milkpoint.

Tinoco, C. M. M., Mendonca, L. J., & Bortolin, J. P. S. (2017). *Aplicação de ferramentas da qualidade em propriedades produtoras de leite. XXIV SIMPEP*. Simpósio De Engenharia De Produção.

Tinoco, C. M. M., Ohishi, T., Lopes, J. D., Freitas, L. C., & Oliveira, P. A. B. A. (2015). Aplicação da metodologia Seis Sigma na melhoria do controle de processos na ordenha em propriedades familiares de Ituiutaba/MG. *V Congresso Brasileiro De Engenharia De Produção*.

Urioste, J. I., Franzén, J., Windig, J. J., & Strandberg, E. (2011). Genetic Variability of Alternative Somatic Cell Count Traits and their Relationship with Clinical and Subclinical Mastitis. *Interbull Bulletin*, *44*, 26–29.

KEY TERMS AND DEFINITIONS

Post-Dipping: Disinfection after milking through chemical protection in order to prevent new infections in the interval between milking sessions, especially contagious mastitis.

Pre-Dipping: A method for disinfecting milking teats through the application of an antiseptic product, usually by immersion of the teats.

Psychrotrophic: These are bacteria capable of proliferating at temperatures below 7°C, and are the main bacteria responsible for the deterioration of refrigerated raw milk and its derivatives (cheese, yogurt) (Holzapfel et al., 2022).

SCC: Somatic Cell Count - cells of the immune system and the epithelium of the mammary gland of cows, which when present in milk, show the level of mastitis in cows, that is, inflammation in the udders normally caused by bacterial infection.

Six Sigma: Methodology used in the systematic improvement of the process in favor of reducing the variability of defects.

TBC: Total Bacteria Count - bacterial contamination in milk normally caused by dust/dirt during the milking process, during collection, which may see an increased volume of bacteria in the milk due to lack of temperature control, which must be below 4 °C.

Trypanosomiasis: Trypanosomiasis or bovine trypanosomiasis, a disease caused by the protozoan Trypanosoma vivax, which can be transmitted by mosquitoes or contaminated needles. Sick animals develop weight loss, anemia, decrease in milk production, reproductive problems, neurological changes and death, resulting in economic losses.

Chapter 2
The Role of Milk Quality in Improvement of Dairy Production

Luiz Carlos Roma Júnior
https://orcid.org/0000-0002-0019-2538
Animal Science Institute, Brazil

Lívia Castelani
https://orcid.org/0000-0001-7801-3256
Animal Science Institute, Brazil

Thatiane Mendes Mitsunaga
São Paulo University, Brazil

EXECUTIVE SUMMARY

Milk is a complete feed, composed of carbohydrates, proteins, vitamins, and minerals of great importance for human nutrition, in addition to playing a representative role in aspect economy. However, milk production is very affected by market variations, such as production costs, influences of exportation activities of the product and international market. To avoid these challenges, every component in the production chain must focus mainly on the sustainability and profitability of dairy farming, especially the producer, the first link in the chain. In this chapter, the relationship between indexes and different areas of the milk production chain is presented, that could generate information able to aid the sustainability and profitability of dairy farms. Also, it will discuss each of these components and its relation to obtaining the maximum amount of information possible to contribute to the dairy industry.

DOI: 10.4018/978-1-6684-5472-5.ch002

INTRODUCTION

Animal Science Institute of Nova Odessa is a research institution operating in São Paulo state, in which the main objective is to develop and apply research focused on sustainable agriculture, for local farmers. However, through partnership with other commercial and research institutes, the results obtained in research projects are extensively used in the country. Another important contribution to agriculture and husbandry is to transmit knowledge and technology to farmers, in order to implement these results for improvement. The Institute has research projects in dairy and beef cattle, sheep farming, swine, poultry and aquaculture.

Our team belongs to The Milk Quality Laboratory, with research focused on improvement of milk quality in production systems, using data from herds for diagnosis of potential problems on farms.

CASE DESCRIPTION

Milk is a product obtained from complete and uninterrupted milking under hygienic conditions of healthy, well-fed, and resting dairy cows. Milk from other animals must be named according to the species they belong. It is considered an easily accessible, complete and one of the most nutritious foods in the world due to its significant amounts of water, proteins and high quality fats, in addition to minerals such as calcium, sodium, potassium, selenium, magnesium and vitamins such as riboflavin, vitamin A and vitamin B12. Due to its complex composition, it is considered of great importance for the health and immunity of neonates, as well as of adult humans.

The chemical classification of milk can be stratified into more than 50 chemical compounds of varying importance, thus defining its quality. Quality is one of the main points for the production chain to achieve sustainability and profitability. In this context, two scenarios can be found: the first scenario is focused on the dairy farm, in which quality can be used for milk production monitoring and planning in order to ensure commercialization and profitability of the activity. In addition to the dairy farm scenario, the industry scenario is also highlighted, which has milk quality as a challenge and incentive for production, industrial yield and variety of products according to the consumer market and microbiological safety.

Involved in both scenarios, legislation and resolutions stand out, which establish minimum acceptable standards of quality and good practices for obtaining the raw material, processing and marketing of the final product. The establishment of these standards aims to guarantee the marketing of safe products for the national and international market.

In terms of legislation, greater progress is observed in countries where dairy farming is more developed. In Brazil, greater progress was observed between 2002 and 2018, with a series of Normative Instructions (INs), which have been improved over these years as a function of the milk quality monitored in Brazil. INs originated from the National Milk Quality Improvement Program (PNMQL), with an initiative by the Ministry of Agriculture, Livestock and Supply (MAPA), support from teaching and research institutes and discussions with the different links in the production chain.

Based on standards present and updated in normative instructions, milk quality, which is determined through laboratory analysis, can be used as a monitoring, evaluation and planning tool for both producer and industry.

Milk Quality Components

Milk is considered a perfect food of nature, as it is linked to the final maternal investment for the survival of the progeny and guarantee of the genetic continuity of different species. Thus, it is necessary to contain a series of components that will guarantee a complete diet for the supplementation of essential nutrients for neonates. Therefore, milk is the result of the combination of water and other solid elements such as proteins, fats, carbohydrates, vitamins and minerals, among which calcium stands out, which is of great importance for human health.

Water

Water is the largest-volume component of bovine milk, with approximately 87% of the total milk (Foroutan et al., 2019). As milk is the primary source of nutrition for mammals, its water concentration is primarily responsible for the hydration of neonates. The factor that most influences the water concentration in milk is the synthesis of lactose, one of the most representative carbohydrates of milk. Therefore, the amount of water in milk can vary, depending on factors such as the concentration of other milk components.

Carbohydrates

The main carbohydrate in milk is lactose, which is produced by the epithelial cells of the mammary gland from glucose, being the main source of energy for neonates. Lactose is a disaccharide composed of glucose and galactose, synthesized by the lactose synthase enzyme, which consists of two protein enzymes, galactosyl transferase and a-lactalbumin, the latter being one of the main proteins in whey (Stelwagen, 2022). This lactose synthesis process is unique to the mammary gland. Lactose plays a fundamental role in milk synthesis, as it is the main osmotic factor during synthesis. After formation, lactose is release through vesicles, called secretory vesicles, which fuse with the apical membrane to release the contents into the lumen of the mammary gland alveoli (Stelwagen, 2022). However, lactose makes the contents of the vesicle hypertonic in relation to the external environment. In this way, water is diffused into these vesicles until it reaches osmotic balance. Therefore, the presence of lactose controls the milk volume produced through the water activity in order to balance the osmotic pressure in the mammary gland. This osmotic control is what gives lactose the characteristics of one of the elements present in milk, with few variations. Thus, the amount of water in the milk and, consequently, the milk volume produced depends on the amount of lactose secreted in the mammary gland, usually at concentration of approximately 5% (4.7 to 5.2%) in bovine milk (Foroutan et al., 2019).

Proteins

Proteins represent between 3% and 4% of solids found in milk. Milk protein originates from the process of absorption of amino acids present in the diet by the intestine, which come mostly from the microbial protein formed in the rumen and from the dietary protein not degraded in the rumen, available in the intestine. Once in the bloodstream, these amino acids are extracted by the secretory cells present in the mammary gland, for the process of synthesizing milk protein, similar to other tissues (Stelwagen, 2022b). There are several types of protein in milk, casein being the main one, with high nutritional quality, essential in the manufacture of cheeses. Casein, when produced by secretory cells of the mammary gland, is secreted in the form of micelles, which are clusters of several casein molecules along with calcium, phosphorus and other salts.

Fat

Fat is an important milk constituent, being the largest in terms of energy. It is responsible for conferring characteristics to dairy products, for their physical and sensory properties. Its structure is in the form of globules suspended in water and is almost entirely composed of triglycerides (above 95%) (Bauman et al., 2022). These triglycerides are synthesized in the cells of the mammary gland from fatty acids extracted from the bloodstream, from three different sources, namely: from acetate and butyrate, which are fatty acids produced in the rumen; from nutrients consumed by the cow via nutrition and that are absorbed in the intestine, and from fatty acids mobilized from body reserves.

Ash

The mineral composition (ash) of milk is relatively constant: 0.7 to 0.8%. Among minerals, calcium and phosphorus stand out, and to a lesser extent magnesium. Calcium and phosphorus in milk are highly available, in part because they are associated with casein. Therefore, milk is considered an excellent source of calcium essential for the development of neonates, in addition to the maintenance of bone integrity in adults.

Vitamins

Milk is also an important source of vitamins, some are associated with fat (A, D, E and K), while others are associated with the aqueous part (B complex and vitamin C). More than ten different B vitamins are found in milk. However, with the exception of vitamin B2 (riboflavin), the others are found in small amounts.

Somatic Cells

Somatic cells are mainly composed of leukocytes, which are the body's defense cells, and some cells of the glandular epithelium (Li et al., 2014). Somatic cell count (SCC) in milk is used as an indicator of udder health and increases in inflammatory and/or infectious processes of the mammary gland is described as mastitis. Although it can appear as a result of physiological disorders or local trauma, infectious mastitis is the most common and of greatest economic importance for milk producers and industry (Gonçalves et al., 2018). SCC in milk is influenced by several factors, such as species, breed, genetics, milk production level, lactation stage, stress, type of infectious agent, sampling time and procedures, as well as individual and environmental factors and management practices. Despite this, SCC is widely used as an indicator of milk quality.

Standard Plate Count (SPC)

This is one of the main methods used to characterize the microbiological quality of milk, also known as total bacterial count (TBC). This parameter is used as an indicator of the hygienic milk quality, in addition to the health of the mammary glands and is based on the quantification of the total number of viable aerobic bacteria in raw milk, including spoilage bacteria and those capable of causing diseases in consumers. SPC can be estimated by traditional methodology, where an aliquot of milk is distributed in plates with culture medium and incubated at 36 °C for 48 hours, and then colonies that grew are counted as a function of the sample volume.

In this methodology, the results are expressed in colony forming units per milliliter of milk (CFU/mL). SPC can also be performed in automated equipment, using flow cytometry as methodology. These procedures estimate the individual bacteria count (IBC) present in the sample, and the results are expressed by number of bacteria in each milliliter of milk (IBC/mL of milk).

With the presentation of the main milk quality components that influence the dairy production chain, the information that can be extracted from these components according to the possible different scenarios will be described below. Regardless of scenario, it is important that the production chain seeks quality through the production efficiency and the promotion of more and better products for the consumer market.

Scenario: Dairy Farm

When approaching the dairy farm-oriented scenario, milk quality includes more factors than just monitoring component data. Scientific research present data that are transformed into information, which are later used by producers as auxiliary tools to maximize production and profitability of the dairy activity. This research relates milk quality to several production factors such as genetics, health, management, well-being, nutrition, milking, human and financial resources, among others (Ruegg, 2017; Bradley, 2002; Cha et al., 2014; Britt et al., 2021). However, which information can be obtained by determining the main milk quality components and how can these results help the producer?

Regarding milk fat, several aspects must be considered, including genetics, management, and production system. Milk fat formation, secretion and profile are highly influenced by the nutrition of cows, being able to decrease in up to 50% of its content or even increase approximately 5% through diet (Wiking et al., 2006). Roughage and the presence of fibers are the diet components responsible for milk fat. This relationship is based on the formation of a wide variety of short- and long-chain fatty acids from the ruminal fermentation of fibrous carbohydrates by rumen microorganisms (Barros, 2001). Acetate and beta-hydroxybutyrate are obtained from this fermentation, captured by the secretory cells of the mammary gland for the synthesis of milk fat, establishing the relationship in which the higher the diet fiber content, the greater the chances of increasing the milk fat content. Therefore, if fat is related to roughage availability and presence of effective fiber in the diet, it is possible to manipulate the milk fat content in different ways, such as altering the concentrate: roughage ratio in confined animals, or modifying the period and range of paddock occupation by animals in a rotational grazing system.

Regarding the effective fiber supply and roughage quality, one of the main factors that affect the rumen fermentation efficiency is the particle size offered to animals, whether from silage or fresh chopped forage. These aspects can influence the use of fibrous carbohydrates by rumen microorganisms, increasing or decreasing the fat content in milk. Considering the diet concentrate, there are several aspects that must be observed, such as type of concentrate supplied and its ingredients, type of grain processing, use of fats and buffering or alkalizing additives.

Another factor related to the fat content is linked to the way in which diet is offered, such as the supply of a mixed diet composed of concentrate and roughage, diets in which the concentrate is supplied separately from the roughage, or even supplying the concentrate distributed over the roughage in feeding troughs (Peres, 2001). These different forms of diet supply interfere with consumption, nutrient availability and rumination, which directly influences the production of fatty acids that will be metabolized to produce milk fat by the mammary gland.

Still on the nutritional management and its effect on the milk fat content, an impacting factor in the consumption of animals is the dominance, in which dominant animals arrive first in the trough and, in diets in which the concentrate is supplied over the roughage, these animals consume more concentrate than the others, interfering with the correct milk fat evaluation, considering all lactating cows. One solution is the distribution of diet with complete mixing, separating dominant animals and dominated animals in different troughs, resulting in less competition for space in the trough.

Fat content can also be influenced by the types of grains that are offered and how they are processed. Another important point is the use of oils and fats in the diet, since the amount offered can result in changes in rumen fermentation, and consequently in the milk fat content (Peres, 2001).

The suckling management in milking with the presence of the calf at foot can also interfere with milk fat. There are three milking management alternatives with calf at foot: 1) milking is carried out with three teats, leaving an exclusive teat for the calf; 2) milking of all four teats is carried out, releasing the cow in the same place as the calf at the end of milking; 3) finally, complete milking, without the presence of the calf, with suckling through bottles and buckets. These managements consider the possibility of a portion of fat that is not milked when there is no complete milking, with part of the milk fat remaining in the final portion of the four teats for the calf (last milking portion), as a function of the fat density in relation to the other components in the residual milk, which can reduce the fat content in the tank.

Furthermore, the use of oxytocin in milk production systems of zebu cows to promote faster milk ejection should be considered (Lammoglia et al, 2016) . Some studies have shown changes in the fat content of cows that received oxytocin in the long term, since it increases the viability of milk-secreting cells with consequent increase in production. However, caution is important with this practice, since oxytocin is applied by needle puncture, which can stress the animal and affect its well-being and health.

Considering milking management, the milking frequency also influences the milk fat content and profile. Between milking procedures, the milk stored in the cisterns, alveoli and ducts can undergo alterations in relation to the fat profile: when milking regularly twice a day, milk has less storage in the cistern and higher content of fat and free fatty acids, important for industrial production.

Milk protein has great relevance for the industry. Milk protein is one of the most valued components in terms of diet balance and feeding management that seek to maximize milk production and protein content. Many studies have been aimed at increasing protein content. However, it is currently known that the best way to increase protein content is through genetic improvement with the use of specific dairy breeds with characteristics of high protein concentration. Furthermore, protein is one of the most heritable genetic traits in breeding programs that select individuals within a breed with higher protein production. Therefore, it is necessary to consider the genetic information of the breed used in the production system for the correct interpretation of the protein results obtained in milk analyses.

In addition, the balance of diets with high levels of amino acids can contribute to increase of protein levels in milk. Some situations such as the type of diet offered, animals under heat stress, dominant animals, and access to the trough, in addition to sudden changes in the diet, can also interfere with milk protein levels.

In relation to the Solids not fat (SNF), this component is related to the consumption of dry matter by animals. Values lower than those recommended may indicate problems mainly related to malnutrition of animals. SNF can be directly or indirectly be influenced by management, nutrition, ingredients used, concentrate and roughage availability. In tropical climates, animal management must consider thermal comfort and increased water intake to ensure milk production and quality.

Another factor that can change the SNF content is related to sudden changes in the diets offered (Peres, 2001). Therefore, animals should be submitted to a period of adaptation to the new diet before this parameter is evaluated. Changing the diet without adaptive management can modify the rumen microbiota, with consequent changes in rumen fermentation and in the synthesis of milk components, including SNF. In addition, the need for dietary changes is related to the availability of ingredients at certain times of the year or regions. Aiming for profitable alternatives, some producers use co-products and by-products with restricted availability according to the seasonality, which results in the need to make changes in the diet of lactating cows.

In short, the analysis of milk composition can generate useful information about the nutritional routine of animals, the way of offering, leftovers, the storage of ingredients, the processing of each ingredient (roughage or concentrate), the ingestion behavior of animals, contributing to a more efficient production.

Regarding the health of the mammary gland, somatic cell count (SCC) can be used as a monitoring tool since animals with mastitis have high SCC. Several factors can affect the udder health of dairy cows and are usually related to poor hygiene practices in milking management and housing conditions (Blowey; Edmondson, 2010). Therefore, it is extremely important to identify the occurrence of mastitis, as it contributes to decision making for its control.

With that in mind, what information can be extracted from a SCC analysis? First, it is important to consider the sampling, whether it was carried out from one teat, from the animal or from the cooling tank. The SCC of teats is important to identify which udder quarters are affected with mastitis, especially in the clinical form. Considering the SCC of lactating cows, it is not possible to identify which teats are affected or which teats are contributing to increase the somatic cells count, but it is possible to obtain information on how many animals are affected with mastitis in the herd. Regarding SCC in the cooling tank, it is possible to obtain information on possible problems related to the control of somatic cells, considering the limits required by legislation and the occurrence of mastitis in the herd.

Therefore, frequent monitoring of these values throughout the herd, with the registration of all cases, is necessary for the establishment of a monitoring and control program for the occurrence of mastitis. It is up to the producer to observe the occurrence of clinical mastitis at the time of milking, together with the observation of the existence or not of changes in the SCC in the tank. In a scenario where the tank presents SCC values above the limit established by legislation, the producer must identify during milking which animals or lots are contributing to the increase in SCC, if there are handling problems in the milking parlor or the need for equipment maintenance. This procedure can help in the elaboration of a mastitis control and prevention plan, fundamental for the reduction of costs with treatment and loss of animals (Santos; Fonseca, 2007).

From the identification of variables that contribute to increase somatic cells count, several actions can be recommended such as the decision making in relation to the need and type of treatment (anti-inflammatory or antimicrobial), management alteration for milking line, early drying of animals with chronic mastitis, drying or inactivation of the affected teat, changes in the diet, among other alternatives that make mastitis control possible.

Finally, widely underestimated by many producers, the Standard Plate Count (SPC) contributes to the monitoring of the hygienic quality in the process of obtaining raw milk, in addition to the storage temperature. This variable is not directly linked to the reduction in milk production, as is the case with SCC. However, high SPC can compromise the entire production stored in the refrigeration tank if milk shows changes in acidity, causing losses with the rejection of the total milk volume.

High SPC mainly indicates failures in milking hygiene procedures and equipment maintenance (Ahmed, 2020). Some situations increase the SPC values in the tank, such as the milking of cows with dirty teats, the occurrence of environmental mastitis, failures in the cleaning of milking equipment, deficiency of rapid milk cooling and the quality of the water used for cleaning.

Therefore, the producer should identify the source of the problem, checking the place where animals remain, the milking parlor and equipment, the cleaning of the udder, the cooling tank, the personal hygiene of the employees, among others. In the cleaning and sanitization process, the water used must be potable or provided by water treatment systems, ensuring its quality through periodical analysis. It is also important to use effective products for cleaning and sanitizing the entire process. Also, the importance of changing milk hoses, teat cup rubbers and milking vacuum maintenance is highlighted, as these are sources of microbial colonization.

For the SPC control, it is also important to guarantee the immediate milk refrigeration, as it is essential to reduce the bacterial growth rate during its storage. In addition, it must be ensured that the milk storage room is isolated, without access by unauthorized persons and animals. Finally, the producer must pay attention to the maintenance of the cooling tank to ensure the ideal temperature for storing the milk until the moment of its removal by the dairy industry.

Far beyond the individual interpretation of each of these components, the producer must consider the combination of them. For example, the relationship between fat and protein to characterize the types of diets (acidogenic or ketogenic), indicate some food management characteristics or even heat stress or other management factors that can interfere with the content of these components. However, the best example of the relationship between components that can result in different information for the producer is the combination of the milk protein content with the milk urea nitrogen (MUN). Although the levels of milk urea nitrogen are not included in the legislation, this analysis is essential to match the protein content and describe the balance between energy and protein in the diet. Based on data from this relationship, it is possible to identify cases of nutrition problems, energy or protein deficiencies, in addition to indicating some excesses that can result in losses in terms of high cost of the diet, or even in the reproduction of animals. Therefore, the relationships among milk components bring important benefits to both producers and technicians in the planning of actions aimed at improving milk production and quality.

In short, the results of milk quality analyses can indicate efficiency or failures in nutritional management, in good hygiene practices, sanitary problems and mastitis control, in the breeding program, among others. All this information contributes to the elaboration and planning of actions aimed at increasing production and improving the quality of milk to be marketed.

Figure 1. Main information extracted from milk quality for milk producers

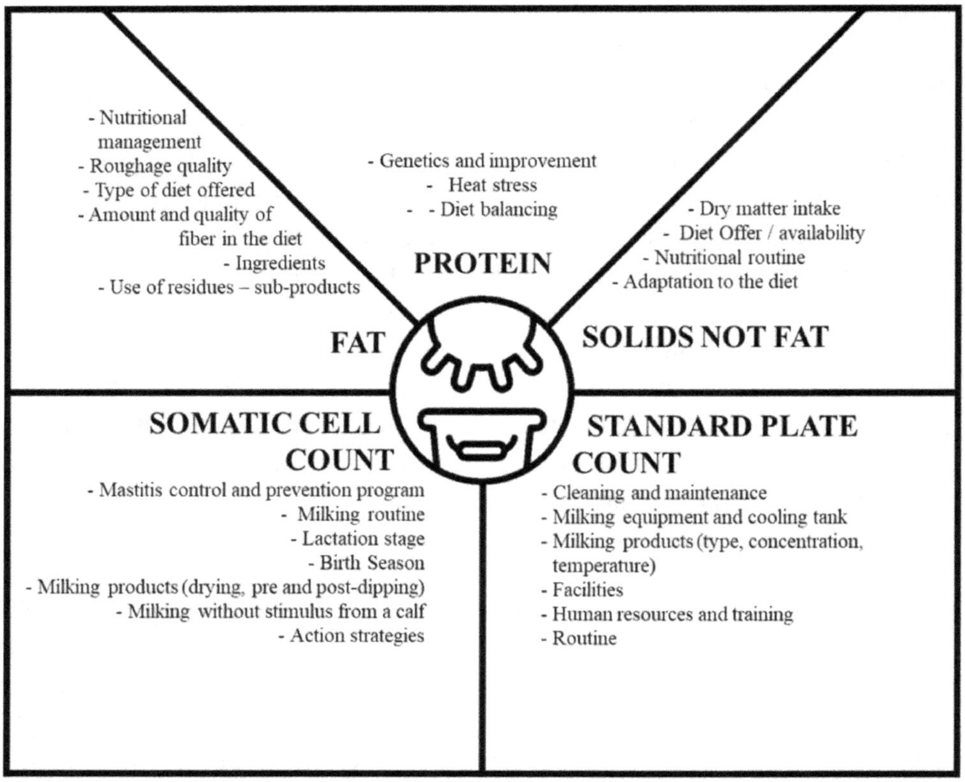

Scenario: Industry

When discussing the industry scenario, two points should be highlighted: milk composition, related to industrial performance and the quality and variety of dairy products and microbiological quality, related to shelf life and control of pathogenic microorganisms.

Regarding milk composition, solids not fat has high relationship with the industrial yield of products such as powdered milk. Lactose, the main milk carbohydrate, is essential in the technology for obtaining milk products, such as in the manufacture of yogurts since it is used by lactic acid-producing microorganisms in the fermentation process. This process gives fermented milks a slightly acidic flavor and favors protein coagulation. The industry also uses lactose in the manufacture of milk jam, condensed milk and ice cream. However, lactose has no influence on the yield of cheese as it is eliminated along with the whey in its manufacturing process (Ribeiro et al., 2018).

Protein is the most economically important element in cheese production. Protein should be evaluated for both quantity and quality. The quality of milk protein determines its stability at high temperatures. Milk with low thermal stability causes problems in processing lines, with consequent damage to industries.

Milk fat is the component that varies the most depending on genetic, physiological, and environmental factors. The first pay-for-quality programs used this variable as one of the main constituents of the milk bonus. Despite the demand for low-fat derivatives by consumers, the industry still highly values this component, since it is crucial to improving the organoleptic characteristics of dairy products. Cheese and butter are good examples of the importance of high fat content to maintain yield and especially the texture of the final product. Fat is also important for the production of whipped cream, cream, ice cream, among others. During the milk centrifugation process, fat is separated in the form of cream from the other milk constituents, resulting in skimmed milk.

Regarding the microbiological milk quality, the industry uses SCC and SPC as a monitoring tool for spoilage and pathogenic microorganisms. High SCC in milk is related to udder health, with consequent reduction in the quality of raw milk, as it reduces the secretory activity of the mammary gland, affecting the volume and composition of the milk produced. The increase in SPC is mainly related to deficiencies in hygiene measures and in raw milk storage temperature, in addition to the occurrence of mastitis in the herd (Ribeiro et al., 2018).

Milk with high SCC shows important changes in its proximate composition, causing numerous damages to the dairy industry. The main changes observed are a decrease in the casein, fat, lactose content and minerals such as calcium and phosphorus and an increase in the concentration of sodium and chlorine, of short-chain free fatty acids and increase in the proteolytic and lipolytic activity of milk due to the action of leukocytes enzymes, which can interfere with industrial processing. In addition, milk with high SCC may contain pathogenic microorganisms that cause mastitis, in addition to the risk of containing antimicrobial residues, which are used for the treatment of this disease. The presence of these residues generates losses to the industry, mainly due to the inhibition of dairy cultures used in the manufacture of derivatives such as cheeses and yogurts, since these microorganisms are sensitive to these residues. In addition, they pose risks to consumer health, such as toxic and allergic reactions (Li et al., 2014).

Regarding cheese making, studies have shown that high SCC can increase milk clotting time and decrease the firmness of the formed clot, which in turn leads to increased moisture content and lower overall cheese yield. High humidity increases the activity of water available to microorganisms, thus reducing the shelf life of cheeses due to early spoilage by these microorganisms. There may also be greater loss of milk components to whey and changes in organoleptic and texture characteristics (Murphy et al., 2016).

In yogurts, high SCC causes a negative effect on the growth and metabolism of lactic acid cultures, compromising quality, coagulation, aroma and texture. The production of antimicrobial substances by leukocytes may be responsible for the inhibition of lactic acid culture.

SPC is one of the main tools used by the industry to predict the conservation quality and shelf life of milk and dairy products. Milk with high bacterial counts can be rejected by the industry due to changes in nutritional value, which make it unsuitable for consumption and industrialization. The excessive presence of microorganisms can compromise the processing of milk and its derivatives due to acidification and coagulation problems, gas production, modification in organoleptic properties, coagulation without acidification and increase in viscosity.

High SPC is related to the presence of psychotropic microorganisms, generally found in soil and untreated water, which contaminate milk mainly through the external surface of contaminated teats and udders. Some species of thermoduric psychotropic bacteria produce thermostable enzymes that remain active even after thermal processing, influencing the yield of milk derivatives. The reduction in cheese yield is directly related to the presence of lipases that hydrolyze lipids and

Figure 2. Main information extracted from milk quality for dairy industries

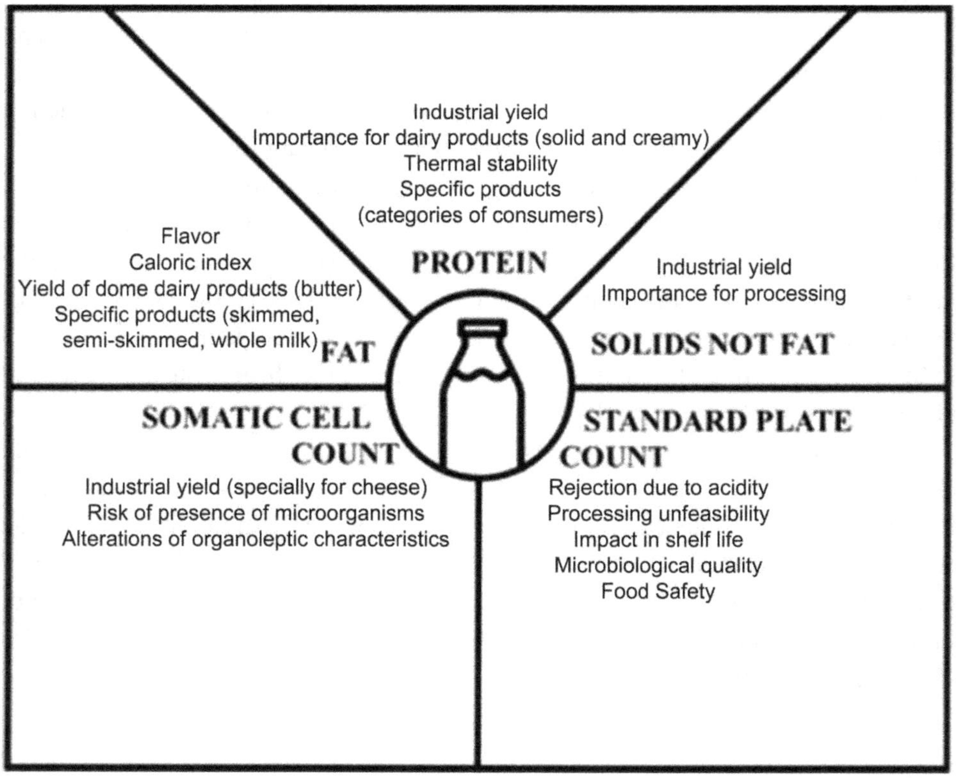

proteases that hydrolyze protein fractions from caseins, causing them to be lost to the whey instead of forming the curd (Ribeiro et al., 2018).

In short, the milk that arrives at the industry must be submitted to quality control tests to avoid technological problems in the production of dairy products. In addition, the results of these analyses contribute to ensuring greater production and diversification of products, always aiming at consumer health and production efficiency.

SOLUTIONS AND RECOMMENDATIONS

Considerations Beyond Analysis

Scenarios for the dairy farm and for the industry, in the extraction of information from the analysis of milk quality, were presented. In practical terms, can the quality of milk be related to financial aspects?

In addition to the damage obtained if the milk is rejected due to high acidity, reduced production, or increased SCC, milk quality is linked to the sustainability of the activity through quality payment programs promoted by some dairy industries. Differentiated payment based on quality criteria is an evolution of the marketing system, but it does not depend exclusively on new rules, but on the agreement between the industry and the producer, who provides a differentiated raw material, generating returns for the industry and the consumer. This practice can encourage producers in the search for maximizing quality production, or in the search for constant improvement of quality indices presented in samples of milk produced. The program can assist the producer in choosing management, nutrition and many other actions, according to the proposed return. However, it is important to emphasize that each

Table 1. Increase in industrial yield of some dairy products according to alteration of some milk components

Product	Altered component	Component variation	Yield (%)
Cheese	Protein	+0.5%(1)	16
Butter	Fat	+0.5%(1)	14
Powdered milk	Lactose	+0.5%(1)	11
Yogurt	Solids Not Fat	+0.5%(1)	6
Condensed milk	Total Solids	+0.5%(1)	4
'Minas Frescal' Cheese	SCC	-300.000 cells/mL	15

(1) increase in percentage points in component content.

industry formulates its program based on the quality challenges for products that will be manufactured. Therefore, the final product influences the valuation of the item.

Some practical examples of how the increase in milk composition interferes with the production of dairy products are given below (Table 1). It is observed, for example, that an increase of 0.5% in milk protein leads to a 16% increase in cheese yield. This can represent a company's financial success or disaster.

In addition to higher industrial yield, the increase in the content of solids also implies in lower water volume, which would determine lower costs with transport, storage and size of facilities required for processing.

In terms of projection, it is important to mention the strengthening of production, with a focus on improving milk quality to meet the consumer market. However, for this projection to be complete, factors that involve all members of the production chain must be included, from producers and intermediaries in processing to the consumer market.

As for the consumer market, there has been an evolution in consumer awareness in relation to food quality, with preference for healthier foods with better characteristics. In some points, this awareness is also related to the production system, such as animal husbandry and welfare, drug use, among others principles, required for example in the organic products market.

Therefore, it is important to consider factors related to efficiency and profitability in the production system, valuing alternative methods to improve milk production and quality, combined with social and ethical responsibility from the dairy farm to the market. It is also important to mention transparency regarding the production method, including attention to the environment, animals and consumers, in addition to the search for the conscious consumption of dairy products.

REFERENCES

Ahmed, I., Kumar, S., & Aggarwal, D. (2020, March). Assessment of Knowledge and Practices of Hygienic Milk Production among Dairy Farmworkers, Southwest Delhi. *Indian Journal of Community Medicine*, *45*(5, Suppl 1), S26–S30. doi:10.4103/ijcm.IJCM_366_19 PMID:32476737

Allen, J. C. (1990). Milk Synthesis and Secretion Rates in Cows with Milk Composition Changed by Oxytocin 1, 2. *Journal of Dairy Science*, *73*(4), 975–984. doi:10.3168/jds.S0022-0302(90)78755-3 PMID:2345208

Bauman, D. E., McGuire, M. A., & Harvatine, K. J. (2022). Mammary Gland, Milk Biosynthesis and Secretion: Milk Fat. In P.L.H. McSweeney (Ed.), J. P. McNamara (Ed.), Encyclopedia of Dairy Sciences (3rd Edition), (pp. 190-197). Academic Press. doi:10.1016/B978-0-12-818766-1.00049-0

Blowey, R. W., & Edmondson, P. (2010). *Mastitis Control in Dairy Herds* (2nd ed.). CAB International. doi:10.1079/9781845935504.0000

Barros, L. (2001). Transtornos metabólicos que afetam a qualidade do leite. In F. H. D. González, J. W. Dürr, & R. S. Fontaneli (Eds.), *Uso do leite para monitorar a nutrição e metabolismo de vacas leiteiras* (pp. 46–60). Ed. UFRGS.

Bradley, A. J. (2002). Bovine mastitis: An evolving disease. *Veterinary Journal (London, England)*, *164*(2), 116–128. doi:10.1053/tvjl.2002.0724 PMID:12359466

Britt, J. H., Cushman, R. A., Dechow, C. D., Dobson, H., Humblot, P., Hutjens, M. F., Jones, G. A., Mitloehner, F. M., Ruegg, P. L., Sheldon, I. M., & Stevenson, J. S. (2021). Review: Perspective on high-performing dairy cows and herds. *Animal*, *xxxx*, 100298. doi:10.1016/j.animal.2021.100298 PMID:34266782

CFIS - Canadian Food Inspection System. (2006). Agriculture and Agri-Food Canada. http://www.cfis.agr.ca/english/indexe.shtml.

FariaA. P. A.PennaC. F. A. M.PintoM. S.EndoÉ. (2020). Influência do leite com elevada contagem de células somáticas sobre características físico-químicas e processo de fermentação de iogurte. *Ciência Animal Brasileira*, v. 21. doi:10.1590/1809-6891v21e-44773

Foroutan, A., Guo, A. C., Vazquez-Fresno, R., Lipfert, M., Zhang, L., Zheng, J., Badran, H., Budinski, Z., Mandal, R., Ametaj, B. N., & Wishart, D. S. (2019). Chemical Composition of Commercial Cow's Milk. *Journal of Agricultural and Food Chemistry*, *67*(17), 4897–4914. doi:10.1021/acs.jafc.9b00204 PMID:30994344

Gonçalves, J. L., Kamphuis, C., Martins, C. M. M. R., Barreiro, J. R., Tomazi, T., Gameiro, A. H., Hogeveen, H., & dos Santos, M. (2018). Bovine subclinical mastitis reduces milk yield and economic return. *Livestock Science*, *210*(January), 25–32. doi:10.1016/j.livsci.2018.01.016

Lammoglia, M. A., Garcez, N., Cabrera, A., López, R. D., Carmen, I., Renteria, D., & Rojas-Ronquillo, R. (2016) Behavior affected by routine oxytocin injection in crossbred cows in the tropics. R. Bras. Zootec. V.45, n.08, 478-482. doi:10.1590/S1806-92902016000800008

Li, N., Richoux, R., Boutinaud, M., Martin, P., & Gagnaire, V. (2014). Role of somatic cells on dairy processes and products: A review. *Dairy Science & Technology*, *1*(94), 517–538. doi:10.100713594-014-0176-3 PMID:25309683

Murphy, S. C., Martin, N. H., Barbano, D. M., & Wiedmann, M. (2016). Influence of raw milk quality on processed dairy products: How do raw milk quality test results relate to product quality and yield? *Journal of Dairy Science*, *99*(12), 10128–10149. doi:10.3168/jds.2016-11172 PMID:27665134

Nostrand, S. D., Galton, D. M., Erb, H. N., & Bauman, D. E. (1991). Effects of Daily Exogenous Oxytocin on Lactation Milk Yield and Composition. *Journal of Dairy Science*, *74*(7), 2119–2127. doi:10.3168/jds.S0022-0302(91)78384-7 PMID:1894807

Peres, J. R. (2001). O leite como ferramenta do monitoramento nutricional. In F. H. D. González, J. W. Dürr, & R. S. Fontaneli (Eds.), *Uso do leite para monitorar a nutrição e metabolismo de vacas leiteiras* (pp. 29–43). Ed. UFRGS.

Ribeiro, B.D., Pereira, K.S., Nascimento, R.P., Coelho, M.A.Z. (2018) Alimentos. *Microbiologia Industrial*. Elsevier, Rio de Janeiro.

Ruegg, P. L. (2017). A 100-Year Review: Mastitis detection, management, and prevention. *Journal of Dairy Science*, *100*(12), 10381–10397. doi:10.3168/jds.2017-13023 PMID:29153171

Santos, M. V., & Fonseca, L. F. L. (2007). *Estratégias para controle de mastite e melhoria da qualidade do leite*. Manole.

Schopen, G. C. B., Heck, J. M. L., Bovenhuis, H., Visker, M. H. P. W., van Valenberg, H. J. F., & van Arendonk, J. A. M. (2009). Genetic parameters for major milk proteins in Dutch Holstein-Friesians. *Journal of Dairy Science*, *92*(3), 1182–1191. doi:10.3168/jds.2008-1281 PMID:19233812

Stelwagen, K. (2022). Mammary Gland, Milk Biosynthesis and Secretion: Lactose. In P.L.H. McSweeney (Ed.), J. P. McNamara (Ed.), Encyclopedia of Dairy Sciences (3rd Edition), (pp. 184-189). Academic Press. doi:10.1016/B978-0-12-818766-1.00025-8

Stelwagen, K. (2022b). Mammary Gland, Milk Biosynthesis and Secretion: Milk protein. In P.L.H. McSweeney (Ed.), J. P. McNamara (Ed.), Encyclopedia of Dairy Sciences (3rd Edition), (pp. 198-205). Academic Press. doi:10.1016/B978-0-12-818766-1.00026-X

USDA - Animal Improvement Programs Laboratory. https://aipl.arsusda.gov

Wiking, L., Nielsen, J. H., Båvius, A.-K., Edvardsson, A., & Svennersten-Sjaunja, K. (2006). Impact of Milking Frequencies on the Level of Free Fatty Acids in Milk, Fat Globule Size, and Fatty Acid Composition. *Journal of Dairy Science*, *89*(3), 1004–1009. doi:10.3168/jds.S0022-0302(06)72166-X PMID:16507695

Wilkinson, J. M., Lee, M. R. F., Rivero, M. J., & Chamberlain, A. T. (2020). Some challenges and opportunities for grazing dairy cows on temperate pastures. *Grass and Forage Science*, *75*(1), 1–17. doi:10.1111/gfs.12458 PMID:32109974

Zacarchenco, P. B., Van Dender, A. G. F., & Rego, R. A. (2017) Brazil Dairy Trends 2020 - Book, (pp. 373). Instituto de Tecnologia dos Alimentos, Campinas, SP, Brazil..

KEY TERMS AND DEFINITIONS

Bacterial Count: estimated quantity of bacteria in milk samples. It is expressed using CFU/mL.

Bulk Tank: Refrigerated container for storage of milk in milking systems.

Milk Quality: is defined as a set of qualities or characteristics that qualify the product. Parameters of milk quality can be about its composition, hygiene, and sanitary condition.

Milk Somatic Cell: Cells produced by the udder, containing desquamation cells from the udder epithelium, as well as immune defense cells, such as macrophages, neutrophils, and lymphocytes

Chapter 3
Quality Requirements Assessed in the Supply of Raw Milk by Brazilian Dairy

Eduardo Guilherme Satolo
https://orcid.org/0000-0002-8176-2423
São Paulo State University, Brazil

Bruna Maria Zumkeller Ferreira
São Paulo State University, Brazil

Priscilla Ayleen Bustos Mac-Lean
São Paulo State University, Brazil

Juliani Elis Duarte Gotardi
São Paulo State University, Brazil

Isabela Garcia Mendes de Araujo Santos
São Paulo State University, Brazil

EXECUTIVE SUMMARY

In Brazil the quality required by dairy is established through regulations 76 and 77. This chapter aims to identify the technical quality requirements evaluated in the supply of raw milk by dairy companies located in the state of São Paulo. Using the voice of the customer (VOC) it is possible to identify compatible internal requirements to drive the improvement process. The results indicate the need for actions to raise awareness of the agents involved in the dairy production chain. First, the role of the dairy, in understanding that the legal parameters must be measured and controlled, given the perishability of milk. Second, producers and dairies must understand

DOI: 10.4018/978-1-6684-5472-5.ch003

that superior milk quality results in better quality products, and to achieve this, the actions carried out on the property and evaluated parameters must be worked out through action plans. Finally, that action plans will only be successful if they are carried out in partnership with the rural producer, understanding that the gain will occur if everyone works collectively and not individually along the production chain.

ORGANIZATION BACKGROUND

Agribusiness represents a production chain, which begins with the rural producer and extends to the final consumer. The agents in this chain correspond to manufacturing of inputs, production in agricultural establishments, technical assistance, transformation and processing by agro-industries, commercialization, distribution, transport, manufacturing, and retail (Gasques, 2004).

It is an economic sector that promote important socio-economic development in Brazil and is essential to food safety of Brazil and world. In view of this, the investments made in agribusiness stand out in relation to other sectors, a reality that can be seen in Brazil and in other countries (Costa et al., 2013).

Therefore, the quality of consumer goods that will be produced by this sector is important for its competitiveness. However, the concept of quality corresponds to a variable concept which allows for different interpretations, once the quality concept required by the consumer may not be compatible with the quality concept proposed for the industry (Telles, 2014). For this reason, For this reason, to create and develop products outside of imitation or copy standards along with total quality requirements, Yoji Akao and Shigeru Mizuno created one method for applying the quality function deployment, originally called Quality Function Deployment (QFD). This theory refers to a way of translating customer needs from their requirements or specifications which is used in the correction, improvement or development of a product or service, so that it is as close to the real needs of the customer (Akao and Mazur, 2003).

According to Wu and Chan (2002), the application of QFD unfolds over product development, quality management and analysis of customer needs. However, the purpose for the application of this method is to give "voice to the customer", because by listening and understanding their desires and needs, the process of developing or improving a product or service becomes assertive, aspects that can provide an increase in product quality and a reduction in development time or production costs. Therefore, its application unfolds over product development, quality management and analysis of customer needs.

The voice of the customer (VoC) transfers what they need and seeks the organization's internal quality requirements (Lorenzo et al., 2004). The most effective way to hear the VoC is to unfold that voice; verify which are the main functions that the customer considers important in the product, identify, and systematize the implicit and explicit desires of the customer, making sure that all needs are identified, analyzed, and transformed into project parameters (Abreu, 2010). As a result, the actions necessary to achieve this satisfaction are planned, which intervene in the design and development of the product or service. In short, it is a system to plan and develop services or products and ensure that they meet or exceed the expectations of its customers (Lorenzo et al., 2004).

The quality requirements in the milk chain are in new Normative Instructions Number 76 and 77, of November 2018, which came into validity in May 2019, and which set new maximum temperature limits, as well as sensory characteristics to be monitored in the reception, conservation, and dispatch of milk as well as its conservation in milk processors (MAPA, 2018 a, b). In this way, new product quality standards started to be demanded by the customers (dairies) and their suppliers (farmers) who sometimes do not know, as well as making it difficult to direct improvements in the production process.

CONCEPTUAL FOUNDATION

Milk is of the five most important agricultural commodities in the world both in volume and in value (GDP, 2016). Furthermore, around the world approximately 150 million families, with an average of 4 members, are committed to dairy production with a positive impact on poverty reduction. And more than 6 billion people consume dairy-based foods (GDP, 2016; GDP, 2018). In this context, Brazil is currently among the top 5 milk producers in the world, occupying the fourth position in the ranking, reaching a margin of 34.8 billion liters in 2019, according to the Brazilian Institute of Geography and Statistics (CONAB, 2018; IBGE, 2020). Still, according to projections, the country may end the year 2030 with a value of more than 42 billion liters produced. And in an optimistic scenario, for the same period, it will reach a margin above 47 billion liters (MAPA, 2020).

To achieve this production volume, improving its production indicators, is important consider to milk has a production chain that involves several agents. A production chain is an organized network of economic agents that collaborate to improve the flow of elements and information in each segment of activities (Govil and Proth, 2002). It is constituted by the link between all the agents involved in the production or transformation of a certain raw material and its proportion is established by the number of members needed to carry out all stages of the production process

(Stein et al., 2019; Talamini et al., 2005). In the milk chain, the main links involved are represented by seven agents (Figure 1). It is worth noting that this number is relatively low compared to the number of agents involved in other supply chains.

Figure 1. Milk production chain

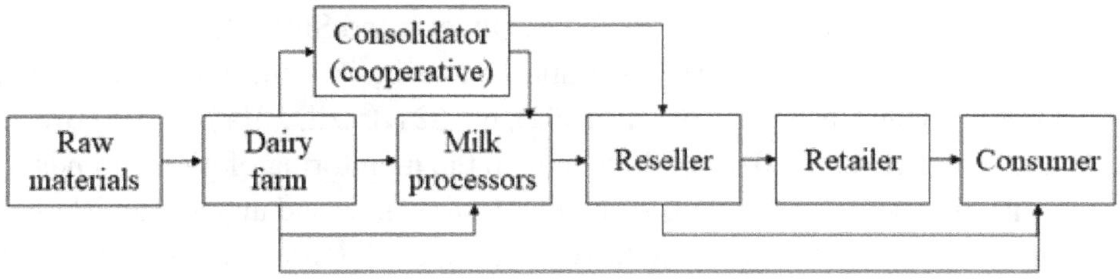

Substantially, raw material suppliers supply food, medicines, seeds, and other inputs necessary for all stages of production within the dairy farms where carry out all the basic techniques for obtaining raw milk. Then milk is assigned to processing, which takes place within the farms themselves, dairies or in cooperatives, which, in addition, perform the function of consolidators of dairy production and assistance to activities and strategies in production. After being processed and properly qualified, raw milk is derived in different forms - cheese, curd, butter, powdered or pasteurized milk - and intended for resellers and retailers, to be sold to consumers (Satolo et al. 2020; Viana and Rinaldi, 2008). In this context, it is worth emphasizing the importance of the private dairies and cooperatives which plays a fundamental role in the milk production chain, once about 60% of milk production in Brazil is acquired and benefited by this link in the milk production chain (Pagnani Netto et al., 2017).

Although in the last years in Brazil dairy products have evolved in terms of revenue and potential, the sector suffers with lower milk production costs arising from Southern Common Market (Mercosur) countries, the high-scale tariff and subsidy policies of countries foreigners and, consequently, the high quality required by them (Morae and Bender, 2017; Scalco and Toledo, 2002). In this way, dairy products ensure the qualification of the milk offered and distributed, through good quality and food safety (Scalco and Toledo, 2002).

To increase the competitiveness of Brazilian milk in the international market in 1996, the Ministry of Agriculture and Livestock and Supply (MAPA), together with universities and research entities, and representative bodies of industry and rural producers, established the Brazilian National Plan for the Improvement of Milk Quality (PNMQL - of the Portuguese term Programa Nacional De Melhoria

Da Qualidade Do Leite), to establish minimum quality parameters for the product (Oliveira and Silva, 2012). Thus, in 2002, was published the normative instruction (ni) n°. 51 that determine the minimum technical and operational rules for the production, identification, classification, and quality of the most varied types of existing milk and, in addition, to define the circumstances for its cooling, storage and transport. The main requirements of this are the refrigeration of milk inside rural properties, collection of bulk milk and the counting of somatic cells and total bacteria through monthly samples (Brasil, 2002; Brito, 2005; Oliveira and Silva, 2012).

Over time it became necessary to improve and implement new milk quality regulations, such as NIs n°. 22 (Brazil, 2002), n°. 62 (Brazil, 2011), n°. 07 (Brazil, 2016) and n°. 31 (Brazil, 2018c). These, due to the monitoring of global trends by the responsible bodies, were revoked and replaced by new and updated legislation. These are NIs n°. 76 and n°. 77, of November 26, 2018 (Brasil, 2018a, b), which came into force at the end of May 2019. Normative Instruction n°. 76 establishes the technical regulations for the identity and characteristics that refrigerated raw milk, pasteurized and type A pasteurized milk must present. While NI n°. 77 determines criteria and procedures for the production, packaging, conservation, transport, selection, and receipt of raw milk in establishments registered with the official inspection service. The parameters established by NI n°. 76 are shown in Table 1.

Table 1. Parameters established by normative instruction n°. 76 of 2018

Compliance with the following parameters at the refrigeration station (producer)		
Parameters	**Minimum acceptable**	**Maximum acceptable**
Temperature	-	£ 4°C
Compliance with the following parameters in the dairy		
Parameters	**Minimum acceptable**	**Maximum acceptable**
Receiving Temperature	£ 7°C	£ 9°C (exceptionally)
Storage temperature		£ 4°C
Fat	3.0 g/100g	
Total protein	2.9 g/100g	
Anhydrous lactose	4.3 g/100g	
Non-Fat Solids	8.4 g/100g	
Total solids	11.4 g/100g	
Titratable acidity	0.14 g/100ml	0.18 g/100ml
Stability to Alizarol	72% v/v	
Density (15°C/15°C)	1,028	1,034
Cryoscopic index	-0.512°C	-0.536°C

For decades, the Brazilian Southeast region currently leads the dairy sector in production volume, representing 34.3% of national production, with 11.9 billion liters per year (IBGE, 2020). The Brazilian states with the highest representation in these items are Minas Gerais, São Paulo, Rio de Janeiro, and Espírito Santo, respectively (Zoccal, 2019). The state of São Paulo, in turn, despite being in second place in the ranking of the largest dairy producer in the Southeast, has suffered stagnation in its production in the last 10 years. The so-called growing deficit of the state is due to several factors such as higher financial income per hectare in other crops, sugar cane and eucalyptus, population aging, changes within the dairy industry, lack of skilled labor and production costs higher than that of other states. Consequently, it is supplied by importing milk from other states or countries (IEA, 2018).

Therefore, technological changes caused a significant increase in the concentration of dairy industries and, consequently, more requirements and quality parameters demanded from production systems (Pagani Netto et al., 2017).

Thus, it becomes necessary to understand in fact what quality requirements are demanded by dairy products to produce high-quality items, so that the producer and final consumers are fully benefited in the production system. In this sense, there is the process of deploying the quality function. According to Abreu (2010), QFD emerged as a tool to ensure that there is quality in products, from conception to final production. This is, therefore, a method used to improve the quality of its products as it understands the real needs of customers (Andrietta and Miguel, 2002). To this end, the first phase of QFD consists of listening to the voice of the customer (VoC) and understanding what their needs are. The term VoC is used to bring customers and the organization closer, so that the requirements demanded by them are applied in the development of products and services, meeting the wishes, expectations and needs exposed by the customer (Costa, 2018). In addition, according to Leonel (2007), the VoC identifies current or possible problems in relation to the quality or particularities of the product. As well as it enables a good relationship between the client and the organization.

As mentioned above, through VoC the needs of customers are identified, and the most efficient way to understand these needs, and carrying out what is called deployment. That is, one must deeply analyze what are the fundamental attributions that the product must have, perceived by the customer, as well as recognize and structure all the customer's wishes - implicit or explicit - so that, after being identified, they are transformed into analysis criteria (Abreu, 2010).Thus, according to Abreu (2010), it is possible to obtain the VoC through techniques that enable the proximity between the company and the customer, such as, with the application of forms and questionnaires - survey type -, individualized and/or group interviews, as well as with direct and directed observation (Figure 2). The second phase, after identifying customer needs, is to produce a table of required qualities. In this sense, the subsequent steps consist of translating the VoC, and then determining the quality functions and/or performance goals and starting the quality matrix (Mazo, 2012).

Figure 2. Techniques and means to obtain the VoC

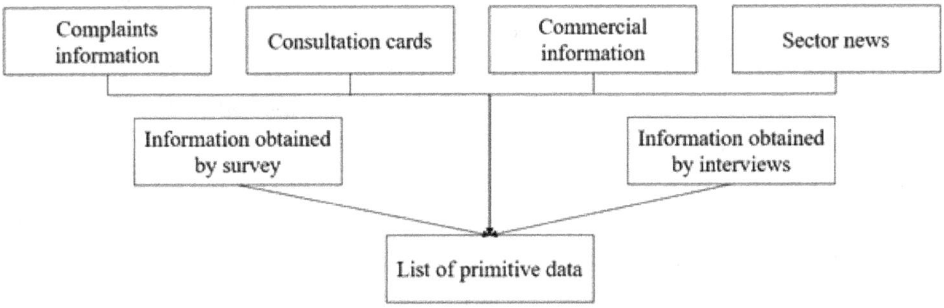

ACTION RESEARCH STEPS

To conduct this research, the survey method was used. Survey method aim to analyze areas and specific problems, of individuals or groups, through field research, questionnaires, telephone interviews, etc. According to Forza (2002), this method has three purposes: exploratory, descriptive, or explanatory.

When it has an exploratory character, it aims to provide greater familiarity with the issue, to make it more explicit. A descriptive characteristic survey intends to describe the characteristics of a given problem. When the survey has an explanatory character, it is concerned with studying and identifying the factors that contribute to the occurrence of the problem. Thus, considering the main objective of this research, the survey adopted was descriptive to verify which occurrences are evident in a population (Forza, 2002). In addition, its main objective is to provide material for the elaboration of theories and/or the deepening of them. Also, regardless of the type, it must be carried out through rigorously structured steps, which begin even before the actual data collection.

The first step is the development of a theoretical model, in which the initial intention is to translate the theoretical domain into the empirical one. In other words, it is necessary to explain in an enlightening way the researcher's expectations regarding the occurrence of relationships and the possible connections with other theories (Miguel and Ho, 2010). The second step is to define how the data will be collected for future analysis. Afterwards, as a third step of the method, it is necessary to collect the data through the chosen research instrument, to finally analyze them and prepare the report (Miguel and Ho, 2010).

Development

To achieve the objectives of this research, data collection was carried out through an online form sent to 143 companies. In this group, all companies that provided a contact telephone number were previously contacted to ensure that they were sent in the correct e-mail address and closer to the customer, with the intention of more effectively engaging these companies to participate in the research.

To define the population, initially, a search for dairy products in the state of São Paulo, Brazil, was carried out via Google platform, and when noticing that there was no website or online magazine dealing with the subject in a consolidated way, the first step was to map all the dairy products in the country state.

The state of São Paulo is made up of 645 municipalities. From there, the search for dairy products was carried out in each of them, letter by letter. For example, the Google Maps tool was used to search for "dairy products in Tupã - SP", and so on. In this way, searches were carried out in all municipalities, to map all existing dairy products. Through this survey, among the 645 cities, only 276 indicated the existence of at least one dairy, according to Google. Based on this, when transferring these dairy products to the control sheet, it was identified that in these 276 cities there were a total of 593 possible dairy products. However, only 419 dairies made up the sample, since, through a filter, it was noticed the repetition of companies in several cities. For example: when searching "dairy products in Adamantina - SP" Google indicated dairy products present in Lucélia-SP, Piquerobi-SP, Herculândia-SP and even in cities of Minas Gerais state.

Hereupon, a new filter was carried out and among the 419 dairy products, it was found that 85 did not have a telephone and, therefore, contact was made directly through the institution's e-mail and/or website. In addition, from the connection, it was found that 83 institutions were not dairy products, but: food distributors, food trade, milk, orange and potato producers, mini-markets, grocery stores and even homes. Among the remaining 336, 25 had no means of contact and 8 had incorrect contact. In addition, it was not possible to contact 48 dairies. Finally, 1 ended its activities and 3 decided not to participate in the research right at this first contact. Thus, the research population, which makes up the mapping of dairy products in the state of São Paulo (Brazil), is composed of 143 research units. Figure 3 visually represents the consolidation of dairy in the state of São Paulo, Brazil.

After completing this objective, the e-mails were sent. This one that contained all the information about the research and the contact of the developers, in addition to the interview form. By the way, the questionnaire sent to the dairies sought to obtain data regarding the forms of organization of the company (appendix A), as well as the parameters evaluated by the dairies in the receipt of raw milk. The main intention was to capture the profile and size of dairy products, using closed and objective questions, in addition to those with a Likert scale from 1 to 5, which represented specific aspects of raw milk processing, referring to quality parameters. In 10 of the 143 dairy products, the e-mail provided and/or made available on the internet was not found or the e-mail was unavailable. In addition, 2 indicated that it was not really a dairy, 4 indicated that they could not participate due to company policy and 15 returned the contact indicating that they would respond to the form soon.

To improve the return rate in the survey, the e-mails and forms were sent back to the dairy, where the deadline for response was extended and the importance of participation was mentioned. At the end of the deadlines, a return of 18 dairy products that answered the form was obtained, and one of them did not perform the processing of milk, so the questionnaire was invalidated for the research and discarded. In this way, the analysis of the results will be made for the 17 dairies responding to the survey, which corresponds to 11.8% of the sample population.

Figure 3. Map of dairy products in the state of São Paulo, Brazil

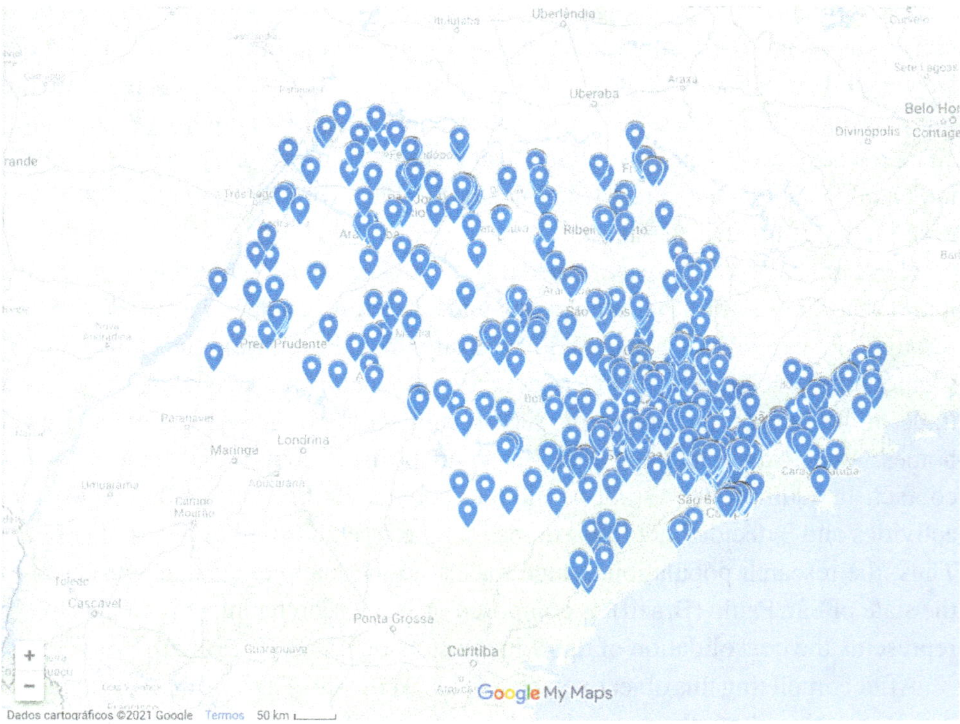

RESULTS

Based on the data obtained in the form applied to the participants, the results are classified into three groups, the first corresponding to the profile of the respondents, the second to the profile of dairy products and the third consistent with the indicators analyzed.

Respondents' Profile

Regarding the profile of the respondents, there are: 6 partner-owners, 4 quality managers, 3 veterinarians, 2 quality leaders, 1 president and 1 analyst. Approximately, 82% have completed higher education with training in the areas of administration, veterinary medicine, chemical engineering, and biological sciences. About experience in the field, 6 of them have more than 20 years of experience, 3 are in the range of 5 years or more and the rest are under 5 years of practice.

Dairy Profile

Regarding the profile of dairy industries, approximately 59% of dairy products with up to 19 employees, 24% with 20 to 99 employees and 17% with 100 to 499 employees (Figure 4). About the size of the company, in terms of its revenue, 59% are classified as micro-companies, 29% are medium-sized companies and 12% are small companies (Figure 5). Still, 82% process less than 1 million liters of milk per month (Figure 6).

Figure 4. Size of the companies participating in the survey (n=17 respondents)

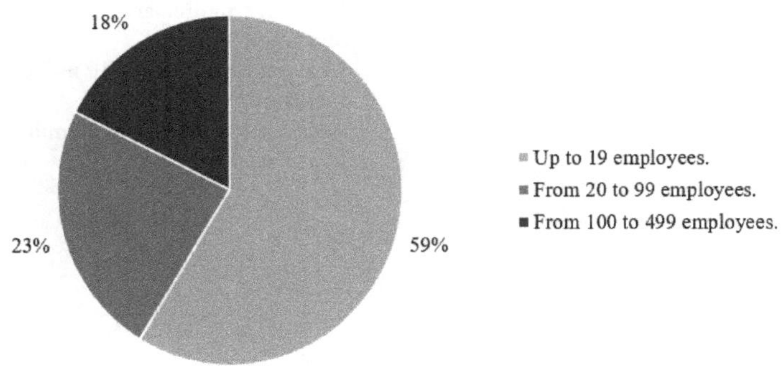

Approximately, 71% is managed in a family management (Figure 7), as well as they do not sell the products to other states. Furthermore, none of the companies interviewed exports and around 53% collect by means of their own transport. In addition, for 53% of dairy products, milk collection is carried out by means of their own transport, while the rest use an outsourced service. Lastly, 71% of the companies analyzed sell their products to other states, but none of them exports.

Figure 5. Monthly invoicing of the companies participating in the survey (n=17 respondents)

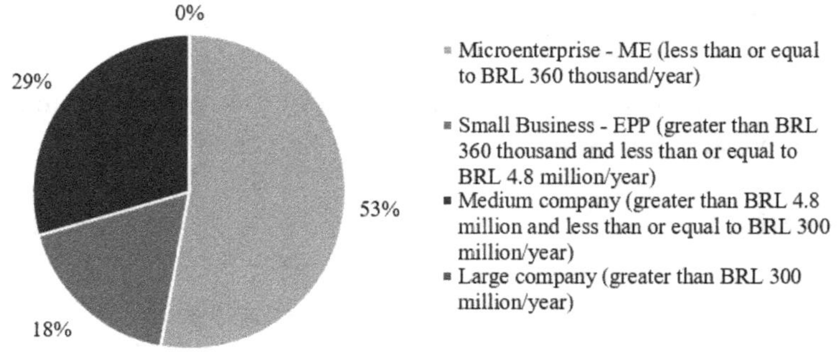

Figure 6. Processing, at the unit, of liters of milk per month (n=17 respondents)

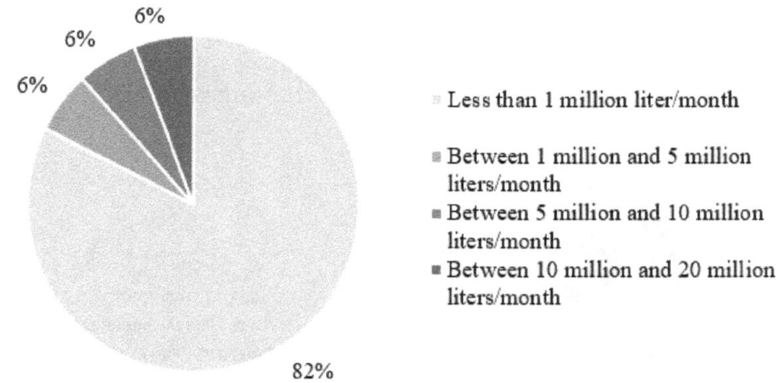

Figure 7. Governance structure (n=17 respondents)

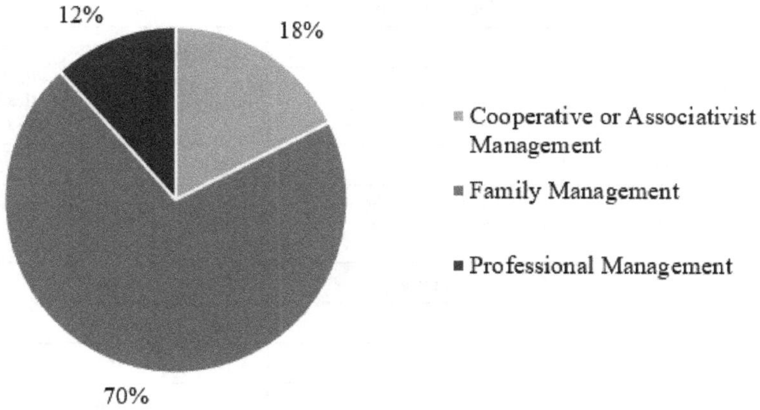

Milk Parameters Evaluated by the Dairy Products Participating in the Research

In this subsection, the results related to the evaluation of milk parameters for the dairy products participating in the research are presented. The indicators analyzed were divided into mandatory and non-mandatory. The demonstrative tables were organized in 3 levels. Level 1 highlights how the parameter is classified according to NI n° 76 while level 2 portrays the purpose of the parameter evaluation and at level 3 the parameter itself. In addition, the tables specify the percentage of dairy products that evaluate and do not evaluate each of the indicators and the average degree of importance, from 1 to 5, which was calculated from the responses of the respondents. In this way, Table 2 presents the results on the evaluation of the mandatory parameters of NI n° 77 for the dairy products participating in the research. As noted, none of the mandatory parameters are evaluated by 100% of dairy products. Some of the parameters even have a very low percentage of evaluation, such as density and somatic cell count, in which only 35.3% and 29.4%, respectively, evaluate. The indicators with the highest percentage of evaluation are the standard plate count - producer and alizarol, which represent, respectively, 94.1% and 82.4%.

Regarding the degree of importance perceived in each of the parameters, it is observed that in none of them the degree is higher than 3.82. For most of them it is extremely low, where the standard plate count stands out - producer (0.24) and truck (1.35). In opposite, the most important are titratable acidity (3.82) and temperature (3.47).

Table 2. Mandatory parameters evaluated by the dairy participating in the research

<table>
<tr><th colspan="6">Voice of the Customer Matrix</th></tr>
<tr><th colspan="3">Levels</th><th colspan="2">% of dairy products</th><th rowspan="2">Degree of importance of the Voice of the Customer</th></tr>
<tr><th>Level 1</th><th>Level 2</th><th>Level 3</th><th>Evaluate</th><th>Do not evaluate</th></tr>
<tr><td rowspan="2">Parameter at the reception of milk on the property</td><td>Acidity</td><td>Alizarol</td><td>94.1</td><td>5.9</td><td>3.35</td></tr>
<tr><td>Temperature</td><td>Temperature</td><td>70.6</td><td>29.4</td><td>3.47</td></tr>
<tr><td rowspan="13">Parameter in the reception of milk at the plant</td><td rowspan="2">Acidity</td><td>Titratable Acidity</td><td>82.4</td><td>17.6</td><td>3.82</td></tr>
<tr><td>Alizarol</td><td>70.6</td><td>29.4</td><td>3.35</td></tr>
<tr><td rowspan="5">Composition</td><td>Fat</td><td>70.6</td><td>29.4</td><td>3.12</td></tr>
<tr><td>Lactose</td><td>41.2</td><td>58.8</td><td>1.65</td></tr>
<tr><td>Protein</td><td>52.9</td><td>47.1</td><td>2.47</td></tr>
<tr><td>Non-Fat Solids</td><td>52.9</td><td>47.1</td><td>2.35</td></tr>
<tr><td>Total solids</td><td>58.8</td><td>41.2</td><td>2.65</td></tr>
<tr><td rowspan="2">Fraud</td><td>Cryoscopy</td><td>64.7</td><td>35.3</td><td>3.18</td></tr>
<tr><td>Density</td><td>64.7</td><td>35.3</td><td>2.88</td></tr>
<tr><td rowspan="3">Microorganism</td><td>Somatic Cell Count</td><td>47.1</td><td>52.9</td><td>1.59</td></tr>
<tr><td>Standard Plate Count - truck</td><td>35.3</td><td>64.7</td><td>1.35</td></tr>
<tr><td>Standard Plate Count - producer</td><td>29.4</td><td>70.6</td><td>0.24</td></tr>
<tr><td>Temperature</td><td>Temperature</td><td>76.5</td><td>23.5</td><td>3.47</td></tr>
</table>

When comparing the evaluation percentages and the degree of importance of the VoC attributed, two important points are observed: the parameter, plate count by the producer, with the highest percentage of evaluation by the dairy, has the lowest importance perceived by the VoC. The parameter, titratable acidity, with the highest degree of importance of the Voice of the Customer does not correspond to the highest percentage of evaluation by the dairy.

About non-mandatory parameters (Table 3), it can be seen that the percentage of dairy products that effectively assess such indicators and the degree of perceived importance in VOC is even lower in this case. The highest percentage of evaluation and degree of importance were antibiotics, 64.7% and 3.24%, respectively, in the group of contaminants.

Table 3. Non-mandatory parameters evaluated by the dairy products participating in the research

Voice of the Customer Matrix					
Levels			% of dairy products		Degree of importance of the Voice of the Customer
Level 1	Level 2	Level 3	Evaluate	Do not evaluate	
Parameter at the reception of milk on the property	Acidity	Ethyl alcohol	35.3	64.7	1.65
		PH	47.1	52.9	1.88
	Contaminants	Antibiotics	64.7	35.3	3.24
		Lactofiltration	5.9	94.1	0.29
		Pus	23.5	76.5	0.94
		Blood	23.5	76.5	1.00
	Fraud	Cryoscopy	64.7	35.3	3.18
		Alkaline	35.3	64.7	1.65
		Starch	23.5	76.5	1.00
		Chloride	23.5	76.5	1.00
		Chlorine and Hypochlorite	29.4	70.6	1.29
		Formalin	23.5	76.5	1.06
		Phosphatase	35.3	64.7	1.59
		Neutralizing	29.4	70.6	1.29
		Peroxidase	47.1	52.9	2.18
		Boiling test	11.8	88.2	0.41
		Sucrose	23.5	76.5	1.12
	Microorganism	Psychrotrophic	17.6	82.4	0.88
		CMT - producer	5.9	94.1	2.24
		Resistant spores	5.9	94.1	0.24
		Total resistant spores	11.8	88.2	0.41
		Reductase	47.1	52.9	1.71

CONCLUSION

From the results obtained, one can see the there is a great incompatibility in the understanding of quality by the intermediaries present in this production system, what was concluded observing very low percentage of the evaluation of milk density (35%) and somatic cell count (29%) evaluated by dairy. In this sense, first, dairy producers have the extremely important role of understanding that, at least, the parameters stipulated by legislation must and need to be measured and controlled, in view of the perishability of milk.

In addition, both producers and dairies need to understand that high quality milk will result in better quality processed products and, consequently, greater competitiveness and reach into other markets.

Finally, to achieve superior quality, the actions to be carried out on the properties and the parameters evaluated must be worked out through action plans. However, such plans will only be successful if carried out in partnership with the rural producer, since the gains will only be achieved if everyone works collectively, and not individually within the production chain.

ACKNOWLEDGMENT

Project funded by the Brazilian agency Conselho Nacional de Desenvolvimento Científico e Tecnológico (CNPq - grant number 424722/2018-6) and to Coordenação de Aperfeiçoamento de Pessoal de Nível Superior (CAPES – grant number CAPES/DS nº 88881.593696/2020-01).

REFERENCES

Abreu, F. D. S. (1997). QFD-desdobramento da função qualidade-estruturando a satisfação do cliente [QFD-deployment of the quality function-structuring customer satisfaction]. *Revista de Administração de Empresas*, *37*(2), 47–55. doi:10.1590/S0034-75901997000200005

Abreu, P. S. (2010). *Desenvolvimento da Voz do Cliente no Desdobramento da Função Qualidade (QFD): Aplicação em Projetos de Cadeira de Rodas [Development of the Voice of the Customer in Quality Function Deployment (QFD): Application in Wheelchair Projects].* [Doctoral dissertation. UNICAMP, Brazil].

Akao, Y., & Mazur, G. H. (2003). The leading edge in QFD: Past, present and future. *International Journal of Quality & Reliability Management*, *20*(1), 20–35. doi:10.1108/02656710310453791

Andrietta, J. M., & Miguel, P. A. C. (2007). Aplicação do programa Seis Sigma no Brasil: Resultados de um levantamento tipo survey exploratório-descritivo e perspectivas para pesquisas futuras [A study on the application of the Six Sigma programme in Brazil: results of an exploratory-descriptive survey and perspectives for future research]. *Gestão & Produção*, *14*(2), 203–219. doi:10.1590/S0104-530X2007000200002

Brasil. (2002). Ministério da Agricultura, Pecuária e Abastecimento. Instrução Normativa n°51, de 19 de setembro de 2002 [Normative Instruction No. 51, of September 19, 2002]. Brasília: Mapa, Brazil.

Brasil. (2011). Ministério da Agricultura, Pecuária e Abastecimento. Instrução normativa n° 62, de 26 de dezembro de 2011 [Normative Instruction No. 62, of December 26, 2011]. Brasília, Mapa, Brazil Brasil. (2016). Ministério da Agricultura, Pecuária e Abastecimento. Instrução Normativa n°07, de 03 de maio de 2016 [Normative Instruction No. 07, of May 03, 2016]. Brasília: Mapa, Brazil.

Brasil. (2018a). Ministério da Agricultura, Pecuária e Abastecimento. Instrução Normativa N° 76, de 26 de Novembro de 2018. [Normative Instruction No. 76, of November 26, 2018]. Brasília: Mapa, Brazil.

Brasil. (2018b). Ministério da Agricultura, Pecuária e Abastecimento. Instrução Normativa N° 77, de 26 de Novembro de 2018. [Normative Instruction No. 77, of November 26, 2018]. Brasília: Mapa, Brazil.

Brasil. (2018c). Ministério da Agricultura, Pecuária e Abastecimento. Instrução Normativa n°31, de 29 de junho de 2018 [Normative Instruction No. 29, of June 29, 2018]. Brasília: Mapa, Brazil.

Brito, L. G., da Silva Netto, F. G., Salman, A. K. D., & da Silva, W. C. (2005). Cartilha para o produtor de leite de Rondônia. Vol. I: Instalações, sanidade animal e a Instrução Normativa 51 [Booklet for the Rondônia milk producer. vol. I: Facilities, animal health and Normative Instruction 51]. Embrapa Rondônia-Documentos (INFOTECA-E).

Companhia Nacional de Abastecimento – CONAB. (2018). O Brasil é responsável por cerca de 7% do leite produzido no mundo [Brazil is responsible for about 7% of the milk produced in the world]. https://www.conab.gov.br/ultimas-noticias/2634-brasil-e-responsavel-por-cerca-de-7-do-leite-produzido-no-mundo

Costa, A. R. V. (2018). Método híbrido de desenvolvimento de produto com utilização de mock-up digital e a voz do cliente no contexto da indústria automotiva [Hybrid method of product development using digital mock-up and the voice of the customer in the context of the automotive industry]. Master's thesis, Universidade Tecnológica Federal do Paraná, Brazil.

Costa, C. C. M., de Almeida, A. L. T., Ferreira, M. A. M., & Silva, E. A. (2013). Determinantes do desenvolvimento do setor agropecuário nos municípios [Determinants of the development of the agricultural sector in the municipalities]. *Revista ADM*, 295–309.

Forza, C. (2002). Survey research in operations management: A process-based perspective. *International Journal of Operations & Production Management*, *22*(2), 152–194. doi:10.1108/01443570210414310

Forza, C. (2002). Survey research in operations management: A process-based perspective. *International Journal of Operations & Production Management*, *22*(2), 152–194. doi:10.1108/01443570210414310

Gasques, J. G., Rezende, G. C. D., Villa Verde, C. M., Salerno, M. S., Conceição, J. C. P. R. D., & Carvalho, J. C. D. S. (2004). Desempenho e crescimento do agronegócio no Brasil [Performance and growth of agribusiness in Brazil]. Instituto de Pesquisa Econômica Aplicada - IPEA, 07–38.

GDP - Global Dairy Platform. (2016). *Global Dairy Platform Annual Review 2016*. GDP. https://www.globaldairyplatform.com/wp-content/uploads/2018/04/2016-annual-review-final.pdf

GDP - Global Dairy Platform. (2018). *Global Dairy Platform Annual Review 2018*. GDP. https://globaldairyplatform.com/wp-content/uploads/2019/02/gdp-annual-report-2018-compressed.pdf

Govil, M., & Proth, J.-M. (2002). *Supply chain design and management: strategic and tactical perspectives*. Elsevier. https://www.sciencedirect.com/science/article/pii/B978012294 1511500023

IEA - Instituto de Economia Agrícola. (2018). Diagnóstico da Produção e Consumo de Leite no Estado de São Paulo [Diagnosis of Milk Production and Consumption in the State of São Paulo]. *Análise e Indicadores do Agrongócio*, *13*, 1–10.

Instituto Brasileiro de Geografia e Estatística - IBGE. (2019). Produção da Pecuária Municipal 2019 [Production of Municipal Livestock 2019]. https://biblioteca.ibge.gov.br/visualizacao/periodicos/84/ppm_2019_v47_br_informativo.pdf

Leonel, S. G. (2007). Um estudo do processo de planejamento tecnológico de uma empresa nascente: alinhando tecnologia, produto e mercado com foco na necessidade do cliente [A study of the technological planning process of a nascent company: aligning technology, product and market with a focus on customer needs.]. Master's thesis, Universidade Federal de Minas Gerais, Brazil.

Lorenzo, S., Mira, J., Olarte, M., Guerrero, J., & Moyano, S. (2004). Análisis matricial de la voz del cliente: QFD aplicado a la gestión sanitária [Matrix analysis of the voice of the customer: QFD applied to health management]. *Gaceta Sanitaria, 18*(6), 464–471. doi:10.1016/S0213-9111(04)72034-8 PMID:15625045

Mazo, S. Z. (2012). Incorporação da voz do cliente nas etapas iniciais do desenvolvimento de sistemas produto-serviço (PSS) [Incorporation of the customer's voice in the initial stages of the development of product-service systems (PSS]. Master's thesis. Universidade Tecnológica Federal do Paraná, Brazil. Miguel, P.A.C., & Ho. L. L. (2010). Levantamento Tipo Survey [Survey Method].São Paulo: Elsevier Editora Ltda.

Ministério da Agricultura. Pecuária e Abastecimento – MAPA. (2020). Brasil Projeções do Agronegócio 2019/2020 a 2029/2030 [Brazil Agribusiness Projections 2019/2020 to 2029/2030]. 11° edição. Brasília: MAPA. https://www.gov.br/agricultura/pt-br/assuntos/politica-agricola/todas-publicacoes-de-politica-agricola/projecoes-do-agronegocio/projecoes-do-agronegocio_2019_20-a-2029_30.pdf

Moraes, B. M. M., & Bender, R. (2017). Mercado brasileiro de lácteos: Análise do impacto de políticas de estímulo à produção [Brazilian dairy market: analysis of the impact of policies to stimulate production]. *Revista de Economia e Sociologia Rural, 55*(4), 783–800. doi:10.1590/1234-56781806-94790550410

Oliveira, L. F. T., & Silva, S. P. (2012). Mudanças institucionais e produção familiar na cadeia produtiva do leite no Oeste Catarinense [Institutional changes and family production in the milk production chain in western Santa Catarina]. *Revista de Economia e Sociologia Rural, 50*(4), 705–720. doi:10.1590/S0103-20032012000400007

Pagani Netto, C., Fontes, J. L., Pimentel, J. C. C., & Martins, S. E. (2017). *Mais Leite, Mais Renda – plano de desenvolvimento da bovinocultura leiteira paulista* [Mais Leite, Mais Renda – development plan for dairy cattle in São Paulo, Brazil]. CATI.

Satolo, E. G., Campos, R. S. D., Ussuna, G. D. A., Simon, A. T., Mac-Lean, P. A. B., & Braga Júnior, S. S. (2020). Sustainability Assessment of logistics activities in a dairy: An example of an emerging economy. *Production*, *30*, e20190036. doi:10.1590/0103-6513.20190036

Scalco, A. R., & De Toledo, J. C. (2002). Gestão da qualidade em laticínios do estado de São Paulo: Situação atual e recomendações [Quality management in dairy products in the state of São Paulo: current situation and recommendations]. *Revista de Administração da Universidade de São Paulo*, *37*(2), 17–25.

Stein, R. T., Malinsk, A., da Silva-Reis, C. M., Soares, B. L. C., & Moura, A. S. (2020). *Cadeias produtivas do agronegócio II* [Agribusiness production chains II]. Grupo A.

Talamini, E., Pedrozo, E. A., & da Silva, A. L. (2005). Gestão da cadeia de suprimentos e a segurança do alimento: Uma pesquisa exploratória na cadeia exportadora de carne suína [Supply chain management and food safety: exploratory research into Brazil's pork export supply chain]. *Gestão & Produção*, *12*(1), 107–120. doi:10.1590/S0104-530X2005000100010

Telles, L. B. (2014). Ferramentas e sistema de custo aplicados a gestão da qualidade no agronegócio [Tools and cost system applied to quality management in agribusiness]. Master's thesis. Universidade Tecnológica Federal do Paraná, Brazil.

Viana, G., & Rinaldi, R. N. (2008). Principais fatores que influenciam no desempenho da cadeia produtiva do leite: um estudo de caso dos produtores de leite do município de Laranjeiras do Sul (PR) [Main factors that influence the performance of the milk production chain: a case study of milk producers in the city of Laranjeiras do Sul (PR)] In *XLVI Congresso da Sociedade Brasileira de Economia, Administração e Sociologia Rural*.

Wu, M. L., & Chan, L. K. (2002). Quality function deployment: A literature review. *European Journal of Operational Research*, *143*(3), 463–497. doi:10.1016/S0377-2217(02)00178-9

Zoccal, R. (2019). Leite nas grandes regiões brasileiras. [Milk in large Brazilian regions.] In *Anuário Leite 2019: Novos produtos e novas estratégias da cadeia do leite para ganhar competitividade e conquistar os clientes finais*. Texto Comunicação Corporativa.

APPENDIX: RESEARCH QUESTIONNAIRE

Part A. Characterization of the research organization and respondent

A.1. Research Unit:	
A.2. Respondent's position:	
A.3. Respondent training:	
A.4. Length of time in the position:	
A.5. Year of company foundation:	
A.6. Company size:	() Up to 19 employees.
	() From 20 to 99 employees.
	() From 100 to 499 employees.
	() More than 499 employees.
A.7. Monthly billing range:	() Microenterprise - ME less than or equal to R$ 360 thousand/year.
	() Small Business - EPP greater than BRL 360 thousand and less than or equal to BRL 4.8 million/year.
	() Medium company greater than BRL 4.8 million and less than or equal to BRL 300 million/year.
	() Large company greater than R$ 300 million/year.
A.8. Processing in the unit of liters of milk per month:	() Less than 1 million liters/month.
	() Between 1 million and 5 million liters/month.
	() Between 5 million and 10 million liters/month.
	() Between 10 million and 20 million liters/month.
	() Above 20 million liters/month.
A.9. Governance structure:	() Family Management.
	() Cooperative/Associative Management.
	() Professional Management.
A.10. Do you sell the products to other states?	() No () Yes. Which one(s):
A.11. Do you export the products to other countries?	() No () Yes. Which one(s):
A.12. Do you have any certification?	() No () Yes. Which one(s):
A.13. How milk is stored:	() In Collective tank. () In Individual tank.
A.14. How milk collection is carried out:	() With own transport. () With outsourced transport.

Part B. Identification of customer needs

B.1. Does the dairy evaluate any parameters at the time of milk collection (on the rural property)? () Yes () No **Which?** **Parameter Acceptable value Awarded value Rejectable value degree of importance** *Acceptable value - accepts the receipt of milk. *Awarded value – accepts the receipt of milk and subsidizes the producer (agreed remuneration + bonus). * Rejectable value - does not accept the receipt of milk. Degree of Importance: 1. Irrelevant / 2. Little relevance / 3. Relevant / 4. Very Relevant / 5. Essential.
B.2. Does the dairy evaluate any parameters when receiving milk at the plant (in the company)? () Yes () No **Which?** **Parameter acceptable value Awarded value Rejectable value Degree of importance** *Acceptable value - accepts the receipt of milk. *Awarded value – accepts the receipt of milk and subsidizes the producer (agreed remuneration + bonus). * Rejectable value - does not accept the receipt of milk. Degree of Importance: 1. Irrelevant / 2. Little relevance / 3. Relevant / 4. Very Relevant / 5. Essential.

Chapter 4
An Outlook on Good Manufacturing Practices in the Dairy Industry

Mohana Priya Rajendran
SNS College of Technology, India

Muthuminal R.
https://orcid.org/0000-0001-9931-2204
SNS College of Technology, India

EXECUTIVE SUMMARY

GMP, termed as good manufacturing practices followed in industries to produce safe and quality product. it was established in the year 1995 by Food and Drug Administration (FDA).Several Rules and Regulations were brought into effect under the FDA in the name of Acts from around 1968. This helps to regulate the cleanliness and product quality in the industrial environment, as it is necessary to maintain the quality of the edible products which will be affected due to the surrounding conditions. The act was implemented in the industry with ten principles which can promote the quality product production which will include the premises & personnel's hygiene, the cleanliness of the raw material section along with processing and storage facilities. As dairy industries plays major role among world population, it is necessary for the industries and the technologist to follow the GMP principles. Hence, at this chapter, the detailed discussion over GMP along with its principles in the fore coming sections.

DOI: 10.4018/978-1-6684-5472-5.ch004

INTRODUCTION

What are Good Manufacturing Practices (GMP)

Federal Food, Drug, and Cosmetic Act promoted regulations named GMP which is referred to as Good Manufacturing Process. It is regulated to maintain cleanliness both in the workplace and with personnel hygiene. It is a quality approach where contamination can be prevented. It enables the customer to purchase a product that is not defective. Violating the GMP ends in seizure, fine, and jail, time.

Why GMP is Important

GMPs explain the method, apparatus, services, and controls for producing processed food. It is the least safety measure followed by the food industry to produce safe and wholesome food and it also plays as the basis for FDP inspections.

History of GMP

In 1903, Harvey W. Wiley, a chemist with the US Department of Agriculture and crusader for safer products, established a volunteer "Poison Squad." The young men who joined him agreed to eat only foods treated with measured amounts of chemical preservatives (including borax, formaldehyde, and salicylic, sulfurous, and benzoic acids) to determine whether they were injurious to health. The Poison Squad became a national sensation and raised awareness regarding the need for food safety[http://themasteryinstitute.org/gmpmastery/history.htm]. After a few years, in 1906, Congress passed the Pure Food and Drugs Act, which was considered to be the first major federal consumer protection law with respect to food processing (Dunkelberger, 1995). This law prevented the misbranding and adulterants in foods, drinks, or drugs, as this can prevent food poisoning. The food products and packs were produced in large quantities there still exist poor-quality food products and defective packaging through loopholes available in the law. Till then the awareness among the consumer about the product was not given. In 1993, Act 1906 was regulated by FDA.

- The regulated act 1906 was replaced by the Food Drug, and Cosmetic Act (FDCA) in 1938t. The FDCA was the major decision maker on the necessary quality standard to avoid Consumer fraud. FDCA is considered the basis for GMPs. Two sections of the FDCA are directly related to conditions in a facility where food has been manufactured (Dunkelberger, 1995).
- Section 402 (a)(3) specifies that food has been manufactured under such conditions that it is unfit for consumption (Dunkelberger, 1995).
- Section 402 (a)(4) considers that food may be adulterated if it is prepared, packed, or held under insanitary conditions whereby it may have become contaminated with filth or rendered injurious to health (Dunkelberger, 1995).

The above provision explains in detail about the facility where the food can be stored and produced. To prove that the food is not adulterant it is mandatory to maintain sufficient sanitary conditions. To monitor the good sanitary condition, FDA started its work towards drafting GMP in the mid of 1960s. The GMP regulations for the food industry was finalized in the year 1968 and the focus point of GMP was to regulate general rules to maintain the sanitary condition that should be also adopted and followed by food industries and must ensure the statutory requirements of Section 402(a)(3) and (4) were met. Three broad categories of interrelated issues arose during the development of the GMPs (Dunkelberger, 1995).

- Concern that the regulations were unduly stringent and especially burdensome for small food companies without necessarily improving the quality or safety of foods (Dunkelberger, 1995).
- Contention that the GMP regulations must prescribe conditions that are "reasonably" related to unsanitary conditions that may contaminate food and render it injurious to health (Dunkelberger, 1995).
- Assertions that the regulations did not have the force of law (Dunkelberger, 1995).

These initial two issues can resolve by using general terms such as adequate and appropriate and ample than unbreakable standards. FDA uses two terms that are "shall" and should". The term "shall" be used when compliance is necessary which is felt by the agency and the term "should" is used when the statutory requirements of the act are less relatable to practices that are followed in the industry. When it was established that the FDA did have the legal jurisdiction to impose the GMP regulations, the third point was deemed irrelevant. In April 1969 the GMP regulations were finalized and published as Part 128 of the Code of Federal Regulations (CFR). Part 128 was remodified and issued as Part 110 of the CFR in 1977.

The last drafted GMP was very extensive and didn't mention the exact facility must do to comply. The uncertainty created by GMPs has been rectified by creating industry specified GMP by FDA in mid of the 1970s. In the late 1970s, the GMP was improved by enhancing the specify GMPs. The revisions were finalized in 1986 and printed in 21 CFR 110 (Dunkelberger, 1995). Specific GMPs were also included and printed in 21 CFR Parts 100 through 169 for:

- Quality control procedures for nutrient content of infant formula (21 CFR 106) (Dunkelberger, 1995).
- Thermally processed low-acid canned foods in hermetically sealed containers (21 CFR 113) (Dunkelberger, 1995).
- Acidified foods (21 CFR 114) (Dunkelberger, 1995).
- Bottled drinking water (21 CFR 129) (Dunkelberger, 1995).

A Food Monitoring group was formed by FDA in July 2002, to determine the effectiveness of the current GMP that has been in prevailing in the food industry from 1986. The research was carried out by the monitoring group to enhance the effectiveness of GMP s and to the impact of it in July of 2002, FDA formed a Food GMP Modernization Working Group to examine the effectiveness of current food GMPs and about the effect of GMPs in Food Safety that also include the economic consequences. Due to the efforts of the monitoring team in June 2004, research is still carried out to determine the components that are critical to retain and components that needs some sort of improvement.

Apart from the above international organization, World Health Organizations also concerned GMP especially about pharmaceutical products. The quality of products has been a concern for WHO. The definition of global standards is required in Article 2 of the Constitution of WHO, which states the duties to "develop, establish and promote international standards with respect to food, biological, and similar products." (Gouveia, 2013, Geneva, 2007, Geneva, 2011, Gouveia et.al, 2015).

Timeline of GMP Carried out for Food Processing

Table 1. GMP timeline

Date	Milestone
1906	The Bureau of Chemistry passes the 1906 Pure Food and Drugs Act, prohibiting interstate commerce in misbranded and adulterated foods, drinks, and drugs
1933	FDA recommends revising the 1906 Pure Food and Drugs Act
1938	FDA passes the 1938 Federal Food, Drugs, and Cosmetics Act, which provides identity and quality standards for food
Mid 1960s	FDA decides to clarify the FDCA through GMP regulations
1968	FDA proposes food GMP regulations
1969	FDA finalizes food GMP regulations
Early 1970s	FDA considers promulgating industry-specific regulations
Late-1970s	FDA decides to revise the general GMPs rather than adopting industry specific GMPs
1986	FDA publishes revised food GMPs
2002	FDA forms Food GMP Modernization Working Group
2004	FDA announces effort to modernize food GMPs

Source: Dunkelberger, 1995; FDA, 1981b

GMP IN DAIRY INDUSTRIES

The Good manufacturing practices in dairy industry helps in controlling the working conditions within their facilities and allow the environmental condition that is suitable to obtain a safe product.

Terms and Definition

Cleaning Work Area

This covers the workspace with high necessity of cleanness. For example, the storage area, filling room, and inner packaging area.

Quasi-Cleaning Work Area

This covers the area next to cleaning work area where pretreatment of raw materials is done.

Commonly Work Area

Milk collecting unit, warehouse of packaging material, finished product warehouse comes under this area.

GMP is overall process of providing suitable condition to produce the safe product such as premises, equipment, hygiene management. The classification of dairy industry is shown in Figure 1.

Figure 1. Classification of dairy industry
Source: Prepared by the authors

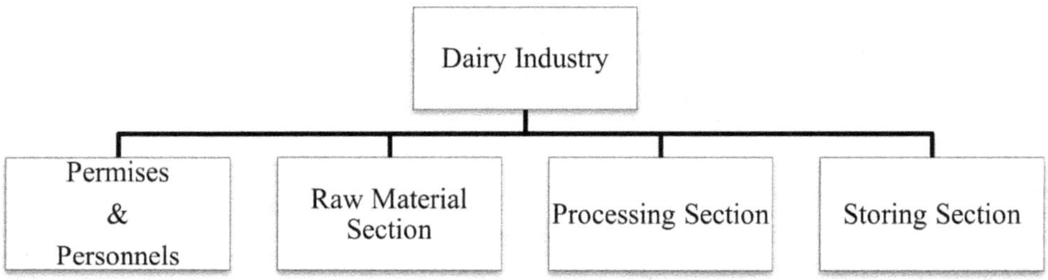

Premises

The premises covers the infrastructure of the plant: building construction and the material used for construction, sanitary amenities, quality of water, design of drainage system, the location surrounding of the plant, waste disposal methods, product flow.

- Building and material used in construction: The building(s) used for the factory shall be planned, constructed, bespoke, and maintained to go with the mechanized operations so as to permit production of milk and milk products under hygienic conditions. They shall conform to the conditions laid down in the Factories Act, 1948 (63 of 1948).
- Sanitary Conditions: The floor should be cleaned using disinfectant often so that contamination from floor can be controlled. The walls of the plant should be designed in such way that prevents contaminations.
- Water System: The water should be potable and should meet the requirement of Bureau of Indian standards or Local Municipality. This could be achieved

by installing a suitable water treatment system for water drawn from own or from other sources. The treated water shall be used only for all operation except washing and cleaning. Water should be stored in such a way that it should be free from microbial affect and shouldn't affect the quality of the water. The water storing container must be cleaned frequently and records of it has to be maintain.

- Location and surroundings: The plant should be in a place which is suitable to control the risk of contamination from external source such as sewage, gas emission, environment, public lavatory, or any other that produces dust, fumes and smoke.
- Disposal of waste: The disposal of sewage and effluents (solid, liquid and gas) from the manufactory shall be in conformity with the requirements of Environmental Pollution Control Board.

Warehousing Area

Sufficient space shall be provided to permit adequate and organized warehouse functioning of different categories of materials and products like preliminary and packaging materials, intermediates, bulk and completed products, products in quarantine, released, rejected, returned, or recalled, machine and equipment spare parts and change items.

- Warehousing shall be designed in as such way that it should ensure good conditions that is adequate to store the goods and it should well be maintained in order to prevent the contamination and spoilage of the raw materials. It should be well maintained in the prescribed temperature and humidity limit, and it should be monitored and recorded regularly. For storing purposes there should be sufficient racks, bins, and platforms. There must be also appropriate housekeeping and rodent, pest control measures and it should be maintained well.
- There should be separate space for storing products that are to be quarantine and it should have clear demarcated.
- Receiving and dispatch bays shall protect materials and products from adverse weather conditions.
- Hazardous and explosive materials such as psychotropic drugs and flammable substances shall be stored in separate safe and secure place. The accessibility to area should be limited. There shall be adequate fire protection measures to ensure the safety of products and personnel.
- There should separate place in warehouse for sampling purpose of the available raw material. If the sampling is done in another place, it should be

done properly in order to prevent the cross contamination of any available product.

- Packaging materials should be placed separately in safe and secure area.

Ancillary Areas

- There should be separate area for changing, storing clothes and other facilities. Toilet should be placed separately from production area. The toilets should be cleaned periodically with disinfection.
- Refreshment area should be provided separately and there should not be any direct connection between these area and manufacturing area.
- Maintenance area shall be separate and should be away from production area. Every time change in spares, it should be placed in well maintained room which should be placed away from the production area. Tools and spare parts that are supposed to be used should be in sterile condition.
- Separate area should be maintained for housing animals and should meet the Drug and Cosmetic rule 150-C (3).

Quality Control Area

- There shall separate area for all sort analysis such physio – chemical, biological, and microbiological. There must be separate instrument room for sensitive and sophisticated result.
- There must separate quality control room with adequate facilities for the operations to be carried out in and to avoid cross contaminations and mix ups. There must be sufficient and appropriate storage facilities to test samples, retained samples. Reference standards, reagents, and record.
- There must separate sections for chemicals and microbiological analysis for microbiology analysis there should be proper air handling system, laminar air flow chamber workstation.
- The laboratory shall be designed in such way that it should be ventilated well and constructed with material suitable properties. The laboratory shall be provided with regular water supply with suitable requirement.

Personnel

The main objective of this topic is to train the employers to ensure safe and sound food production process.

- At initials stages, there should separate orientation program to employs regarding product handling and self-maintenance. This orientation should be carried over thought their journey in industry at interval times. The provided training should be adequate suitable to operation they are involved.
    - Train employees to understand the importance of the critical control points for which they are responsible, the critical limits, the procedures for monitoring, the action to be taken if the limits are not met, and the records to be kept.
    - Train employees responsible for maintenance of equipment impacting on food safety to identify deficiencies that could affect product safety and to take the appropriate corrective action.
    - Ensure employees and supervisors responsible for the sanitation program understand the principles and methods required for effective cleaning and sanitizing.
    - Train employees handling hazardous chemicals on safe handling techniques.
- It is also mandatory to provide additional knowledge about other operations as it is necessary to maintain the continuous operation of specific processing without any breakthrough.
- Maintain separate record that has all the activities conducted regarding the training is given to the employees.

Flow and Practices

It includes operations or flow of both equipment and people throughout the facility and processing practice.

People

The main source of contamination or spoilage next to air and water is through employs who have direct contact with food. So, care must be taken to prevent the contamination of product by following simple rules and regulations.

- To prevent the contamination of products in processing and packaging area there must controlled movement of employs as these are the crucial step of any food processing unit.
- This shall include lab employees, delivery, and maintenance employees as well as employees performing various processing activities. For example:

- Employees in raw receiving areas do not move into any other areas of the facility unless strict procedures are followed (for example, foot baths or sprays, hand dips).
- Employees responsible for the receiving of raw milk do not move throughout the facility as their clothing and shoes may be a source of contamination in clean areas.
- The flow of employees and equipment between wet areas of the facility is controlled to minimize the risk of spreading contamination.

- There must be Control over the access of all visitors to the facility, such as farmers, salespersons, and field personnel.
- Monitor the flow of people in the facility.
  - For example, use color-coded clothing.

Equipment

To reduce the contamination of products movement of employers should be controlled as well as the movement of equipment from one area to another. The process flow must simple straight in order to prevent this contamination.

- Restrict and control the movement of portable equipment (for example, pallets, carts, etc.) from one area to another.
- Do not move equipment to or back from cleaner areas after being in less clean areas without appropriate controls.

Hygiene and Health

This includes the employee's hygiene as well as their personal behavior to ensure safe food protection. There are certain parameters followed by many industries and it also considered to be GMP practices for personnel's. This included clean and suitable clothes for the processing followed by few selfcare to personnel's as well as employee health monitoring which are considered to be mandatory in milk and milk products processing unit. As employs are having direct contact with milk the inappropriate clothing, not grooming according the regulations of GMP are consider to be great threat to milk industry.

Clothing

Clothing should be sound and clean. For example:

- Clothes must be washed or disposed after completing the shift.

- There should not be any buttons or pockets above the waist as buttons can accidently get into the final product.
- Washing of clothes can be done separately in industry or by third party.
- The clothes should wash using disinfectant and dried thoroughly before wearing.
- Strict monitoring should made to Ensure hair and beard coverings are worn properly and provide complete coverage. After completing the shift head caps and mask should dispose and for next day new head cap and mask should be given.
- Disposal gloves should be used, and reputation of same gloves has to be avoided.
- Separate clothes and foot wares must be used for certain areas. Especially when they meet allergy material such as Latex which pose high threat.
- Avoid personal things in food handling area. Should have wear in facility foot ware and clothes to avoid contamination to the product.

Grooming and Behaviors

- The jewelry has to covered and secured, if they aren't removed.
- Forbid any unhygienic activity as it can contaminate the food.
- Nails in hand should remove periodically and kept clean, in case of nail polish it should be removed.
- Activities such as smoking, chewing, spitting, and sneezing and coughing inside processing area should be avoided as it causes contamination of product.
- Instruct employees to clean and sanitize their hands:
    - When starting and returning to work.
    - After using toilet facilities.
    - After hands become soiled or contaminated.

Health

- Periodic health checkup must be carried out to employees those who come in contact with food.
- If there is any illness, symptom of a diseases instructs the employees to report it to the supervisor and immediate care must be given to them and should not come in contact with product.
- If employees have any open cut wounds it must treated well and should be covered properly and assign them work that has no contact with product.

Raw Material Section

Incoming Material (Ingredients and Packaging)

This topic covers all the incoming materials that are being used in processing as well as packaging. The incoming materials are classified into raw product acceptability and nonfood chemical. Milk and creams come under the Raw product acceptability and the materials that nonfood materials will cover in Non-Food Chemicals.

Raw Product Acceptability

As discussed earlier this topic covers the acceptance, receiving, sampling, and monitoring the ingredients such as milk, cream, and other dairy products that are considered to be raw ingredients, and the products are included from all species.

Acceptance

- The incoming milk/cream shall be accepted with a few criteria.
- Each load should be checked separately to ensure the procured materials should meet the criteria such as
    - The incoming milk should be at a temperature 6 °C or less.
    - Antibiotic residues test should be run to verify the product is free from residues.
- Any load that doesn't meet the requirement should be segregated and disposed.
- For instance, if the incoming milk is at a higher temperature than 6 °C, some rectification can be done in the industry. For example:
    - The milk or cream can be cooled to 6 °C or less.
    - Process the milk as earlier as possible so that milk can be preserved.
    - Microbiology should be carried out to check the presence of S. aureus.

Receiving

- Hazardous material higher than 2 mm can be eliminated by using filters on receiving equipment.
- Test should be carried out from each tanker by obtaining a sample of raw milk to meet the requirements.
- The milk should be stored at the appropriate temperature in safe and sound sanitary containers.

- ◦ Most provincial governments have regulations stating the length of time and temperature that milk and cream samples must be held prior to testing and under what conditions they must be held.

Raw Product Monitoring

The raw material such as milk and cream should be examined at frequent intervals to verify that raw materials meet the quality standard in accordance with federal authority such as antibiotics, microbiological content, sediment, and treatable acidity.

Apart from above mention activity GMP covers the following points which must be followed strictly inside the industry.

- Sufficient space shall be provided to permit adequate and organized warehouse functioning of different categories of materials and products like preliminary and packaging materials, intermediates, bulk, and completed products, products in quarantine, released, rejected, returned, or recalled, machine and equipment spare parts and change items.
- Warehousing shall be designed in such a way that it should ensure good conditions that are adequate to store the goods and it should be well maintained in order to prevent the contamination and spoilage of the raw materials. It should be well maintained within the prescribed temperature and humidity limit and it should be monitored and recorded regularly. For storing purposes there should be sufficient racks, bins, and platforms. There must be also appropriate housekeeping and rodent, and pest control measures and they should be maintained well.
- There should be separate space for storing products that are to be quarantined and it should have clear demarcated.
- Receiving and dispatch bays shall protect materials and products from adverse weather conditions.
- Hazardous and explosive materials such as psychotropic drugs and flammable substances shall be stored in separate safe and secure place. The accessibility to area should be limited. There shall be adequate fire protection measures to ensure the safety of products and personnel.
- There should separate place in warehouse for sampling purpose of the available raw material. If the sampling is done in another place, it should be done properly in order to prevent the cross contamination of any available product.
- Packaging materials should be placed separately in safe and secure area.

Non-Food Chemicals

- A list of non – food chemicals should be maintained separately.
- They should be placed in safe and secure place as it can deteriorate the product.
- Identification of them should easy so that the containers which they are kept should be colored or labeled.
- Containers has to be kept closed between uses.
- Only trained personals must use these nonfood chemicals and the usage should be in accordance with manufacturer's instruction.

Processing Section

As mentioned in the flow chart, the processing section consists of a few equipment such as a pasteurizer and homogenizer. They are designed in accordance with GMP Compliance:

Design and Installation

- The design of equipment shall have accessible to sanitize, maintain and examine.
- The equipment designed shall be made of food-grade material that is nontoxic, and noncorrosive with a smooth surface.
- The design of the equipment shall have provisions for drainage to drain the materials used whenever necessary.
- The design and installation of the equipment shall prevent the cross-contamination of products during operation.
- The design of the equipment shall prevent the condensation.
- The parts of equipment shall have coating or lubricant but that should not affect the quality and safety of the product.
- Keep a detailed plan showing the types and locations of equipment used in the facility.

Preventative Maintenance

- There shall be a preventive maintenance program periodically to maintain the equipment in good condition. For instance, the equipment should be calibrated often. Gaskets and filters should be replaced, and there must be periodic conduct routine testing on all significant processes such as High temperature and short-time, batch pasteurizers.

- The equipment shall maintain to ensure there is no physical and chemical hazard and records for that should be maintained.
- The maintenance shall take place in a such way that it will not contaminate the food.
- Whenever faults occur, analysis the preventive program and update it as mandatory.

Seals

- It is mandatory to monitor the use and replacement of seals on important equipment.
- There shall be the usage of unique code to each seal so that it becomes easier to identify the seal that can be replaced.
- There shall be the maintenance of records for the usage of seals and the list of used seals with their location, when the seals are broken, the reason why the seal was broken, that the equipment is re-sealed and the person responsible.

Air Quality

This topic covers all the gases such as air and inert gases that are added directly or by any other means. It includes both ambient air and compressed air sources. Air that is directed onto any process such as agitation, drying process, or incorporated into the products considered to be one of the main sources of contamination. in order to prevent the contamination or spoilage of product there set of requirements to be met by any company they are:

- Processing air can be used which can reduce the potential contamination of the product.
- Air can be processed using filters.
- Sanitary check valves are mandatory in order to avoid product back up into the airline.
- Dismantling the agitation equipment, and sanitary check valves should be done periodically to reduce the contamination and spoilage of the product.
- If air is produced, then it should be from compressing equipment where oil-free air is produced.
- Air intakes should be from clean space.

Metal Detector

Metal detectors are equipment to remove specific product to ensure the safety of the product that is processed and also helps to remove particles that are associated with hazard. Install, maintain and use metal detectors in accordance with the equipment manufacturer's manual.

- Adjust for product effect, selection of target metal and size, and timing of the reject mechanism.
- Test metal detectors at a regularly.
- Keep records that demonstrate the equipment can identify the test material.

Storage Section

This section gives a wide range of knowledge on how to store the finished goods. To prevent spoilage of finished product it must be stored under sterile conditions.

- The products that are stored at ambient temperature must be stored carefully against external agents such as direct sun, excessive heating, moisture, external contaminants, and rapid temperature changes which could adversely affect the integrity of the product container or the safety or suitability of the product.
- Finished Product should store with the required refrigeration temperature at 4oC or less.
- Storage room should be monitored properly.
- Care must be given to humidity sensitive products while storing.
- To prevent deterioration, stock rotation must be carried out.
- To prevent contamination, the wooden shelves should be maintained, clean and periodic inspections must be carried out. The activities carried out should be recorded.

Temperature and Humidity Control

Temperature and humidity play a major role in storing dairy products to prevent the development of condensation and mold growth. It also plays an essential role in the salting and curing stages of the manufacturing process, also important in enzymatic activity and microbial growth in the aging process.

- Use non-breakable thermometers and other automatic devices, such as hygrometers and thermometers, that do not offer a risk of contamination.

- Keep refrigerated units free of excessive frost buildup and maintain them in a clean and hygienic condition.
- Monitor and record temperature and humidity levels to ensure that the product is being held adequately.
- Join the refrigerator unit to a drainage system to ensure proper cooling condensate is removed.

IMPLEMENTATION OF GMP IN INDUSTRIES

To implement the GMP in industries there are set of 10 Principles which helps industry to follow up the GMP which are shown in Figure 2.

Figure 2. The ten principles of GMP

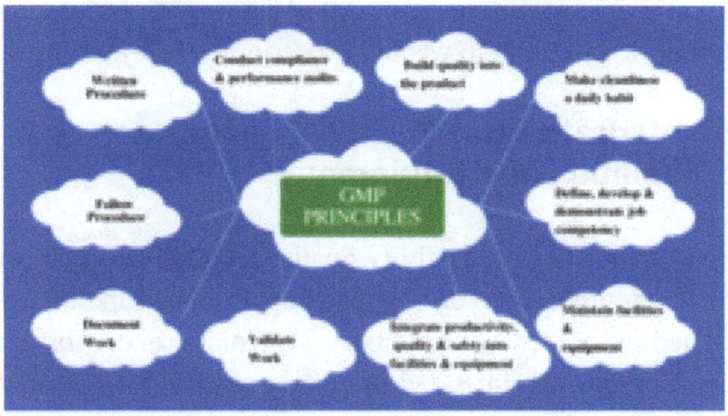

Principle 1: Step-by-Step Written Procedures

In order to make employees, easily understand all the regulations and working instructions, it can be written down and can paste in important areas such as the entrance and near each operation unit. It also includes Standard operating procedures which are methods that have to be followed to perform the designated function regularly.

Principle 2: Follow Procedures

To have controlled and consistent performance, procedures that are to be followed should be to the point and reasonable and can easily be followed to reduce the hazard there should be no shortcuts as it leads to deviation and it should be written clearly.

Principle 3: Document Work

Documentation is essential as it provides certified information which is helpful in traceability and conformity. It also helps fill in the investigation in case of any error occurrence.

Principle 4: Validate Work

The evidence for the documentation is mandatory as it maintains conformity at all stages. Validating and maintaining are essential for regular performance. This helps the industry to check whether the written procedure is followed up.

Principle 5: Integrate Productivity, Quality, and Safety into Facilities and Equipment

It is mandatory to incorporate productivity, quality & employee safety while designing the company and equipment. According to the operational function of equipment, it should be designed and placed in the industry. To avoid error or mix up of segregation of materials according to the intended use is mandatory. Assigning the equipment in the right place can maintain consistent operation at all stages.

Principle 6: Maintain Facilities and Equipment

To reduce the potential risk, it is mandatory to maintain the facility and the equipment with proper documentation. It also helps in reducing the contamination and to maintain the quality of the product. Equipment breakdown can be prevented by proper maintenance.

Principle 7: Define, Develop, and Demonstrate Job Competency

A brief explanation and training must be given to the employees about the procedure of GMP. Training of employees should include both Theory and Practical of GMP Each employee should be provided with training whose activities could affect product quality. Training should be inclusive of basic training on the theory & practice of GMP as well as role-related training. By creating quality goods in a safe and effective manner, employees should show that they are up to the task at hand.

Principle 8: Make Cleanliness a Daily Habit

Good Hygiene practices must be carried out in order to prevent the product from being contaminated and this principle ensures the operation mentioned above is carried out properly. Good hygiene is achieved by maintaining cleanliness in the workplace regularly. The dirtiness at a workstation can cause contamination which contaminates the product by physical, chemical, or microbiological. So, it is important to maintain and sanitize the workplace.

Principle 9: Build Quality into the Product

The quality of the product can be obtained by various efficient controls such as control components, controlling the manufacturing process, packaging & labelling controls, and holding and distribution controls as it becomes mandatory in all stages of product development. Each component used in the processing of products should ensure that they meet defined specifications. This can be achieved only by storing them in a quarantine area for sampling and testing. Maintaining a master record of all the processes can help with error rectification. A batch number is mandatory for all products as it is helpful in traceability. Measures are mandatory to prevent mix-ups and errors.

Principle 10: Conduct Compliance and Performance Audits

Periodic audits must be carried out to ensure GMP is followed regularly. Audits can be internal or external bodies such as FDA to ensure conformity at all stages.

REFERENCES

Dunkelberger, E. (1995). The statutory basis for the FDA's food safety assurance programs: From GMP, to emergency permit control, to HACCP. *Food and Drug Law Journal*, *50*(3), 357–383. PMID:10343005

Gouveia, B. G. (2013) Good Manufacturing Practices for Medicinal Products for Human Use [Dissertation, Lisbon: Lusófona University of Humanities and Technologies.]

Gouveia, B. G., Rijo, P., Gonçalo, T. S., & Reis, C. P. (2015). Good manufacturing practices for medicinal products for human use. *Journal of Pharmacy & Bioallied Sciences*, *7*(2), 87. doi:10.4103/0975-7406.154424 PMID:25883511

The Mastery Institute. (2011). *History of GMP.* http://themasteryinstitute.org/gmpmastery/history.htm

US Food and Drug Administration. (2017). *Good Manufacturing Practices for the 21st Century for Food Processing (2004 Study) Section 1: Current Food Good Manufacturing Practices*. FDA. https://www.fda.gov/food/current-good-manufacturing-practices-cgmps-food-and-dietary-supplements/good-manufacturing-practices-21st-century-food-processing-2004-study-section-1-current-food-good

World Health Organization. (2007). *Quality assurance of pharmaceuticals: a compendium of guidelines and related materials. Good manufacturing practices and inspection* (Vol. 2). World Health Organization.

World Health Organization. (2011). WHO Expert Committee on Specifications for Pharmaceutical Preparations. Forty-fifth report. *World Health Organization Technical Report Series*, (961), 1. PMID:21699061

Chapter 5
Adaptive Management Practices for Robust and Profitable Dairy Operations in the Tropics Upon the Uncertainty of Climate Change

Guillermo Ortiz-Colón
https://orcid.org/0000-0002-4602-2964
University of Puerto Rico at Mayagüez, Puerto Rico

EXECUTIVE SUMMARY

One of the greatest challenges of dairying in tropical regions of the world is heat stress. Upon the uncertain effects of climate change might have in dairy cattle productivity in the tropics, intrinsic advantages exhibited by heat tolerant breeds, in combination with adaptive nutritional management for heat stress, and modifications to the tropical environment, could be valuable in the quest to assure a robust and profitable dairy operation in our region. In the American continent, farmers have selected bovines better adapted to the region, like the hot and humid environment of Puerto Rico, where we can find "purebred" Holstein cattle that phenotypically have a very short and slick hair. This chapter intends to present the main managements practices carried out in dairy herds in the tropics upon the climate changes.

DOI: 10.4018/978-1-6684-5472-5.ch005

INTRODUCTION

One of the greatest challenges of dairying in tropical regions of the world is heat stress (HS). Currently, but worsening in the future, climate changes may also expose the temperate regions of the world to hotter and longer hot seasons, perhaps with higher relative humidity (West, 2003; Berry et al., 2016; Assenget al., 2021). Because the principal dairy breeds have been mostly selected from temperate regions, climate change might increasingly expose dairy cows in all parts of the world to chronic and acute HS, negatively affecting dairy profitability. To minimize the effects of HS, three main management approaches have been identified: 1) selection for heat-tolerance; 2) dietary approaches specifically developed for the heat stressed cow, 3) alterations of the dairy cow's environment (Beede and Collier, 1986; Asseng et al., 2021). Upon the uncertain effects of climate change might have in dairy cattle productivity in the tropics, intrinsic advantages exhibited by heat tolerant breeds, in combination with adaptive nutritional management for HS, and modifications to the tropical environment, could be invaluable in the quest to ensure robust and profitable dairy operations in our region.

GENETIC DEVELOPMENT AND USE OF HEAT-TOLERANT BREEDS

A valuable strategy to decrease the severity of heat stress (HS) on dairy cattle is the selection of animals with improved tolerance to tropical conditions. For European cattle breeds localized in the tropics, HS is chronic and usually there is only small heat relief at night, if any (Berman, 2011). In addition, year-round intense radiant energy significantly adds to the heat load to these animals. The tropics are particularly harsh to European lactating dairy cows that generate a large amount of metabolic heat (Wheelock et al. 2010). In tropical conditions, the high environmental temperature, combined with augmented humidity, added to the metabolic heat associated to lactogenesis and galactopoiesis, results in greater rectal temperatures in cattle susceptible to HS (West, 2003; Berry et al., 2016). Indeed, once the bovine corporal heat exceeds 25°C, the capacity of dairy cattle to dissipate heat diminishes significantly (Kadzere et al., 2002; Bernabucci et al., 2010; Belhadj et al., 2016), which further diminishes dry matter intake (DMI) and negatively affect galactopoiesis (West, 2003; Berry et al., 2016). As the prospect of global warming threatens extended periods of aggravated HS across the world, severe negative effects on the productivity, reproduction, and welfare of dairy cattle would be a dire consequence of climate change worldwide (Berman, 2011; Mylostyvyi and Sejian, 2019).

Unfortunately, heat tolerance is negatively correlated with milk production (−0.3; Ravagnolo and Misztal, 2000). Therefore, if selection for higher milk production continues while ignoring heat tolerance, the result would be a progressive decrease in heat tolerance in dairy breeds (West, 2003; Berry et al., 2016). However, because the negative correlation between milk production and heat tolerance is small, a combined selection for milk production and heat tolerance is possible (West, 2003; Berry et al., 2016). Upon the uncertain effects of climate change might have in dairy cattle productivity, research of intrinsic differences exhibited by heat tolerant breeds would be invaluable in the quest to genetically select animals for thermal tolerance, with the goal to maintain high milk production efficiency under the foreseeable global warmer climate.

SLICK (SL) GENE BOVINES

A valuable strategy to decrease the severity of HS impact on dairy cattle is the selection of animals with improved thermo-tolerance, a management practice that has been practiced in our American continent for centuries. Concomitantly with the European conquest of the American continent, European cattle arrive to America more than five centuries ago. Therefore, farmers in America have been selecting the bovines better adapted to their regions, and in the particular case of Puerto Rico, to the heat and humid environment of the national archipelago (Molina-Fernández, 2001). In 1898, Puerto Rico was militarily invaded by the United States of America (USA), and subsequently there was a large influx of European dairy breeds (primarily Holstein cattle) into Puerto Rico from the USA (Molina-Fernández, 2001). Eventually farmers in Puerto Rico started cross breeding their creole cattle; product of centuries of selection in the tropics; with the "novel" Holstein dairy cattle that although more productive in terms of milk production were non-adapted to HS. Presently, after decades of Puerto Rico's farmers selecting these crossbreed animals, in Puerto Rico we can find "purebred" Holstein cattle that phenotypically have a very short, sleek, and mostly glossy coat very similar to the phenotype described by Olson et al. in 2003, after introducing the Slick gene from the Senepol into Holstein cows. We will call these short hair Holstein cows Slick (SL) Holstein cattle.

At the molecular level, specific single nucleotide polymorphisms (SNP) have been associated with the thermo-tolerance in these Holstein SL cattle (Dikmen et al., 2008; Dikmen et al., 2014). In Lajas, Puerto Rico, lactating SL Holstein cows have been documented to have lower rectal temperatures under HS conditions, and therefore are regarded as more tolerant to high temperatures and humidity than Wildtype (WT) Holstein cattle (Curbelo et al., 2016; RíosSolís et al., 2019a). Moreover, in Florida, USA; SL Holstein dairy cows were documented to have lower respiration rates and lower vaginal temperatures than WT Holstein cows (Dikmen et al., 2008). Dikmen et al., 2008 also documented that SL Holstein dairy cows showed greater sweating rates than WT cattle, a difference that was vanished if WT animals have their hair clipped (Dikmen et al., 2008). Therefore, it was suggested that the main thermo-tolerance advantage of SL cows could be predominantly a result of an enhanced capability to dispel heat through perspiration, resulting from the shorter of hair length in the SL phenotype, as clipped hair cows showed a decrease in rectal temperature and an increase in milk production that paralleled that of the SL phenotype (Dikmen et al., 2008). The short hair length in SL bovines therefore decreases insulation to conductive and convective heat loss in the hair coat, causing in an enhanced tolerance to HS (Berman, 2011). Moreover, Dikmen et al. (2014), evaluated HS tolerance of lactating SL Holstein cows, non-SL Holstein cows (but genetically related to the SL group), and WT Holstein cows (completely genetically unrelated to the SL group). Body temperature (vaginal and rectal) were inferior in SL cows than in non-SL relatives and WT cows (Dikmen et al. (2014). Furthermore, as HS increased during the day, respiration rate increased more in non-SL relatives or WT cows than in SL dairy cows, and in concordance, perspiration rates were greater in SL dairy cows (Dikmen et al. 2014). At the end, the superior thermoregulatory ability of Holsteins with slick hair compared with non-slick animals also resulted in a less drastic depression in milk yield during the summer (Dikmen et al., 2014). Curiously, Dikmen et al. (2014) reported that non-SL cows that were genetically related to the SL group still demonstrated an improved tolerance to HS than the WT dairy cows, keenly proposing that other mutations besides the SL one could be associated with adaptations to HS (Dikmen et al., 2014). Unidentified inherent dissimilarities in HS tolerance between SL and WT could be the result of other intricate mechanisms in addition to merely hair length. In this same experiment, seasonal variations in milk components (lactose, protein, fat) and mastitis indicators (i.e., somatic cell counts) were larger in SL Holstein cattle (Dikmen et al., 2014), what powerfully denote that SL bovines may have differences associated with the regulation of lactogenesis, galactopoiesis and in the immune system.

It has been well established that under HS, differences exist between SL and WT dairy cows in terms of body temperature (Curbelo et al., 2016; RíosSolís et al., 2019a), milk production (Dikmen et al. 2014), and perspiration (Dikmen et al. 2008). However, to the best of our knowledge, only one study has compared the feed efficiency (FE) between SL and WT dairy cows. Feed efficiency evaluates the relative ability of dairy cows to turn feed nutrients into milk or milk components. Basically, FE is the kg of milk produced per kg of feed (dry matter basis) consumed. This selection criterion must always be contemplated when selecting dairy animals, but it is extremely important when feed prices are high and milk prices are low. Therefore, an important tool to improve profit margins for dairy producers in Puerto

Rico and the rest of Latin America is to increase the FE of the dairy animals. Moreover, dairy cattle with higher FE will excrete less nutrients per unit of product, minimizing the environmental impact of dairy farms in our region. Feed efficiency can be enhanced by two mechanisms: increasing milk yield with the same DMI or decreasing DMI while maintaining the same milk yield.

Field evaluations, using official data from the Dairy Records Management System (DRMS - https://www.drms.org/) of dairy cows in Puerto Rico, insinuated that SL Holsteins had higher milk production than WT Holsteins in tropical conditions (Pantoja et al., 2005; OrtizCol6n et al., 2018). However, when data from SL and WT dairy cows was used to compare the milk production of both genotypes throughout the years, factors such as stage of lactation and number of lactations per cow (i.e. cows 'maturity) were not considered. Moreover, it was unfeasible to record DMI in these preliminary evaluations (Pantoja et al., 2005; OrtizCol6n et al., 2018).

Therefore, Ríos-Solís et al., (2019b) conducted a controlled study in Tai South Farm, Lajas, Puerto Rico, to determine how HS conditions could affect milk yield (MY), milk composition, DMI, energy corrected milk (ECM), feed efficiency (FE) and some physiological responses in SL and WT Holstein dairy cows. The physiological responses evaluated were rectal temperature (RT), respiration rate (RR) and heart rate (HR). Slick and WT Holstein dairy cows were evaluated in mid lactation. Contrary to our expectations, no differences in MY and milk composition between genotypes were found (MY: SL 20.49 $\pm$ 0.94 kg vs. WT 19.49 $\pm$ 0.88 kg). In addition, no differences in DMI were documented (SL 18.72 $\pm$ 0.3773 kg vs. WT 19.11 $\pm$ 0.3840 kg). Accordingly, there was no difference in FE between genotypes (FE: SL 1.22 $\pm$ 0.05 vs. WT 1.11 $\pm$ 0.0780). Although we were expecting to see marked differences between SL and WT Holstein dairy cows, both genotypes presented equivalent milk production capabilities. These results, however, were in direct conflict with the field data that supported that SL Holsteins presented a superior ability to manage heat stress and produce milk in the dairy farms in Puerto Rico (Pantoja et al., 2005; OrtizCol6n et al., 2018).

In light of the results discussed above, we revisited the official data from the Dairy Records Management System (DRMS - https://www.drms.org/) of dairy cows in Puerto Rico during the "hot" season (March to August) but this time considering lactation number (1st, 2nd, and [3] 3rd lactations and lactation stage (i.e. early lactation, peak lactation, midlactation, and latelactation) (Ortiz-Uriarte et al., 2020). Under these stricter conditions of data analysis, no differences (P = 0.766) were found in MY during the hot period between SL and WT cows, in congruence with the study of Ríos-Solís et al. (2019b). We hypothesized that as a consequence that in Puerto Rico dairy cows are under HS constantly, both genotypes (SL and WT) can be acclimatized to HS, a different situation to experiments done in other latitudes (i.e. Florida, USA), and this acclimatized status could explain why no differences in milk yields were found between SL and WT in our studies (Ríos-Solís et al., 2019b; Ortiz-Uriarte et al., 2020). Conversely, in this same study, the calving interval (CI) of SL Holsteins was 1.64 months shorter than WT Holsteins, what suggest a strong reproductive advantage for SL animals.

It is well established the negative impact that HS has on bovine's reproductive efficiency by hampering the development of gametes and embryos (Hansen, 2009). Moreover, Hansen (2019) indicated that HS can negatively modify embryos' attributes via alterations in folliculogenesis. Bovines' gonadotropin secretion, steroidogenesis, ovarian follicular growth, oocyte competence, development of the corpus luteum, embryonic development, and the uterine endometrial environment are all deleteriously affected by HS (Hansen, 2019; Wolfenson and Roth, 2019). Indeed, under HS, embryo viability during the first week is diminished resulting in a lower pregnancy rate (Ryan et al.,1993; Bertipaglia et al., 2018). Moreover, HS has been linked with briefer and/or silent heats, leading to longer CI (Younas et al., 1993; De Rensis and Scaramuzzi, 2003; ElTarabany and El Tarabany, 2015). Indeed, in WT bovines a temperature humidity index (THI) exceeding 72 has been related with a diminished 1st service conception frequency (Bohmanova, et al., 2007; Morton et al., 2007; Dash et al., 2016;).

Fortunately, HS adapted SL Holstein consistently exhibit shorter calving interval than their wild-type (WT) cohorts (Pantoja et al. 2005; Ortiz-Uriarte et al. 2020). Conceivably, under HS, SL Holstein dairy cows can maintain balanced hormonal secretions, more noticeable estrous, oocytes with characteristic that enhanced the likelihood of initial impregnation, what when combined, result in enhancements in conceptus development that allows SL cows to maintain their pregnancies under HS, what eventually result the observed phenotype of a shorter calving interval.

It is suggested that the reputed superior milk production advantage of SL dairy cows under HS is the indirect result of a shorter CI, what will result in lower average of days in milk (DIM) in SL cow herds. A herd with lower CI and lower DIM at any given time will have a higher milk yield as a group. Since in HS, SL Holstein lactating cows seem to perform reproductively better than WT Holsteins, the SL gene could be a suitable adaptive approach to maintain a proficient dairy industry within a hotter worldwide weather. Slick cattle ability to maintain the expression of their inherited reproductive advantage under HS is of paramount importance as an adaptive management practice for a robust and profitable dairy production in the tropics upon the uncertainty of climate change.

IMPROVED NUTRITIONAL MANAGEMENT PRACTICES FOR THE HEAT STRESSED COW

Under heat stress (HS) the dry matter intake (DMI) can be decreased up to 30%, what consequently limits the energy and protein intake of the lactating dairy cow. We did not detect differences in DMI between Slick (SL) and Wild type dairy cows (Ríos-Solís et al., 2019b). Consequently, it seems that even if we used dairy cows tolerant to HS, certain diet modification must be considered in order to provide the required nutrients for an optimum milk production.

It is typical in tropical grazing dairy productions systems that dairy cows under HS present lower DMI throughout the day causing the grain portion of the diet to be a higher proportion of the dairy cow DMI (Stone, 2004;). Moreover, those cows on pasture are generally supplemented with concentrated grain feeds during the milking time, obliging the dairy farmer to split the concentrate ration in mostly two (seldomly three) short (i.e., £ 10 minutes) feeding periods. However, to promote the stability of the rumen pH, we should not feed more than 2.26 kg of concentrate in a single cow feeding period (i.e., at milking) (Stone, 2004; Alatas et al., 2015). Therefore, dairy farmers that desire to offer daily 9 kg of concentrated feed to each cow, should divide the concentrate ration into 4 feeding periods within a 24-hour period, that is, we must provide 2.26 kg of feed to the cow every 6 hours, instead of 4.5 kg in each one of two milking periods. A viable alternative is to use automatic feeders that distribute concentrated feed to the cow during the day and night in portions £ 2.26 kg.

To avoid ruminal acidosis under HS, it is imperative that forage intake should be ^{3} 1.8% of cow's live weight (LW). That is, for a 544 kg cow, it should consume per day 9.8 kg of forage (DM basis), equivalent to 49 kg of fresh forage. On chemical terms, the neutral detergent fiber (NDF) intake of the forage should be ^{3} 0.9% of the LW or 19% of the DMI (Stone, 2004; Alatas et al., 2015) and the total NDF in the diet should be ^{3} 28% of total DM (1.1-1.2% of the animal's LW). Moreover, according to the National Research Council (NRC) acid detergent fiber (ADF) in the diet should be 19-21% of total DMI (NRC 2001).

Beyond the chemical definitions of fiber (NDF and ADF), in the diets of dairy cattle we should assure a minimum level of physically effective NDF (peNDF), also known as physically effective fiber (NRC 2001). The peNDF stimulates rumination and saliva production (Van Soest, 1994). Ruminal contractions are also stimulated by peNDF, which in turn stimulate volatile fatty acid uptake (VFA) and saliva distribution throughout the rumen, buffering VFA and thus stabilizing rumen pH (Van Soest, 1994). Providing dairy cows with 0.23-0.45 kg of straw/cow/day can help maintain rumination (Stone, 1994; Alatas et al., 2015).

When feeding Totally Mixed Rations (TMR), forages used should be processed to a length of 2.5-5 cm to prevent cows from selecting grains over forages (Stone, 2004; Alatas et al., 2015). Moreover, the TMR must always be available to the cows, what requires the TMR to be pushed towards the cows 8-10 times a day, targeting 3% of residual feed to ensure that all cows ate ad libitum and to avoid over-consumption at the next feeding (Stone, 2004; Alatas et al., 2015). The critical minimum linear space per feeder should be 0.2 linear meters per cow. Additionally, although the NRC 2001, recommends that the effective neutral detergent fiber (peNDF) in the diet should be 22% of DMI, other authors have recommended that dairy cows should be fed peNDF 1-2% above the NRC 2001 recommendations (Stone, 2004; Alatas et al., 2015).

Under normal conditions, the rumen of dairy cows is slightly acidic, with a pH of $\cong$ 6.0 (Van Soest, 1994). Under these conditions, forage fiber-degrading cellulolytic bacteria and grain starch-degrading amylolytic bacteria present in high-energy diets commonly fed to dairy cattle can coexist (Van Soest, 1994). Although a pH < 5.2 represents acute rumen acidosis and the animal is clinically ill, and with a pH < 5.6 dairy cows are considered to have subclinical ruminal acidosis (SARA); with a pH <6.0 the rumen microflora is negatively affected causing inhibition of fiber fermentation and negative alterations in the biohydrogenation of fatty acids in the rumen, what predisposes the dairy cows to milk fat depression (Stone, 2004; Alatas et al., 2015). Under HS, dairy cows are predisposed to ruminal acidosis when fed typical dairy cattle diets high in non-structural carbohydrates (Stone, 2004; Alatas et al., 2015). Therefore, under HS, to minimize ruminal disturbances and enhance milk production dairy managers should take the proactive actions discussed below.

Under HS the dairy cow ruminates less. With less rumination, fewer saliva gets produced. Thus, decreasing bicarbonate (HCO3-) influx into the rumen (Stone, 2004; Alatas et al., 2015). In addition, the heat stressed cow has a lower number of ruminal contractions, which minimizes the mixture of bicarbonate with ruminal acids and increases the digest retention time in the rumen, producing more organic acids per unit of DM (West, 2003; Berry et al., 2016). To top it all, under HS, dairy cows use panting to dissipate the excess heat but, in doing so, they lose HCO3- in the process of exhaling CO2 (Stone, 2004; Alatas et al., 2015). Consequently, cows suffering from HS will be more predisposed to suffer rumen acidosis.

The addition of sodium bicarbonate and potassium bicarbonate to cows receiving high-grain rations can help stabilize rumen pH, prevent a decline in milk fat, and help optimize fiber digestion (NRC, 2001). However, bicarbonate is not palatable enough to expect cows to consume the amount of bicarbonate needed per day when offered ad libitum. Therefore, the dairy nutritionist must ensure that dairy cows consume between 0.8-1.0% of its total DMI as bicarbonate. Hence, for a dairy cow consuming 18 kg of DM, it should consume between 144 and 180 grams of sodium bicarbonate and/or potassium bicarbonate per day. Assuming the cow consumes 9 kg of grain concentrate dry feed per day, the feed should contain between 1.6 to 2.0% bicarbonate. Moreover, magnesium (i.e., magnesium oxide) at a rate of 0.36 to 0.4% of the DMI has been shown to help stabilize the rumen pH (NRC 2001; Stone, 2004; Alatas et al., 2015).

Besides assuring a minimum level of NDF and peNDF in the dairy cow diet, we must limit the amount of non-fibrous carbohydrates (NFC) to 38% of dry matter intake (DMI) (NRC, 2001). Among the NFC, we find organic acids, fructans, starch, sugars, and pectin. Under HS we should limit the starch level to 26% of the DMI (Koch and Lascano 2018). Moreover, we should limit the sugar level to 4-6% of DMI. It has been shown that substituting some of the starch energy with molasses (£ 5% of DMI) can help stabilize the rumen pH (Koch and Lascano 2018). The addition of molasses can increase butyrate production, thus stimulating blood flow and potentially increasing the rate of absorption of VFA through the ruminal epithelium (Koch and Lascano 2018). Sugars may also increase the competitive uptake of the VFA's acetate and propionate in the rumen, suggesting that their regulated transport is stimulated, which could increase the flow of bicarbonate (HCO3-) into the rumen and could attenuate a drastic drop in rumen pH when the cow is fed sugars (Chibisa et al. 2015). Another valuable strategy to stabilize the rumen pH and prevent low milk fat is to replace part of the dietary starch with pectin (Voelker and Allen 2003). However, the maximum total NFC in the diet should remain 38%. Various pectin sources include legumes; soybean hulls; beet pulp and citrus pulp (Van Soest, 1994).

The heat stressed cow has an impaired appetite. With a lower feed consumption, the heat stressed cow is predisposed to have a greater negative energy balance that a cow in a thermoneutral environment. One of the strategies that have been use with success to improve the energy balance of dairy cows under heat stress is fat supplementation (Harvatine and Allen; 2006; Palmquist and Jenkins, 2017). Under HS fat supplementation offers the advantage that is does not add fermentation heat, and improvements in milk production of up to 1 kg/milk/cow/day have been reported (Harvatine and Allen; 2006; Palmquist and Jenkins, 2017). Scientific studies have documented increases of 1.5 L milk/cow/day (Harvatine and Allen; 2006; Palmquist and Jenkins, 2017). The supplementation with rumen inert fat at the rate of 226-453 grams/cow/day is a common recommendation (Harvatine and Allen; 2006; Palmquist and Jenkins, 2017). An example of a product that delivers rumen inert fat is Energy Booster H™ (Hubbard Feeds® an Alltech® Company [Nicholasville, Kentucky, USA]). However, because dairy cows under HS are susceptible to rumen acidosis, they are also predisposed to the milk fat depression (MFD) phenomenon. Consequently, we need to watch closely our fat supplementation sources and limit the amount of total unsaturated fatty acids (UFA) in the diet in order to avoid MFD (Lock et al., 2008; Palmquist and Jenkins, 2017). Because good quality forages can have significant amounts of poly unsaturated fatty acids (PUFA), but this type of forage should be the basis of the dairy cow's diet (³ 1.8% of the cow's live weight); then we must manage the other sources of PUFA in the diet (Lock et al., 2006; Palmquist and Jenkins, 2017). Indeed, the total amount of lipids in the diet should be limited to 5-5.5% of total DMI (Lock et al., 2008; Palmquist and Jenkins, 2017). If we are using rumen inert fats, we could use lipids £ 6.5% of total DMI (Koch and Lascano 2018; Palmquist and Jenkins, 2017). Indeed, under HS we must carefully monitor the amount of UFA that the diet provides to the rumen (RUFAL) (Koch and Lascano 2018; Palmquist and Jenkins, 2017). This burden of UFA to the rumen have been defined as RUFAL (Palmquist and Jenkins, 2017).. Standing for Rumen Unsaturated Fatty Acid Load, RUFAL) "is a number meant to show the total unsaturated fatty acid level entering the rumen from all feeds consumed each day and it was designed to be a better indicator of fermentation disruption in the rumen than just relying on the percentage of fat added to the diet" (Palmquist and Jenkins, 2017). When we are referring to RUFAL, we are considering the summation of oleic (C18:1), linoleic (C18:2), and linolenic (C18:3) acids (Koch and Lascano 2018; Palmquist and Jenkins, 2017). To prevent MFD, we must limit RUFAL to £ 3.5% of DMI. For a cow consuming 18 kg of DM, the RUFAL should be less than 630 grams. To achieve a RUFAL £ 3.5% we must limit the addition of the following dietary ingredients: fish oil, Distillers Dry Grains (DDG), and whole oily grain seeds (i.e., soy and cotton) (Koch and Lascano 2018).

Under HS, the dairy cow may decrease feed consumption between 15-30%. Therefore, the dairy nutritionist must increase the crude protein (CP) content of the diet to comply with the protein requirements of a high producing dairy cow. Some reports have determined the need of increasing the diet CP from 16.3% 21.2% to maintain a similar level of production under HS (Stevens et al., 2021). However, to minimize the energy cost of urea production from "excess" protein, we must supplement the dairy cows with high quality protein, not degradable in the rumen like blood meal (226-453 grams/cow/day); fish meal (<453 grams/cow/day); soybeans (Extruded (£1.81 kg/cow/day) (Stevens et al., 2021). Moreover, because the efficiency of converting CP from feed to milk protein is £ 30%, but this efficiency increases as milk productivity per cow increases, the protein supplement must be precisely fed in the amount of CP that the cow needs for a given amount of milk (VandeHaar and St-Pierre, 2006). Crucial for this is the use of cow milk production data that indicate the quantity and quality of milk produced and that we use scientifically formulated feed rations specifically formulated to cover the nitrogen (and another nutrients) needs of the cow. In this way we optimize the utilization of nitrogen by the ruminant and, consequently, limit the amount of excreted nitrogen and N2O emissions (VandeHaar and St-Pierre, 2006).

The heat stressed dairy cow has different minerals requirements when compared to similar animals under thermoneutral conditions because of the decrease in DMI and increased perspiration. Under HS, bovines suffer larger losses of sodium (Na+) and we should increase the Na+ concentration in the diet from 0.18% to 0.4-0.5% (NRC 2001). Moreover, bovine sweat is high in potassium (K+) and HS significantly increase K+ losses and we should increase the % of K+ in the diet from 0.18% to 0.4-0.5% (NRC 2001). Importantly, when we increase the K+ concentration in the diet, we will interfere with the absorption of magnesium (Mg+) in the rumen, and consequently we should increase the Mg concentration in the diet from 0.25% to 0.35% (NRC 2001). Indeed, we should aim for a K: Mg ratio < 5:1 (NRC 2001; Stone, 2004). Magnesium oxide is a common source of Mg supplementation that also acts as a buffer (NRC 2001). Under HS we must refrain from offering salt ad libitum. Salt (NaCl), in terms of mass, is 39% Na and 61% Cl. Dairy cows only required about 0.30% of DM (NRC 2001). Because salt is palatable for dairy cows, they might end of consuming amounts of Cl detrimental for the performance of dairy cows under HS (Sánchez et al., (1994).

In terms of the dietary cationic anionic differences (DCAD), under HS we should aim for a positive DCAD 250-350 mEq/kg, although benefits have been found with DCAD of up to 520 mEq/kg (NRC 2001). This desirable DCAD can be achieve by supplementing dairy cows with a combination of sodium bicarbonate, potassium carbonate, and/or potassium bicarbonate at a rate of 113-151 grams/cow/day combined (NRC 2001).

PHYSICAL MODIFICATION OF THE ENVIRONMENT

Lower DMI of dairy cows will partially explain lower milk production under HS (Mylostyvyi, and Sejian, 2019; Asseng et al., 2021). Moreover, HS will also hamper reproduction efficiency in the dairy cow (Mylostyvyi, and Sejian, 2019; Asseng et al., 2021). This negative HS effects can be seen with ambient temperatures as low as 25°C (Collier et al., 2008; Asseng et al., 2021). Moreover, the tropical conditions of high temperatures with concomitant high relative humidity, makes the HS strenuous for dairy cows (West, 2003; Berry et al., 2016; Asseng et al., 2021). To account for the influence of ambient humidity on heat stress, the Temperature Humidity Index (THI) is utilized to evaluate the level of HS that dairy cows are exposed to. The THI is computed with the formula: THI = (Dry bulb temperature C°) + (0.36 x dew point temperature C°) + 41.2 (Mylostyvyi, and Sejian, 2019; Asseng et al., 2021). When dairy cows are under a THI of ^{3} 65, we should take measures to ameliorate HS and avoid its detrimental effects on milk production and fertility (Mylostyvyi, and Sejian, 2019; Asseng et al., 2021).

Mylostyvyi, and Sejian, 2019 have summarized the welfare of dairy cattle in conditions of global climate change. Briefly, at the bare minimum, dairy farmers must provide their cows with adequate shade (³2.61 m2/cow) to minimize the amount of heat gain through radiation. Appropriate shade has been shown to increase milk production 13.6%, decrease respiratory rate in 34%, increase calf birth weight and colostrum IgG concentration. Moreover, water just be available ad libitum, and on pasture cows must have available water troughs at £ 213.36 m from any point of the pasture to optimized pasture use. In pasture and at the exit of the milking parlor as cows can drink 200-250 liters of water/cow/day under heat stress. Under extreme heat, dairy farmers should pasture cows exclusively at night, while providing a partial TMR under shaded structures equipped with fans, water sprinklers and or showers to aid dissipate heat when indoors, and while waiting to be milked. Indeed, cows should only be milk in the cool hours of the day (6:00 and 18:00) and fresh food should be added overnight or in the early morning. Dairy farmers in the tropics should also consider stablishing silvopastoral systems as it has been reported temperatures 12°C cooler under the tree shades. Trees should be planted at a proportion of 30-50 ha (14-18 x 14-18 m) and tree of different heights should be combined (Tall: 25-30 m and Low height: 9-10 m) while shade tolerant grasses should be used (Orozco et al., 2012).

CONCLUSION

To manage profitable dairy operations in the tropics upon the uncertainty of climate change, dairy farmers must utilize animals adapted to the tropical environment like slick (SL) dairy cattle, provided them diets specially formulated for the heat stressed cow (i.e., with supplemental inert fats; positive DCAD); and protect them from excessive heat gains from the environment.

REFERENCES

Alatas, M. S., Citil, O. B., Kahraman, O., & Ozbilgin, A. (2015). Causes of milk fat depression in dairy cows. Scientific Papers. *Series D. Animal Science*, 80-85. http://animalsciencejournal.usamv.ro/pdf/2015/Art11.pdf

Asseng, S., Spänkuch, D., Hernandez-Ochoa, I. M., & Laporta, J. (2021, June). The upper temperature thresholds of life. [Erratum in: Lancet Planet Health. 2021 Sep;5] [9] [: e578. PMID: 34119012. Doi: 10.1016/S2542-5196] [21]. *The Lancet. Planetary Health*, *5*(6), e378–e385. doi:10.1016/S2542-5196(21)00079-6 PMID:34119012

Beede, D. K., & Collier, R. J. (1986). Potential nutritional strategies for intensively managed cattle during thermal stress. *Journal of Animal Science*, *62*(2), 543–554. doi:10.2527/jas1986.622543x

Belhadj Slimen, I., Najar, T., Ghram, A., & Abdrrabba, M. (2016). Heat stress effects on livestock: Molecular, cellular and metabolic aspects, a review. *Journal of Animal Physiology and Animal Nutrition*, *100*(3), 401412. doi:10.1111/jpn.12379 PMID:26250521

Berman, A. (2011). Invited review: Are adaptations present to support dairy cattle productivity in warm climates? *Journal of Dairy Science*, *94*(5), 2147–2158. doi:10.3168/jds.2010-3962 PMID:21524505

Bernabucci, U., N. Lacetera, L.H. Baumgard, P.P. Rhoads, B. Ronchi, and A. Nardone, (2010). Metabolic and hormonal acclimation to heat stress in domesticated ruminants. *Animal: an international journal of animal bioscience, 4*, 1167. doi:10.1017/S175173111000090X

Berry, D. P., Friggens, N. C., Lucy, M., & Roche, J. R. (2016). Milk production and fertility in cattle. *Annual Review of Animal Biosciences*, *4*(1), 269290. doi:10.1146/annurev-animal-021815-111406 PMID:26526546

Bertipaglia, E. C. A., Silva, R. G., & Maia, A. S. C. (2018). Fertility and hair coat characteristics of Holstein cows in a tropical environment. *Animal Reproduction (AR)*, *2*, 187194.

Bohmanova, J., Misztal, I., & Cole, J. B. (2007). Temperaturehumidity indices as indicators of milk production losses due to heat stress. *Journal of Dairy Science*, *90*(4), 19471956. doi:10.3168/jds.2006-513 PMID:17369235

Chibisa, G. E., Gorka, P., Penner, G. B., Berthiaume, R., & Mutsvangwa, T. (2015). Effects of partial replacement of dietary starch from barley or corn with lactose on ruminal function, short-chain fatty acid absorption, nitrogen utilization, and production performance of dairy cows. *Journal of Dairy Science*, *98*(4), 2627–2640. doi:10.3168/jds.2014-8827 PMID:25704977

Collier, R. J., Collier, J. L., Rhoads, R. P., & Baumgard, L. H. (2008). Invited Review: Genes involved in the bovine heat stress response. *Journal of Dairy Science*, *91*(2), 445–454. doi:10.3168/jds.2007-0540 PMID:18218730

CurbeloRodríguez, J. E., V. RodríguezCruz and A. AlmeidaMontenegro. (2016). Evaluación de la capacidad termoregulatoria en bovinos lecheros Holstein pelona puertorriqueña, Holstein normal y Jersey. *The Journal of Agriculture of the University of Puerto Rico*, *100*, 112. doi:10.10071058401721101

Dash, S., Chakravarty, A. K., Singh, A., Upadhyay, A., Singh, M., & Yousuf, S. (2016). Effect of heat stress on reproductive performances of dairy cattle and buffaloes: A review. *Veterinary World*, *9*(3), 235–244. doi:10.14202/vetworld.2016.235-244 PMID:27057105

De Rensis, F., & Scaramuzzi, R. J. (2003). Heat stress and seasonal effects on reproduction in the dairy cow—A review. *Theriogenology*, *60*(6), 11391151. doi:10.1016/S0093-691X(03)00126-2 PMID:12935853

Dikmen, S., Alava, E., Pontes, E., Fear, J. M., Dikmen, B. Y., Olson, T. A., & Hansen, P. J. (2008). Differences in thermoregulatory ability between slick-haired and wild-type lactating Holstein cows in response to acute heat stress. *Journal of Dairy Science*, *91*(9), 3395–3402. doi:10.3168/jds.2008-1072 PMID:18765598

Dikmen, S., Khan, F. A., Huson, H. J., Sonstegard, T. S., Moss, J. I., Dahl, G. E., & Hansen, P. J. (2014). The SLICK hair locus derived from Senepol cattle confers thermotolerance to intensively managed lactating Holstein cows. *Journal of Dairy Science*, *97*, 55085520. doi:10.3168/jds.20148087

ElTarabany, M. S., & ElTarabany, A. A. (2015). Impact of maternal heat stress at insemination on the subsequent reproductive performance of Holstein, Brown Swiss, and their crosses. *Theriogenology*, *84*(9), 15231529. doi:10.1016/j.theriogenology.2015.07.040 PMID:26324113

Hansen, P. J. (2009). Effects of heat stress on mammalian reproduction. *Philosophical Transactions of the Royal Society of London. Series B, Biological Sciences*, *364*(1534), 33413350. doi:10.1098/rstb.2009.0131 PMID:19833646

Hansen, P. J. (2019). Reproductive physiology of the heatstressed dairy cow: Implications for fertility and assisted reproduction. *Animal Reproduction*, *16*(3), 497507. doi:10.21451/1984-3143-AR2019-0053 PMID:32435293

Harvatine, K. J., & Allen, M. S. (2006). Effects of fatty acid supplements on milk yield and energy balance of lactating dairy cows. *Journal of Dairy Science*, *89*(3), 1081–1091. doi:10.3168/jds.S0022-0302(06)72176-2 PMID:16507705

Kadzere, C. T., Murphy, M. R., Silanikove, N., & Maltz, E. (2002). Heat stress in lactating dairy cows: A review. *Livestock Production Science*, *77*(1), 5991. doi:10.1016/S0301-6226(01)00330-X

Koch L.E. and G.J. Lascano, (2018). Milk Fat Depression: Etiology, theories, and soluble carbohydrate interactions. *J. Anim. Res. Nutr 3*. doi:10.21767/2572-5459.100046

Lock, A. L., Bauman, D. E., & Jenkins, T. C. (2008). Understanding the biology of milk fat depression: from basic concepts to practical application. *Proc. Intermountain Nutr. Conf.* (pp. 27-44). IEEE.

Lock, A. L., Overton, T. R., Harvatine, K. J., Giesy, J. G., & Bauman, D. E. (2006). Milk fat depression: impact of dietary components and their interaction during rumen fermentation. *Proc. Cornell Nutr. Conf.* (pp. 75-85). IEEE.

Molina-Fernández, J. F. (2001). Origin and development of the Dairy Industry of Puerto Rico. [Origen y desarrollo de la industria lechera en Puerto Rico. Página 30 en La industria lechera en Puerto Rico]. NUPRESS of Miami, Inc. Miami, FL.

Morton, J. M., Tranter, W. P., Mayer, D. G., & Jonsson, N. N. (2007). Effects of environmental heat on conception rates in lactating dairy cows: Critical periods of exposure. *Journal of Dairy Science, 90*(5), 22712278. doi:10.3168/jds.2006-574 PMID:17430927

Mylostyvyi, R. V., & Sejian, V. (2019). Welfare of dairy cattle in conditions of global climate change. *Theoretical and Applied Veterinary Medicine.*, *7*(1), 47–55. doi:10.32819/2019.71009

National Research Council. (2001). *Nutrient Requirements of Dairy Cattle. 7th rev. ed.* Natl. Acad. Sci., Washington, DC. https://www.nap.edu/catalog/9825/nutrient-requirements-of-dairy-cattle-seventh-revised-edition-2001

Olson, T. A., Lucena, C., Chase, C. C. Jr, & Hammond, A. C. (2003). Evidence of a major gene influencing hair length and heat tolerance in Bos taurus cattle. *Journal of Animal Science, 81*(1), 8090. doi:10.2527/2003.81180x PMID:12597376

Orozco, A. J., Angulo, L. M., Pérez, A. P., & Liodoro, J. H. (2012). Aspectos fisiológicos y bromatológicos de Brachiaria humidicola. *CES Medicina Veterinaria y Zootecnia, 7*, 87–98. http://www.scielo.org.co/scielo.php?script=sci_arttext&pid=S1900-96072012000100008

Ortiz-Colón, G., Fain, S. J., Parés, I. K., Curbelo-Rodríguez, J., Jiménez-Cabán, E., Pagán-Morales, M., & Gould, W. A.OrtizColón. (2018). Assessing climate vulnerabilities and adaptive strategies for resilient beef and dairy operations in the tropics. *Climatic Change, 146*(1-2), 4758. doi:10.100710584-017-2110-1

OrtizUriarte. (2020). B. P., N. L. RosaPadilla, R. LópezLópez, J. CurbeloRodríguez, V. M. NegrónPérez and G. OrtizColón, (2020). Comparison of milk production and calving intervals between Slick and WildType Holsteins in a tropical grazing production system. *Archivos Latinoamericanos de Producción Animal, 28*(34), 145–153. https://ojs.alpa.uy/index.php/ojs_files/article/view/2817

Palmquist, D. L., & Jenkins, T. C. (2017). A 100-Year Review: Fat feeding of dairy cows. *Journal of Dairy Science, 100*(12), 10061–10077. doi:10.3168/jds.2017-12924 PMID:29153155

Pantoja, J. T. Olson, T. Ruiz, Á. Custodio, B. Vallejo and M. Pagán, (2005). Evaluation of factors that influence dairy productivity in dairy cows with short hair in the tropics. *Proceedings of the Annual Meeting of the Puerto Rican Society of Agricultural Sciences*. IEEE.

Ravagnolo O. and I. Misztal, (2000). Genetic component of heat stress in dairy cattle, parameter estimation. *J Dairy Sci. 83*, 2126–2130. doi: . S0022-0302(00)75095-8 doi:10.3168/jds

RíosSolís, C. G, Valencia, E. CurbeloRodríguez, aJ., & Ortiz, C. (2019). Feed efficiency of slick hair and wildtype dairy cows under heat stress. *J. Dairy Sci. 102*(1), 311 (Abstract).

RíosSolís[a], C. G., N. R. CidHernández, E. RuizCortés, E. Valencia, J. E. CurbeloRodríguez, and G. Ortiz Colón, (2019). Rectal temperature, respiration rate, and heart rate of slickhair and wildtype lactating Holstein cows under heat stress. *J. Dairy Sci.* Vol. 102, Suppl. 1: 11 (Abstract).

Ryan, P. D., Prichard, J. F., Kopel, E., & Godke, R. A. (1993). Comparing early embryo mortality in dairy cows during hot and cool seasons of the year. *Theriogenology, 39*(3), 719737. doi:10.1016/0093-691X(93)90257-6 PMID:16727249

Sánchez, W. K., McGuire, M. A., & Beede, D. K. (1994). Macromineral Nutrition by Heat Stress Interactions in Dairy Cattle: Review and Original Research. *Journal of Dairy Science, 77*(7), 2051–2079. https://www.journalofdairyscience.org/article/S0022-0302(94)77150-2/pdf. doi:10.3168/jds.S0022-0302(94)77150-2 PMID:7929965

Stevens, A. V., Karges, K., Rezamand, P., Laarman, A. H., & Chibisa, G. E. (2021). Production performance and nitrogen metabolism in dairy cows fed supplemental blends of rumen undegradable protein and rumen-protected amino acids in low-compared with high-protein diets containing corn distillers' grains. *Journal of Dairy Science, 104*(4), 4134–4145. doi:10.3168/jds.2020-19697 PMID:33612237

Stone, W. C., (2004). Nutritional approaches to minimize subacute ruminal acidosis and laminitis in dairy cattle. *J. Dairy Sci. 87*, E13-E26 doi:10.3168/jds.S0022-0302(04)70057-0

Van Soest, P. J. (1994). *Nutritional ecology of the ruminant* (2nd ed.). Cornell University Press. https://www.cornellpress.cornell.edu/book/9780801427725/nutritional-ecology-of-the-ruminant/#bookTabs=1 doi:10.7591/9781501732355

Vande Haar, M. J. and N. St-Pierre, (2006). Major Advances in Nutrition: Relevance to the Sustainability of the Dairy Industry. *J. Dairy Sci.* 89:1280–1291. doi:. S0022-0302(06)72196-8 doi:10.3168/jds

Voelker, J. A., & Allen, M. S. (2003). Pelleted beet pulp substituted for high-moisture corn: 1. Effects on feed intake chewing behavior, and milk production of lactating dairy cows. *Journal of Dairy Science*, *86*(11), 3542–3552. doi:10.3168/jds.S0022-0302(03)73959-9 PMID:14672184

West, J. W. (2003). Effects of Heat-Stress on Production in Dairy Cattle. *Journal of Dairy Science*, *86*(6), 2131–2144. doi:10.3168/jds.S0022-0302(03)73803-X PMID:12836950

Wheelock, J. B., Rhoads, R. P., VanBaale, M. J., Sanders, S. R., & Baumgard, L. H. (2010). Effects of heat stress on energetic metabolism in lactating Holstein cows1. *Journal of Dairy Science*, *93*(2), 644–655. doi:10.3168/jds.2009-2295 PMID:20105536

Wolfenson, D., & Roth, Z. (2019). Impact of heat stress on cow reproduction and fertility. *Animal Frontiers*, *9*(1), 3238. doi:10.1093/af/vfy027 PMID:32002237

Younas, M., Fuquay, J. W., Smith, A. E., & Moore, A. B. (1993). Estrous and endocrine responses of lactating Holsteins to forced ventilation during summer. *Journal of Dairy Science*, *76*(2), 430436. doi:10.3168/jds.S0022-0302(93)77363-4 PMID:8445097

Chapter 6
Mastitis in Dairy Species:
Identification, Control, Prevention, and Milk Quality

Bruna Samara dos Santos Rekowsky
https://orcid.org/0000-0002-2161-2714
Bahia Federal University, Brazil

Matheus Carvalho Freire de Oliveira
Lactpremium Technical Consultancy in Food, Brazil

Uiara Moreira Paim
Bahia Federal University, Brazil

José Givanildo da Silva
Bahia Federal University, Brazil

Marion Pereira da Costa
https://orcid.org/0000-0002-3003-6763
Bahia Federal University, Brazil

EXECUTIVE SUMMARY

The milk composition and quality are related to the animal's health (cow, buffalo, goat, and sheep). The mammary gland disease negatively impacts agricultural economies by reducing milk production and quality, treatment costs, discarded milk, and reduced longevity of animals. Therefore, the chapter reports how and which anatomical, physiological, and behavioral differences inherent to each species can influence the occurrence, treatment, and prevention of mastitis. Thus, early detection and antibiotic resistance are essential factors to exploit new detection methods and more natural remedies associated with reducing antimicrobial therapies such as

DOI: 10.4018/978-1-6684-5472-5.ch006

antioxidants, nanoparticles, or essential oils base-plant. Furthermore, healthy animal selection and being genetically resistant to mastitis associated with reasonable hygienic procedures, training programs, and periodic equipment maintenance are crucial steps to ensure the supply of good quality milk and dairy products without risks to the population. In addition, these measures must respect the specificities of each milk-producing species.

INTRODUCTION

The constant growth of dairy products consumption world-wide, turn the dairy sector into an essential choice to supply the nutritional demands of the population and justifies the intensification of milk production, with emphasis on bovine (cows and buffalos), goats, and sheep milk. The nutritional characteristics of milk and dairy products largely depend on the herd's health, which directly reflects on the quality of the milk produced. Thus, the efficiency in detecting diseases in the dairy herd becomes crucial to minimize economic losses, while ensuring milk and dairy products quality, and preventing risks to consumers health (Puppel et al., 2020). Mastitis, an inflammation of the mammary gland, is a primary disease of dairy herds. Mastitis promotes adverse effects on agricultural economies, mainly associated with reduced milk production, poor milk quality, treatment costs, discarded milk, and reduced longevity of animals (Kim et al., 2019). In this scenario, significant research is carried out to identify the main etiological agents involved in the disease, such as Staphylococcus aureus and Staphylococcus spp. (Silva et al., 2022; Merz et al., 2016) Streptococcus uberis, Streptococcus dysgalactiae, Corynebacterium bovis, Escherichia. Coli and others (Vakkamäki et al., 2017).

Clinical and Recently subclinical mastitis have become much more prevalent than clinical mastitis in herds while it is difficult to detect (Krishnamoorthy et al., 2021). Consequently, early detection and diagnosis of mastitis allows producers to provide quick remediation and control, like timely medical treatment of animals, avoiding severe damage. The diagnostic methods can be subdivided into direct and indirect tests, that are considered faster and cheaper. Diagnostic methods can be done during the initial stages of milking, such as California Mastitis Test (CMT), Wisconsin Mastitis Test (WMT), and black bottom mug test, or they can be more accurate using Somatic Cell Count (SCC), and detection of biomarkers such as lactate dehydrogenase, lactose, and proteins associated with inflammation (Antanaitis et al., 2021; Boora et al., 2021; Dore et al., 2016; Holko et al., 2019; Gómez-Gascón et al., 2022). Bacterial culture, Polymerase Chain Reaction (PCR), and Matrix Associated Laser Desorption-Ionization - Time of Flight (MALDI-TOF) are direct tests that are usually slower and more expensive. However, these tests present more detailed information at the bacterial strain level, such as virulence factors, circulating antibiotic resistance genes, antibiogram, and new species involved in the occurrence of mastitis (Aragão et al., 2021; Gelasakis et al., 2016; Oliveira et al., 2022a; Oliveira et al., 2022b).

Advancement in research alert to the spread of bacterial strains resistant to antibiotic therapy, and the risks of this phenomenon increasing the probability of finding antibiotic residues in dairy products have been taken seriously. The development of nanoparticles and the effectiveness of natural products used in homeopathy and phytotherapy are presented as healthier and secure alternatives since they are effective in the treatment of diseases and reduce the use of antibiotics, also reducing antibiotic residues in products and in the environment (Kala et al., 2022; Mimoune et al., 2021; Silva et al., 2020; Sousa et al., 2020). Training workers for hygienic dairy management, data organization, periodic herd testing, and appropriate veterinary treatment procedures are fundamental measures of prevention and control to minimize losses in the production and quality of milk offered to the population (Mimoune et al., 2021; Silva et al., 2020). Therefore, this chapter aims to present recent scientific data on clinical and subclinical mastitis in different dairy species approaching detection methods, etiological agents, primary and new therapies, control trends, and prevention methods. Moreover, we also report how and which anatomical, physiological, and behavioral differences inherent to each species can influence the occurrence, treatment, and prevention of mastitis.

METHODOLOGY

For the development of this chapter, peer-reviewed publications and conference proceedings were selected from 2004 to 2022 using electronic search engines such as PubMed/Medline, Science Direct, Lilacs, Scielo, and Google Academic to review new trends regarding the detection, treatment, and prevention of mastitis. The research was carried out with an emphasis on the main milk-producing species: cow, buffalo, goat, and sheep.

ASPECTS OF MASTITIS IN THE DIFFERENT DAIRY SPECIES

Mastitis is one of the leading causes of economic losses in dairy farms. It can be classified into subclinical, clinical, and chronic mastitis, and it depends on different factors related to the animals, their environment, and management (Krishnamoorthy et al., 2021). Clinical mastitis is characterized by an inflammatory response of the mammary gland tissue with tumescence, redness, and pain of the udder; milk-containing flakes, clots, and/or watery; and ultimately a gross reduction in milk production and quality is observed (Chinnappan et al., 2017). According to (Ferreira & Ribeiro, 2022), the visible signs can be scored in: (i) Grade 1: only changes in milk (clots, lumps, pus, blood, and yellowish coloration); (ii) Grade 2: changes in the milk and the udder (swollen, painful and red) and (iii) Grade 3: milk, udder, and systemic changes (fever, dyspnea, hypotension, prostration, and anorexia).

In subclinical mastitis, the animals show no clinical signs, and it can only be identified in milk through laboratory analyses, which causes a significant loss in milk production due to difficulty in detection (Krishnamoorthy et al., 2021). Good water quality, proper milking procedures, adequate hygiene (Dore et al., 2016) well maintained equipment, and clean workers' hands may preclude the multiplication of the microorganisms (Youssif et al., 2020). As a result, multiple environmental and management factors can affect the epidemiological characteristics of mastitis.

The etiological agents that cause mastitis are classified into two groups: contagious and environmental, where an ample spectrum of microorganisms is involved in disease occurrence. Environmental mastitis can be defined as those Intra-Mammary Infections (IMIs) caused by pathogens found primarily in the environment (Youssif et al., 2020). In the environmental form, the contact of animals, preferably in the environment, in places with the presence of clay, mud, manure, and organic bedding, is the primary source of contamination to them once the entire animals' habitat disseminates the pathogens involved in the cases of environmental mastitis. Unlike contagious mastitis, this type of mastitis is characterized by a high rate of clinical cases, usually in well-managed herds and with low SCC (Massote et al., 2019).

The contagious mastitis is provoked by pathogens that live and multiply on or within the mammary gland and are spread from cow to cow, primarily during milking and tend to result in subclinical infections with flareups of clinical form, where udders of infected animals serve as the main reservoir (Abdalhamed et al., 2018). The contagious mastitis type is the biggest threat to the transmission of the disease in the herd, often due to inadequate handling during the milking of the animals. The pathogens are preferably located inside the mammary gland or on the teet's surface (Oliveira et al., 2019).

Dairy Cows

In dairy cows, the primary pathogens of contagious mastitis are Staphylococcus aureus (Zafalon et al., 2005) and Streptococcus agalactiae (Souza et al., 2009), and among the secondary pathogens, Corynebacterium bovis (Langoni et al., 2012). Staphylococcus aureus has the capacity to invade deep parts of the mammary gland and form reservoirs in fibrous tissue, which is crucial to prevent antibiotic access and results in a long-duration infection. Thus, infections by Staphylococcus aureus tend to cornify and a low healing rate, which leads to loss of secretory function, and consequently, a significant reduction in milk production (Zalafon et al., 2008).

The microorganisms most often isolated from environmental samples were Escherichia coli, Enterococcus species, Staphylococcus aureus, and Pseudomonas aeruginosa (Youssif et al., 2020). In clinical mastitis of dairy cows, the animals can be recognized by physical and visual examination and by the appearance of the milk (Figure 1). Still, it is not possible to infer the etiological agent involved.

The main representative microorganisms of the coliforms group involved in mastitis are Escherichia coli, Klebsiella sp., and Enterobacter sp. (Ribeiro et al., 2006; Hogan & Smith, 2012). Furthermore, according to these authors, the main characteristics are self-limiting, low prevalence, and generally less than 10% of infected quarters in the herd. Streptococcus uberis and Streptococcus agalactiae play a vital role in spreading environmental mastitis. However, recent studies indicated that other species, such as S. dysgalactiae, S. hyovaginalis, and S. pluranimalium, are also related to environmental mastitis (Oliveira et al., 2022).

Figure 1. A daily routine of dairy cows at the time of milking
Legend: A) Positioning and visual inspection of animals before milking; B) Suspected animal with mastitis due to clinical signs of edema C) Visible presence of flakes in milk.

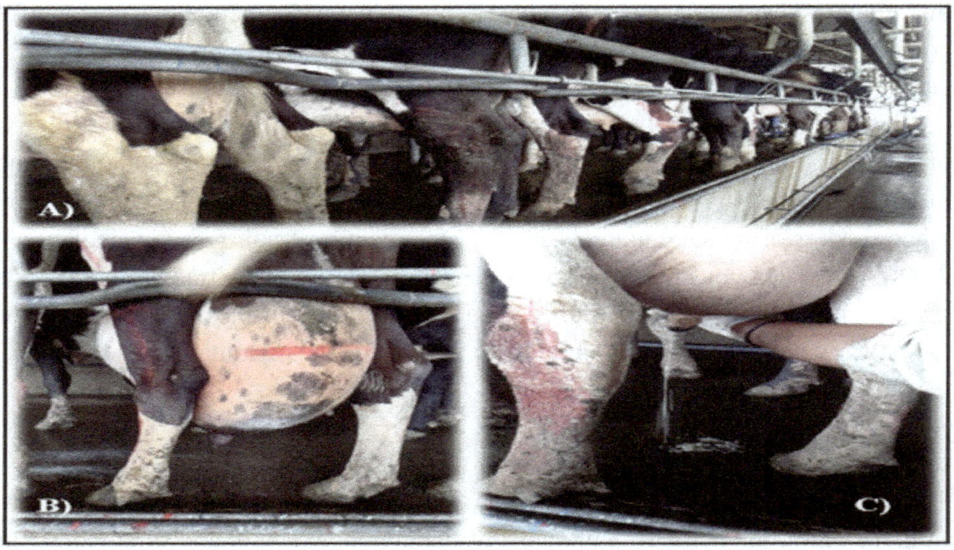

Dairy Buffaloes

In buffaloes, mastitis most commonly affects herds that present inadequate management, that has unregulated milking equipment, or dissemination by other vectors such as the milker's hands, clothes, and other utensils (Cunha et al., 2006; Sales et al., 2018; Patel et al., 2019). Staphylococcus epidermidis S. aureus, Bacillus spp.; Acinetobacter spp.; Pseudomonas aeruginosa; Shiguela. flexneri; Streptococcus spp.; Corynebacterium spp.; E. coli and Streptococcus agalactiae often causes mastitis cases (Boora et al., 2021; Vásquez-García et al., 2017). The bacterial pathogens responsible for subclinical mastitis in lactating buffaloes were Staphylococcus aureus, Streptococcus agalactiae, E. coli, and other streptococci, where this scenario may reflect several factors such as poor milking parlor hygiene, herd size, variation in feeding and handling systems (Ali et al., 2015). The most recurrent agent in buffalo clinical mastitis was coagulase-negative Staphylococcus, Escherichia coli, Staphylococcus aureus, Streptococcus spp., (Boora et al., 2015).

Small Dairy Ruminants

In clinical cases, abnormalities of the udder and milk with some form of abnormal appearance has been used as a visual identification of mastitis (Balemi et al., 2021). Pseudomonas aeruginosa is considered an opportunistic pathogen that compromises animal health when the environment, such as water, teat dip, feed, bedding, and the hoses used to fill the water troughs and wash udders is contaminated (Kelly & Wilson, 2016). Clinical cases with systemic signs in sheep and goat species are mainly provoked by Gram-negative bacteria such as Escherichia coli, Klebsiella pneumoniae, and Pseudomonas spp. Among pathogens isolated from goat milk with subclinical mastitis, Escherichia coli, Pseudomonas aeruginosa, Citrobacter koseri, and Klebsiella oxytoca, Staphylococcus non-aureus, and were directly associated with reduced daily milk yield (Gelasakis et al., 2016). Pathogenic agents identified in sheep and goat herds seem similar, except for Pseudomonas spp. and Streptococcus uberis, which are most frequently detected in sheep than in goat herds (Krishnamoorthy et al., 2021).

DETECTION METHODS OF MASTITIS

Despite its higher prevalence, the subclinical disease can only be detected by a more specific method in each suspicious animal, such as SCC or microbiological culture analysis (Antanaitis et al., 2021). Many methods can be used for the detection of mastitis, such as physical examination of animals (due to inflammatory process),

CMT, SCC, WMT and black bottom mug test for the presence of flakes in milk for clinical mastitis, for example. On livestock farms, mastitis is traditionally detected by visual inspection of the condition of the animals. However, because this visual inspection is less specific for subclinical phases, relaying on it results in a delay in starting the treatment period due to the identification of the disease only after its symptoms are visible (Kim et al., 2019). So, early detection by direct or indirect methods can prevent the aggravating deterioration of udder health due to tissue damage while facilitating the efficacy of treatment, reducing the spread (Kim et al., 2019) in cows, buffaloes, and small ruminants (Table 1).

The most common pathogens in mastitis cases in buffaloes, cows, and small ruminants are Escherichia coli, Streptococcus uberis, Streptococcus agalactiae, Staphylococcus non-aureus, Pseudomonas aeruginosa and the leading pathogen representative of contagious for all species studied is Staphylococcus aureus (Table 1). Several direct and indirect methods for detecting mastitis (clinical or subclinical) are recognized, and the sensitivity and specificity of the tests vary widely (Kala et al., 2021). Regardless of the greater specificity of indirect methods, the traditional use of some indirect techniques stems from the practicality and speed easily used in the routine protocols of the dairy farm, and improvement or new methods constantly prospect.

Indirect Methods

Physical examination and screening tests are essential methods to detect mastitis. Recently, Antanaitis et al. (2022) proposed a study in which animal behavior, for example, rumination time, eating time, drinking time, rumination chews, and drinking gulps could be used to identify mastitis in its early stages (subclinical). They observed that reduced rumination time, rumination chews, and bolus formation could be used as behavior biomarkers for subclinical mastitis identification. They suggest that changes in the time a cow spends feeding and ruminating can reflect a state of contentment and well-being.

Body temperature is another parameter commonly used in physical examination. A real-time body temperature monitoring with ingestible biosensors seems to be easily used to monitored many animals anytime and anywhere; using a mobile application that require less workforce and time (Kim et al., 2019). Researching a system to develop a real-time, lightweight multiscale detection algorithm, based on thermal images of the eyes and udder, with temperature measurements might allow us to automatically recognize cows with mastitis (Xudong et al., 2020). In this context, infrared thermography (IRT) equipment is a portable tool that requires specific software for analysis and is more accurate to the forequarters of the udder (Machado et al., 2021).

Table 1. Microorganisms involved in clinical and subclinical mastitis in dairy species

Mastitis type	Detection method	Sample	Microorganisms detected	References
Dairy Species: Bovine				
Clinical	CMT and PCR	Milk; Teat cup Milkers swab	*Staphylococcus* spp.; *Sta. aureus MRSA*	Silva et al., 2022
	CMT; Bacteriological culture; Biochemical tests	Milk	*E. coli*; *Sta. aureus*; *Str. uberis*; *Str. agalactiae*	Holko et al., 2019
Subclinical	SCC; evaluation of milk lactose	Milk	*Staphylococcus* non-*aureus*; Sta. *aureus*	Antanaitis et al., 2021
	CMT; Bacteriological culture; Biochemical tests	Milk	*E. coli*; *Sta. aureus*; *Str. uberis*; *Str. agalactiae*	Holko et al., 2019
	CMT; Bacteriological culture; Biochemical tests; PCR	Milk	*Sta. aureus; Sta. chromogenes; Sta. epidermidis; Sta. sciuri; Sta. warneri; Sta. haemolyticus; Sta. hominis; Sta. arlettae; Sta. capitis; Sta. gallinarum*	Oliveira et al., 2022a
	CMT; Bacteriological culture; Biochemical tests; PCR	Milk	*Enterococcus faecalis; Str. Agalactiae; Enterococcus faecium; E. coli; Lactococcus garvieae; Acinetobacter baumannii; Bacillus subtilis; Kocuria marina; Macrococcus caseolyticus; Microbacterium resistens; Micrococcus luteus; Str. Dysgalactiae; Str. hyovaginalis and Str. pluranimalium*	Oliveira et al., 2022b
	Bacteriological culture	Milk	*Sta. aureus; Enterococcus* spp.; *E. coli; Str. agalactiae*	Youssif et al., 2020
Dairy Species: Buffalo				
Clinical	CMT; Bacteriological culture; Biochemical tests	Milk	SNA; *E. coli*; *Sta. aureus*; *Pseudomonas* spp.	Boora et al., 2021
	Physical examination, CMT, Bacteriological culture; Biochemical tests;	Milk	*Staphylococcus* spp.; *Streptococcus* spp; *Bacillus* spp.; *Corynebacterium* spp.; *E. coli Klebsiella* spp.; *Micrococcus* sp.; *Enterobacter* sp.	Cunha et al., 2006
	SCC; CMT, and PCR	Milk	*Staphylococcus, Enterococcus, Escherichia, and Streptococcus*	Patel et al., 2019
	Bacteriological culture; Biochemical tests; PCR	Milk	*S. aureus MRSA*	Panahi and Saei, 2019

continues on following page

Table 1. Continued

Mastitis type	Detection method	Sample	Microorganisms detected	References
Subclinical	Physical examination; SCC; Bacteriological culture; Biochemical tests and PCR	Milk	*S. epidermidis*; *S. aureus*; *Bacillus* spp; *Acinetobacter* spp.; *P. aeruginosa*; *Shigella. flexneri*; *Streptococcus* spp; *E. coli; Corynebacterium* spp.	Vásquez-García et al., 2017
	SCC; CMT; Bacteriological culture	Milk	*S. aureus*; *Str. agalactiae*; *E. coli*	Ali et al., 2015
	Physical examination, CMT, Bacteriological culture; Biochemical tests;	Milk	*Staphylococcus* spp.; *Streptococcus* spp.; *Bacillus* spp.; *Corynebacterium* spp.; *E. coli Klebsiella* spp.; *Micrococcus* sp.; *Enterobacter* sp.	Cunha et al., 2006
	SCC; CMT, and PCR	Milk	*Lactococcus; Ralstonia, Enterococcus; Pseudomonas and Bacillus*	Patel et al., 2019
Dairy Species: Goats				
Clinical	CMT; Bacteriological culture; Biochemical tests	Milk	SNA; *Sta. aureus; Sta. Epidermidis; E. coli; Klebsiella pneumoniae; Pseudomonas* spp.	Abdalhamed et al., 2018
	MALDI- TOF; PCR	Milk	*B. cereus; B. licheniformis; B.; Enterobacter cloacae; E. coli; Micrococcus luteus; Morganella morganii; Aerococcus viridans; C. ammoniagenes; C. efficiens; C. falsenii; Sta. aureus; Sta. capitis; Sta. caprae; Sta. epidermidis; Sta. haemolyticus; Sta. hominis; Sta. lugdunensis; Sta. pasteuri; Sta. pettenkoferi; Sta. simulans; Sta. warneri; Sta. xylosus; Turicella otitidis*	Aragão et al., 2021
	Bacteriological culture and PCR	A swab of ears and milk	*Mycoplasmas auris*	García-Galán et al., 2020

continues on following page

Table 1. Continued

Mastitis type	Detection method	Sample	Microorganisms detected	References
Subclinical	CMT; Bacteriological culture; Biochemical tests	Milk	SNA; *Sta. aureus; Sta. Epidermidis; E. coli; Klebsiella pneumoniae; Pseudomonas* spp.	Abdalhamed et al., 2018
	Physical examination; SCC; Bacteriological culture and Multiplex PCR	Milk	*E. coli; P. aeruginosa; Citrobacter koseri; Klebsiella oxytoca; Mycoplasma agalactiae*	Gelasakis et al., 2016
	CMT and Bacteriological culture	Milk	*Sta. epidermidis; Sta. aureus; Str. agalactiae; Candida albicans* *Corynebacterium bovis; Pasteurella multocida; Bacillus* spp. *E. coli; Acinetobacter calcoaceticus*	Langoni et al., 2006
	MALDI- TOF; PCR	Milk	*B. cereus; B. licheniformis; B.; Enterobacter cloacae; E. coli; Micrococcus luteus; Morganella morganii; Aerococcus viridans; C. ammoniagenes; C. efficiens; C. falsenii; Sta. aureus; Sta. capitis; Sta. caprae; Sta. epidermidis; Sta. haemolyticus; Sta. hominis; Sta. lugdunensis; Sta. pasteuri; Sta. pettenkoferi; Sta. simulans; Sta. warneri; Sta. xylosus; Turicella otitidis*	Aragão et al., 2021
Dairy Species: Sheep				
Clinical	CMT, Bacteriological culture, and Biochemical tests	Milk	SNA; *Sta. aureus*; *E. coli*; *Pseudomonas* spp.; *Streptococcus uberis*	Dore et al., 2016
	CMT; Bacteriological culture; Biochemical tests	Milk	SNA; *Sta. Aureus; Sta. Epidermidis; E. coli; Klebsiella pneumoniae; Pseudomonas* spp.	Abdalhamed et al., 2018
Subclinical	CMT, Bacteriological culture, and Biochemical tests	Milk	SNA; *Sta. aureus*; *E. coli*; *Pseudomonas* spp.; *Streptococcus uberis*	Dore et al., 2016
	CMT; Bacteriological culture; Biochemical tests	Milk	SNA; *Sta. Aureus; Sta. Epidermidis; E. coli; Klebsiella pneumoniae; Pseudomonas* spp.	Abdalhamed et al., 2018
	Whiteside test; Bacteriological culture; Biochemical tests	Milk	*Staphylococcus* spp.; *Sta. aureus; Sta. epidermidis; Acinetobacter* spp.; *Enterobacter aerogenes; E. coli; Aeromonas* spp.; *Streptococcus* spp.; *Pseudomonas* spp.; *Corynebacterium* spp.; *Klebsiella* spp.; *Alcaligenes faecalis*	Drescher et al., 2010

Legend: B. = Bacillus; C. = Corynebacterium; Sta. = Staphylococcus; Str. = Streptococcus; SNA = Staphylococcus non-aureus; P.= Pseudomonas; CMT= California Mastitis Test; SCC= Somatic Cell Count; MALDI- TOF = Matrix Associated Laser Desorption-Ionization - Time of Flight.

The black bottom mug test is a rapid test where milk from all quarters of the mammary glands is ejected onto a plate with a black background and screened for detection of gross abnormalities in color, consistency, presence of flakes, purulence, and even blood (Kandeel et al., 2018). In the CMT, indirect measures of leukocyte concentration can be estimated by aggregation of nucleic acids in the milk samples (Chinnappan et al., 2017). In this test (Figure 2), equal parts of milk samples from each mammary quarter and an anionic detergent (sodium alkyl aryl sulfonate) disrupts the cytoplasm and nucleus of cells, are mixed in shallow cups of the paddle by gentle circular rotation. The results are interpreted based on the density of the gel formed (Abdalhamed et al., 2018) and widely used in developed countries to detect subclinical mastitis (Muhammad et al., 2010). The WMT, commonly referred to as Somaticell®, estimates the number of somatic cells through the increase in viscosity developed in the reaction medium in volumetric tubes. According to Rossi et al. (2018), the WMT seems to have a similar level of accuracy when compared with CMT but can be more sensitive in general.

Figure 2. A positive result (+3) for the California Mastitis test

Somatic cell counts (SCC) represent a relatively constant proportion of mammary epithelial cells and various immune cell populations, mainly lymphocytes, polymorphonuclear neutrophils, and macrophages (Bisutti et al., 2022). Somatic cell counts is a robust indicator of the intensity of the mammary gland immune response (Oliveira et al., 2019). It cannot indicate the stage of the inflammatory response or the progression of the disease so a differential SCC (DSCC) method able to differentiate the profile of SCC in milk and indicate the proportions of macrophages, polymorphonuclear neutrophils, and lymphocytes of epithelial cells (Bisutti et al., 2022).

Due to many cytoplasmic particles released in the apocrine milk secretion process in goats (Gómez-Gascón et al., 2022; Kuchtík et al., 2021), healthy goats have a higher milk SCC compared with other dairy ruminants.

Specific proteins detection such as haptoglobin, lactoferrin, or changes in the enzyme activity such as N-acetyl-D-glucosaminidase, lactate dehydrogenase has been studied as biomarkers directly related to SCC (Gómez-Gascón et al., 2022). These authors found that Lactate dehydrogenase (LDH) activity in raw goat milk presented excellent sensitivity and specificity compared with SCC and Total Bacteria Count (TBC) values. Another study tested a biosensor formed by immobilizing a plasmin-specific peptide coupled to magnetic nanobeads for qualitative and quantitative measurements (Chinnappan et al., 2017).

The milk miRNAs have been proposed as biomarkers of different mammary gland conditions and for subclinical mastitis size and concentration of vesicles seems not to be altered, but the cargo of vesicles has changed (Saenz-de-Juano et al., 2022). miRNAs expression (MIR29B-2, MIR146A, and MIR155) were good candidates for determining infected/non-infected milk statuses and appeared to have potential as bovine mastitis biomarkers with high accuracy, specificity, and sensitivity (Srikok et al., 2020). The immune system compounds such as IL-8, IFN-γ, and IP-10 could be biomarkers for mastitis-prone animals' adverse selection (Castro et al., 2019).

The primary function of whey proteins (i.e., β-lactoglobulin, α-lactalbumin, serum albumin, immunoglobulins and glycomacropeptide) is to maintain the host's immune system and these protein components can lyse the cell wall of potentially pathogenic bacteria. Puppel et al. (2020) found that the presence of Enterobacteriaceae spp. Considerably affect the synthesis of lysozymes and lactoferrins, while infection by Staphylococcus spp. Influenced the level of β-lactoglobulin, demonstrating that they can be used as strain-specific inflammatory biomarkers. In addition to whey proteins, lactose has also been explored as a biomarker of subclinical mastitis in dairy cows since the reduction of lactose levels in cow's milk has a positive correlation with the prevalence of pathogens such as S. agalactiae, S aureus, and other streptococci (Antanaitis et al., 2021).

Direct Methods

Direct tests are based on identifying the etiological agent by demonstrating the presence of microorganisms in the milk samples sent to the laboratories (Peixoto et al., 2010). Bacteriological examinations are considered a standard method for confirming subclinical mastitis despite being time-consuming and laborious (Kala et al., 2021) and by identifying the pathogenic agent it is possible to improve therapeutic strategies such as the application of antibiograms that optimize the use of antibiotic therapy if necessary.

Culture techniques to selectively identify microorganisms involved in mastitis may reveal the cause of mastitis (Kayitsinga et al., 2017). Although this technique has little demand for equipment, it requires skilled personnel and certified laboratories. Furthermore, culturing is time-consuming, taking minimally 24 h before results are delivered (Chinnappan et al., 2017). For more specific information about the species involved, it is necessary to carry out complementary tests such as colorimetric and biochemical methods.

Polymerase chain reaction (PCR) is an accurate direct method for bacterial DNA detection from bovine milk (Soares et al., 2022), buffaloes (Panahi & Saei, 2019) and widely used in other species as small ruminants (Silva et al., 2015). From milk samples, PCR amplifies a fragment of the region of the bacterial 16S ribosomal RNA gene (Castro et al., 2019). According to these authors, PCR products from the samples were pooled at approximately equimolar DNA concentrations and run on an agarose gel. As sequence fragments are amplified, bacterial DNA samples can be taxonomically analyzed using the software.

Immunosensors are commonly produced with nanostructured films and a biomolecule interacting with the analyte of interest. The immunosensor tested by Soares et al. (2022) was made with a nanostructured film formed from a chitosan matrix, carbon nanotubes, and a layer of specific antibodies to detect S. aureus, which can be produced at low cost. According to these authors, the detection limits can be as low as 2.6 CFU/mL in milk samples, which makes it sufficient to detect mastitis even in the early stages when using impedance spectroscopy. The main characteristics are the high sensitivity and selectivity obtained using this technique.

MASTITIS TREATMENT ON THE DIFFERENT DAIRY SPECIES

Antibiotic Therapy Applied in the Treatment of Mastitis

Virulence factors are related to bacterial pathogenicity, which includes structural components (penicillin-binding protein, collagen, fibrinogen, β-lactamase, proteases, capsules), enzymes (coagulase, staphylokinase, DNase, phospholipase), and toxins. These factors help the bacterium against the host immune system and allow the colonization of the mammary glands of ruminants (Abril et al., 2020) and must be taken into account to formulate the most appropriate antibiotic protocol. The antimicrobial drug's effects in mastitis treatment depend on its administration route and pharmacokinetics. Moreover, the therapeutic effects vary on disease severity, drug choice, reasonable drug usage and dosages (Algharib et al., 2020).

According to Oliveira et al. (2022), from strains tested in antibiogram (E. faecalis, S. agalactiae, E. faecium, E. coli, L. garvieae, Acinetobacter baumannii, B. subtilis, Kocuria marina, Macrococcus caseolyticus, Microbacterium resistens, Micrococcus luteus, S. dysgalactiae, S. hyovaginalis, and S. pluranimalium) gentamicin and vancomycin were less than 30% of resistance cases proving to be a good option of treatment. Despite some pathogens showing frequent resistance to aminoglycosides, overall susceptibility to potentiated penicillin, cephalosporins of the second generation, sulfonamides, and quinolones can be considered satisfactory (Holko et al., 2019).

In buffalo, isolates of Staphylococcus spp. it was characterized by the presence of genes associated with β-lactam resistance (mecA and Blaz), showing the lowest resistance in a drug combination of tetracycline, neomycin, and bacitracin suggesting potential for mastitis treatment (Medeiros et al., 2011). S. aureus and S. agalactiae revealed significant sensitivity to ciprofloxacin and enrofloxacin (Ali et al., 2015).

The use of antioxidants as a complement to antibiotic therapy presented better treatment efficacy than only antibiotics (Kala & Rani, 2022), and the non-specific immune modulators like amino nitrogen, minerals, vitamins, protein hydrolysate and lactobacilli maximize host response against bacterial invasion.

Occurrence of Antibiotic Resistance in Dairy Species

Bacterial resistance to different antibiotics is due to the emergence of multidrug-resistant bacteria that cause negative economic impacts, either through the difficulty of increasingly fewer effective treatments or through the spread of resistant bacteria from animal to animal (Abdalhamed et al., 2018). The multi-drug resistant isolates, especially of E. coli, point to the trend of increasing environmental resistance in livestock farming (Holko et al., 2019), and many studies bring the alert of the spread of multi-drug resistant strains of S. aureus, known mainly as Methicillin Resistant Staphylococcus aureus (MRSA).

An important factor in bacterial resistance is the production of enzymes that inactivate antimicrobials. Beta-lactamase is included among the main mechanisms of resistance to beta-lactams where this enzyme acts, causing hydrolysis on the beta-lactam ring and leading to antibiotic inactivation in drugs frequently used to treat intramammary infections (GarinoJunior et al., 2011). In about 50% of S. aureus strains isolated from mastitis cows, the ability to produce beta-lactamase, abscess formation, and alteration of glandular tissue around the infected site was observed (Algharib et al., 2020). Another well-known resistance mechanism is the multi-drug efflux systems that comprise cytoplasmic proteins involved in the expulsion of toxic agents from inside the bacteria to the external environment preventing the drug concentrations inside the bacterial cell have an inhibitory effect (Silva et al., 2020).

Genes of antibiotic resistance (norA, norC, tet-38, and msrA) were detected in dairy cattle farms, possibly indicating the circulation of strains with mechanisms to reduce the efficacy of quinolones, tetracyclines, erythromycin, and macrolides (Silva et al., 2020). In a detailed investigation, (Silva et al., 2022) found MRSA and methicillin-resistant Staphylococcus non-aureus (MRSNA) in milk, milking utensils, milker's hands, milker's nostrils proving the importance of farmer workers in the maintenance and spread of farm mastitis. In Brazilian dairy cattle, the genotypic resistance test in Staphylococcus spp. was revealed by blaZ, tetL, tetM, mecA, and vanB gene, and resistance profiles were identified for penicillin, tetracycline, oxacillin, cefoxitin and to vancomycin (Oliveira et al., 2022).

In buffaloes, multi-drug resistance was observed among Staphylococcus spp. isolates with prevalence as mecA and blaZ genes and efflux pump as determinants of antimicrobial resistance mainly to penicillin, oxacillin, cefquinome, tetracycline, erythromycin, doxycycline, and others; warning for greater caution in the choice of therapy to minimize the risk to public health (Medeiros et al., 2011). Less efficacy of penicillin G and amoxicillin against Staphylococcus aureus and Streptococcus agalactiae may be due to the production of resistance factors such as a β-lactamase enzyme (Ali et al., 2015).

According to antimicrobial susceptibility testing, 73.1% of the S. aureus from bovine and small ruminants were resistant to at least one antimicrobial, and 25.4% were multi-drug resistant (Dastmalchi Saei & Panahi, 2020). According to these authors, 90% of S. aureus isolates presented resistance to penicillin, followed by tetracycline, streptomycin, erythromycin, and clindamycin. S. aureus isolates from goats were resistant to chloramphenicol, penicillin G, polymyxin B, and spectinomycin but sensitive to vancomycin, ceftriaxone, doxycycline, and nitrofurantoin (Balemi et al., 2021).

New Therapeutic Strategies

The group of formulated homeopathic products vary according to their properties such as anti-inflammatory, anti-infectious, complementary analgesic effect, strengthening, and restorative properties, with applicability in different stages of mastitis (Mimoune et al., 2021). Few studies have focused on treating mastitis with homeopathy and natural products based on a holistic approach to stimulating the animal's immune system and treating diseases (Table 2). The antimicrobial activity of bioactive molecules obtained from essential oils of several plants points to a positive potential against several bacteria due to a chemical composition with a large variety of substances and their mechanism that involve several targets in the bacterial cell (Furlani et al.,

2021). These therapeutic strategies can be relevant considering the use of endemic plants, the low cost of phytotherapy, and the activity against causative pathogens of mastitis in dairy species (Peixoto et al., 2016).

Table 2. New approaches to Mastitis treatments

Components	MIC (mg/ml)	Microorganisms tested	Isolated/ Tested	References
Silver – Copper nanoparticles complex	0.003 to 0.012	*S. dysagalactiae; Salmonella* spp.; *E. faecalis; E. cloacae. C. albicans; E. coli; S. agalactiae; S. aureus*	In vitro	Lange et al., 2021
Polypyrrole nanoparticles	0.062 to 0.125	*Staphylococcus* spp.	Cow milk	Silva et al., 2020
EO *Lippia origanoides* EO *Lippia alba* EO *Cymbopogon citratus*	0.560 to 1.280	*Staphylococcus* spp. *Staphylococcus aureus*	Goat and Sheep milk	Furlani et al., 2021
AE *Moringa oleifera*	0.190 to 6.100	*Staphylococcus* spp.	Cow milk	Silva et al., 2020
EO Salvia officinalis	0.120	*Staphylococcus aureus*	Sheep milk	Alekish et al., 2017
EO Mimosa tenuiflora *EO Commiphora leptophloeos*	1.680 to 2.730	*Staphylococcus* spp.	In vitro	Sousa et al., 2020

Legend: EO= Essential Oils; AE = Aqueous extract; MIC = Minimal Inhibitory Concentration.

The prepared dose with essential oil of Salvia officinalis (12.5%) administered intramammary for three days had a positive effect reducing the SCC and increasing milk fat and lactose from sheep with subclinical mastitis (Alekish et al., 2017).

The use of chromatography with mass spectrometer is a standard method to describe the profile of essential oil components such as Lippia organoids (pepper rosemary) which mainly present carvacrol, p-cymene, and terpinene, which conferred better antibacterial activity against several strains of Staphylococcus spp. and S. aureus (Furlani et al., 2021).

Despite the antimicrobial activity, Sousa et al. (2020) highlight the necessity of studies to investigate other relevant aspects of these extracts, such as toxicity and better efficacy, to establish a therapeutic modality without causing unpredictable effects or injury to the host. Mimosa tenuiflora and Commiphora leptophloeos are tested for phytotherapeutic properties, and both presented myricetin as the main compound in the extracts and biological activity against Staphylococcus spp. (Sousa et al., 2020). Despite their potential effect against mastitis pathogens, the authors confirm that only Commiphora leptophloeos present low toxicity and could be considered safe to use in treatments.

Recently, nano drugs have been used as a substitute measure to solve the multi-drug resistance and intracellular persistence of S. aureus. So, these new nanocarriers provide a new strategy to combat S. aureus mastitis problems (Algharib et al., 2020). Both silver nanoparticles and the silver-copper complex could inhibit microbial growth by disrupting the cell membrane at a relatively low concentration with greater efficacy than Salmonella spp. and E. coli (Lange et al., 2021). Antioxidant therapy and antibiotics significantly improved treatment efficacy which could be accredited to the presence of non-specific immune- modulators such as vitamins, minerals, amino nitrogens, protein hydrolysate, Trisodium citrate dehydrate, and lactobacillus (Kala et al., 2022).

PREVENTION AND CONTROL

In addition to the various studies applied to improve the safety and efficiency of traditional treatments and new, more natural alternatives, prevention and control measures still occupy a prominent position in reducing mastitis in dairy herds. Despite the individual characteristics of each farm and production system, some aspects must be considered essential for efficient milk production and minimization of mastitis.

Genetic selection programs for dairy animals are usually related to improvement of economically important traits such as fertility and milk production yield. The intensification of the production system leads to a greater predisposition to physiological, behavioral, and immunological disorders, where a higher incidence of clinical mastitis leads to lower reproductive performance (Oltenacu & Broom, 2010). For mastitis, somatic cell counts are an essential tool in genetic selection due to their association with clinical and subclinical forms of mastitis, ease of measurement, and moderate heritability. However, other aspects, such as the genetic correlation between clinical mastitis and udder height, can also be explored (Miglior et al., 2017), aiming at selecting healthier animals and reducing mastitis cases on the property.

The use of direct methods allows a better understanding of the disease and the causative pathogen. The significant correlation between hygiene and the prevalence of Staphylococcus aureus, Escherichia coli, Pseudomonas aeruginosa, and Enterococcus (Yousiff et al., 2020) and finding of those bacteria in swab from the hands and nostrils of milkers; milking bucket swab and teat cup (Silva et al., 2022) demonstrate the importance of complete training program for farm workers by consultants or veterinarians in charge of the herd and allowing conscious decision-making regarding the treatment and, mainly, regarding the antimicrobial protocol. The workers must be aware of incorrect handling of animals and in their role in the spread of diseases and how primary conditions of hygiene and cleanliness of the environment can reduce the prevalence of sick animals. Especially in extensive and semi-extensive systems, where the animals are more exposed to climatic conditions like rain and mud (Figure 3).

Figure 3. Challenging situations in the hygienic milking of animals are managed in extensive systems

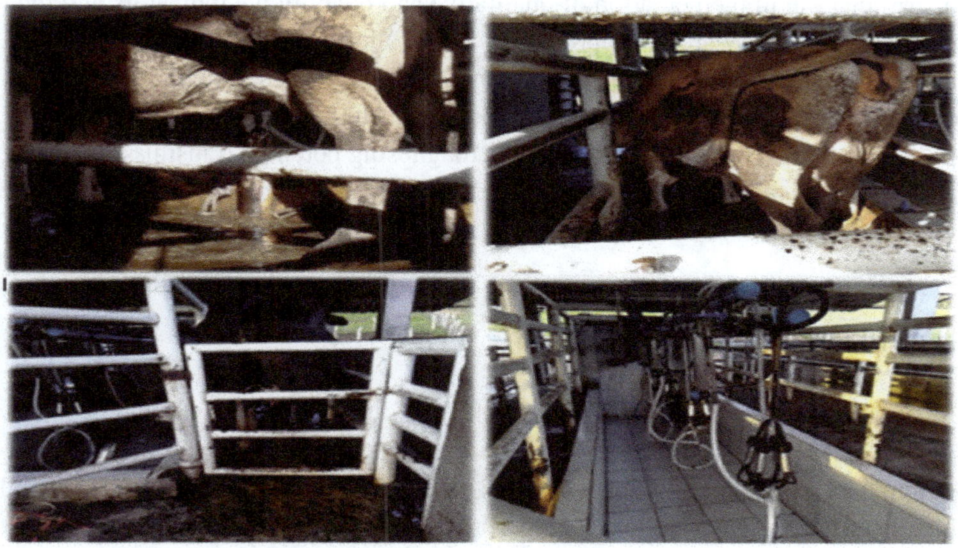

With the animals in the milking line, a physical examination of the udder should first be carried out through inspection and palpation, followed by the removal of more considerable dirt; only teats should be washed with good-quality running water or chlorinated water and dried with disposable paper towels (Cunha et al., 2006; Silva & Nogueira, 2010). The milking order of the animals must be performed according to the severity of the symptoms observed, leaving the most severe animals for last to be milked in a separate set (Ferreira & Ribeiro, 2022). The use of rapid screening tests such as CMT, WMT and SCC are relevant to identify animals especially those with subclinical mastitis and are used as a basis for the establishment of the milking line, as it avoids the transmission of pathogens between animals during milking.

Pre-dipping consists of the disinfection of teats that occurs before milking by immersing the teats in an antiseptic product, aiming to reduce the resident microbial load and consequently reducing the possibility of infection and transmission of microorganisms between the animals by letting the product act in the ceiling between 15 and 30 seconds before drying (Silva & Nogueira, 2010). After the hygienic milking of the animals, post-dipping is another immersion procedure that aims to protect the teats and reduce the bacterial load. The choice of teat disinfectant product is according to the need, cost-benefit, and safety for animals and milkers, and iodine, lactic acid, chlorine, hydrogen peroxide, chlorhexidine, and phenolic compounds are typically used.

According to Dore et al. (2016), the interaction of different laboratories in obtaining information can favor the homogenization of the methodology and evaluate the effectiveness of control programs across the country. In this context, the authors affirm that the implementation of qualitative and quantitative indicators and their constant evaluation are an approach to better allocation of economic resources and, when necessary, the application of individual strategies. For example, data on antimicrobial virulence and resistance genes and clonality of S. aureus detected in milk from small ruminants are critical to diagnosing potential transmission routes, improving the management of affected herds, and developing effective therapeutic interventions (Merz et al., 2016).

MILK QUALITY VS. MASTITIS

Mastitis alters the milk composition which magnitude depends on the pathogenicity of the microorganisms involved and the extent of the mammary gland infection (Chinnappan et al., 2017). Differentiated payment for milk, based on raw material quality criteria, is an evolution of the product marketing system and an improvement in the relationship between the industry and producers (Zafalon et al., 2005).

In general, there is a reduction in the synthesis of fat, casein, and lactose and an increase in blood components such as albumin, immunoglobulins, sodium, and chlorine during inflammation (Ferreira & Ribeiro, 2022). The volume of milk produced by cows with healthy mammary glands was significantly higher than those milked by sick animals (Leitner et al., 2019). Lactose is a significant bovine milk solid, and a reduced level of milk lactose is commonly observed during mastitis and might be able to serve as a potential biomarker for early identification of subclinical mastitis caused by S. agalactiae, S. aureus and other Streptococci (Antanaitis et al., 2021; Wang et al., 2021). The higher content of chlorides can be explained by the change in the ionic concentration of milk caused by the increase in capillary permeability in the mammary gland (Zalafon et al., 2008). Mastitis may lead to increased ions, proteins, and phagocytic cells resulting from leakage or migration, respectively (Chinnappan et al., 2017).

A decrease in casein fraction followed by an increase in whey protein (mostly β-LG and α-LA) content is probably related to the proteolytic activity of milk native enzymes, such as plasmin, and directly influences the levels of endogenous peptides serving as potential biomarkers (Addis et al., 2020; Bisutti et al., 2022). Lysozyme and lactoferrin may be markers of udder inflammation caused by Enterobacteriaceae and Streptococcus spp., whereas β-lactoglobulin may prove helpful in diagnosing Staphylococcus spp. induced mastitis (Puppel et al., 2020).

In buffalos milk, there is also an increase in the somatic cell count (Catozzi et al., 2017). Moreover, as observed in other species with mastitis, in buffalo's milk through the lipidome assay it was observed that both milk from healthy buffaloes and animals affected with mastitis had a composite lipid fraction, mainly from Triacylglycerols (TAG), Free Fatty Acids (FFA), and Ceramides (Cer), however, an increase in FFA (such as 40:7, 20:3 and 30:0) and Cer 25:1;2O/18:5 while a decrease of FFA (such as 34:0 and 34:7) was observed in milk of affected buffaloes (Ceciliani et al., 2022) and the susceptibility of milk fat to some microorganisms' lipases and lipolytic activity can explain this.

In small ruminants, it can be observed that a significant decrease in lactose characterizes the evolution of subclinical mastitis in goats compared to milk from a healthy mammary gland (Leitner et al., 2019). Regarding the proteome, Pisanu et al. (2020) observed that mastitis milk showed an increase in the differential proteins counting almost three times, offering an essential change in the protein profile of the milk. Besides, the coagulation properties, rennet coagulation time, and curd stability were influenced not only by the infection but also by the etiology of this infection (Leitner et al., 2019).

The values of sodium, magnesium, and chlorine were higher in mastitic milk samples (Sramek et al., 2018) and t. Moreover, the concentration of medium and short-chain fatty acids is reduced what directly influences the taste of goat milk because the characteristic flavor of goat milk is due to the short-chain fatty acids caproic, caprylic, and capric. From sheep, Martí-De Olives et al. (2020) concluded that the losses in dairy production could surpass 43%, but depends on the infection severity and etiology of the mastitis. They also stated that there is a decrease in the values of lactose and whey proteins. In the case of the impacts on production, Alba et al. (2019) observed a significant decline in the volume milked by sick sheep, such as in the protein levels, fat, and total solids.

CONCLUSION

Mastitis is a disease that affects different animal species, and although the clinical form is easily diagnosed, the subclinical phase of the disease has been a silent danger. Despite the various possible causes, the high load of microorganisms poses a more significant challenge to the immune system of dairy animals. From environmental mastitis, the main microorganisms identified are E. coli, S. uberis, and S. agalactiae, while Staphylococcus aureus is undoubtedly the most frequent of the pathogens that cause contagious mastitis. Indirect methods such as CMT and SSC are considered the fastest and most practical for detecting mastitis early. Rapid tests such as infrared thermal imaging and biomarkers are new options that consider speed and cost-effectiveness.

Nevertheless, when we think about the presence of contagious pathogens with multi-drug resistant gene expression, direct methods such as bacteriological associated with antibiogram or PCR being indispensable for the detection and choice of the best therapy. With the spread of resistant pathogens and the concern about the presence of antimicrobials residues in dairy products, the possibility of reducing doses by associating the use of antioxidants, the use of nanoparticles and plant extracts has been explored as alternative therapies. The basis to contain the spread of mastitis in dairy herds is still found in selecting resistant animals with adequate food management, training workers, and maintaining environmental hygiene and good equipment.

ACKNOWLEDGMENT

The authors are thankful for the financial support provided by the Conselho Nacional de Desenvolvimento Científico e Tecnológico (CNPq) - grant number [303074/2021-3], and the Coordenação de Aperfeiçoamento de Pessoal de Nível Superior (CAPES) Brazil.

REFERENCES

Abdalhamed, A. M., Zeedan, G. S. G., & Zeina, H. A. A. A. (2018). Isolation and identification of bacteria causing mastitis in small ruminants and their susceptibility to antibiotics, honey, essential oils, and plant extracts. *Veterinary World*, *11*(3), 355–362. doi:10.14202/vetworld.2018.355-362 PMID:29657429

Abril, A. G., Villa, T. G., Barros-Velázquez, J., Cañas, B., Sánchez-Pérez, A., Calo-Mata, P., & Carrera, M. (2020). Staphylococcus aureus Exotoxins and Their Detection in the Dairy Industry and Mastitis. *Toxins*, *12*(9), 1–18. doi:10.3390/toxins12090537 PMID:32825515

Addis, M. F., Maffioli, E. M., Ceciliani, F., Tedeschi, G., Zamarian, V., Tangorra, F., Albertini, M., Piccinini, R., & Bronzo, V. (2020). Influence of subclinical mastitis and intramammary infection by coagulase-negative staphylococci on the cow milk peptidome. *Journal of Proteomics*, *226*(June), 103885. doi:10.1016/j.jprot.2020.103885 PMID:32645476

Alba, D. F., da Rosa, G., Hanauer, D., Saldanha, T. F., Souza, C. F., Baldissera, M. D., da Silva dos Santos, D., Piovezan, A. P., Girardini, L. K., & Schafer Da Silva, A. (2019). Subclinical mastitis in Lacaune sheep: Causative agents, impacts on milk production, milk quality, oxidative profiles, and treatment efficacy of ceftiofur. *Microbial Pathogenesis*, *137*(September), 103732. doi:10.1016/j.micpath.2019.103732 PMID:31505264

Alekish, M. O., Ismail, Z. B., Awawdeh, M. S., & Shatnawi, S. (2017). Effects of intramammary infusion of sage (Salvia officinalis) essential oil on milk somatic cell count, milk composition parameters and selected hematology and serum biochemical parameters in Awassi sheep with subclinical mastitis. *Veterinary World*, *10*(8), 895–900. doi:10.14202/vetworld.2017.895-900 PMID:28919679

Algharib, S. A., Dawood, A., & Xie, S. (2020). Nanoparticles for treatment of bovine Staphylococcus aureus mastitis. *Drug Delivery*, *27*(1), 292–308. doi:10.1080/10717544.2020.1724209 PMID:32036717

Ali, Z., Dimri, U., & Jhambh, R. (2015). Prevalence and antibiogram of bacterial pathogens from subclinical mastitis in buffaloes. *Buffalo Bulletin*, *34*(1), 41–44.

Antanaitis, R., Juozaitienė, V., Jonike, V., Baumgartner, W., & Paulauskas, A. (2021). Milk lactose as a biomarker of subclinical mastitis in dairy cows. *Animals (Basel)*, *11*(6), 1–11. doi:10.3390/ani11061736 PMID:34200862

Antanaitis, R., Juozaitienė, V., Malašauskienė, D., Televičius, M., Urbutis, M., Rutkaukas, A., Šertvytytė, G., & Baumgartner, W. (2022). Identification of Changes in Rumination Behavior Registered with an Online Sensor System in Cows with Subclinical Mastitis. *Veterinary Sciences*, *9*(9), 454. doi:10.3390/vetsci9090454 PMID:36136670

Balemi, A., Gumi, B., Amenu, K., Girma, S., Gebru, M., Tekle, M., Rius, A. A., D'souza, D. H., Agga, G. E., & Dego, O. K. (2021). Prevalence of mastitis and antibiotic resistance of bacterial isolates from cmt positive milk samples obtained from dairy cows, camels, and goats in two pastoral districts in southern ethiopia. *Animals (Basel)*, *11*(6), 1530. doi:10.3390/ani11061530 PMID:34073967

Bisutti, V., Vanzin, A., Toscano, A., Pegolo, S., Giannuzzi, D., Tagliapietra, F., Schiavon, S., Gallo, L., Trevisi, E., Negrini, R., & Cecchinato, A. (2022). Impact of somatic cell count combined with differential somatic cell count on milk protein fractions in Holstein cattle. *Journal of Dairy Science*, *105*(8), 6447–6459. doi:10.3168/jds.2022-22071 PMID:35840397

Boora, A., Yadav, S., Devi, P., Singh, K. P., Cheema, P. S., Muley, V., Sehrawat, V., Dhamanaskar, K., & Singh, I. (2021). an Open Label Study To Assess the Efficacy of Ceftiofur in Treatment of Clinical Mastitis in Buffaloes. *Buffalo Bulletin*, *40*(4), 545–555.

Castro, I., Alba, C., Aparicio, M., Arroyo, R., Jiménez, L., Fernández, L., Arias, R., & Rodríguez, J. M. (2019). Metataxonomic and immunological analysis of milk from ewes with or without a history of mastitis. *Journal of Dairy Science*, *102*(10), 9298–9311. doi:10.3168/jds.2019-16403 PMID:31421883

Catozzi, C., Sanchez Bonastre, A., Francino, O., Lecchi, C., De Carlo, E., Vecchio, D., Martucciello, A., Fraulo, P., Bronzo, V., Cuscó, A., D'Andreano, S., & Ceciliani, F. (2017). The microbiota of water buffalo milk during mastitis. *PLoS One*, *12*(9), 1–20. doi:10.1371/journal.pone.0184710 PMID:28926595

Ceciliani, F., Audano, M., Addis, M. F., Mitro, N., Lecchi, C., Ghaffari, M. H., Albertini, M., De Carlo, E., Vecchio, D., Di Vuolo, G., Cappelli, G., Tangorra, F., Piccinini, R., Bronzo, V., & Caruso, D. (2022). Changes in the lipidome of water buffalo milk during intramammary infection by non-aureus Staphylococci. *Scientific Reports*, *12*(1), 1–12. doi:10.103841598-022-13400-0 PMID:35690599

Chinnappan, R., Al Attas, S., Kaman, W. E., Bikker, F. J., & Zourob, M. (2017). Development of magnetic nanoparticle based calorimetric assay for the detection of bovine mastitis in cow milk. *Analytical Biochemistry*, *523*, 58–64. doi:10.1016/j.ab.2017.02.009 PMID:28219684

Cunha, A. Silva, L. B. G. da, Pinheiro Júnior, J. W., Silva, D. R. da, Oliveira, A. A. da F., Silva, K. P. C. da, & Mota, R. A. (2006). Perfil De Sensibilidade Antimicrobiana De Agentes Contagiosos E Ambientais Isolados De Mastite Clínica E Subclínica De Búfalas. *Arquivos do Instituto Biológico*, *73*(1), 17–21. doi:10.1590/1808-1657v73p0172006

da Silva, J. G., Alves, B. H. L. S., Melo, R. P. B., Kim, P. C. P., Neto, O. L. S., Bezerra, M. J. G., Sá, S. G., & Mota, R. A. (2015). Occurrence of anti-Toxoplasma gondii antibodies and parasite DNA in raw milk of sheep and goats of local breeds reared in Northeastern Brazil. *Acta Tropica*, *142*, 145–148. doi:10.1016/j.actatropica.2014.11.011 PMID:25438258 da Silva, J. G., Camargo, A. C., de Melo, R. P. B., Aragão, B. B., de Oliveira, J. M. B., de Sena, M. J., Nero, L. A., & Mota, R. A. (2022). Meca positive staphylococcus spp. In bovine mastitis, milkers, milking environment, and the circulation of different MRSA clones at dairy cows' farms in the northeast region of Brazil. *Ciência Rural, 52*(3), e20210008. doi:10.1590/0103-8478cr20210008

Dastmalchi Saei, H., & Panahi, M. (2020). Genotyping and antimicrobial resistance of Staphylococcus aureus isolates from dairy ruminants: Differences in the distribution of clonal types between cattle and small ruminants. *Archives of Microbiology, 202*(1), 115–125. doi:10.100700203-019-01722-z PMID:31501949

Dore, S., Liciardi, M., Amatiste, S., Bergagna, S., Bolzoni, G., Caligiuri, V., Cerrone, A., Farina, G., Montagna, C. O., Saletti, M. A., Scatassa, M. L., Sotgiu, G., & Cannas, E. A. (2016). Survey on small ruminant bacterial mastitis in Italy, 2013–2014. *Small Ruminant Research, 141*, 91–93. doi:10.1016/j.smallrumres.2016.07.010

Drescher, G., Mattiello, S. P., Peixoto, R. M., Vargas, A. C., Maciel, M. N., & Costa, M. M. (2010). Caracterização Bioquímica E Perfil De Sensibilidade Aos Antimicrobianos De Agentes Bacterianos Isolados De Mastite Subclínica Ovina Na Região Oeste De Santa Catarina. *Ciência Animal Brasileira, 11*(1), 188–193. doi:10.5216/cab.v11i1.5189

Ferreira, B. H. A., & Ribeiro, L. F. (2022). Mastites causadas por Escherichia coli, Klebsiella spp. e Streptococcus uberis relacionadas ao sistema de produção Compost Barn e o impacto na qualidade do leite. Https://Medium. Com/, 1–18. https://medium.com/@arifwicaksanaa/pengertian-use-case-a7e57 6e1b6bf

Furlani, R. DE SOUSA, M. M., Rocha, G. N. da S. A. O., Vilar, F. C. R., Ramalho, R. C., & Peixoto, R. D. M. (2021). Antibacterial activity of essential oils against pathogens of importance in caprine and ovine mastitis. *Revista Caatinga, 34*(3), 702–708. doi:10.1590/1983-21252021v34n322rc

García-Galán, A., García-Romero, E., Sánchez, A., Corrales, J. C., Contreras, A., & De la Fe, C. (2020). Isolation of Mycoplasma auris from milk of goats with clinical mastitis. *Small Ruminant Research, 185*(March), 106089. doi:10.1016/j. smallrumres.2020.106089

GarinoJunior, F., Camboim, E. K. A., Neves, P. B., Sá, A. V. V, & Almeida, A. P. (2011). Suscetibilidade a antimicrobianos e produção de betalactamase em amostras de staphylococcus isolados de mastite caprina no semiárido paraibano. *Arquivos do Instituto Biológico, 28*(1), 103–107.

Gelasakis, A. I., Angelidis, A. S., Giannakou, R., Filioussis, G., Kalamaki, M. S., & Arsenos, G. (2016). Bacterial subclinical mastitis and its effect on milk yield in low-input dairy goat herds. *Journal of Dairy Science*, *99*(5), 3698–3708. doi:10.3168/jds.2015-10694 PMID:26898280

Gómez-Gascón, L., Galán-Relaño, A., Cardoso-Toset, F., Barrero-Domínguez, B., Astorga, R. J., Luque, I., Tarradas, C., & Gómez-Laguna, J. (2022). Lactate dehydrogenase: Detecting high bacterial and somatic cells counts in goats from whole milk samples. *Small Ruminant Research*, 208. doi:10.1016/j.smallrumres.2022.106632

Hogan, J., & Smith, K. L. (2012). Managing Environmental Mastitis. *The Veterinary Clinics of North America. Food Animal Practice*, *28*(2), 217–224. doi:10.1016/j.cvfa.2012.03.009 PMID:22664204

Holko, I., Tančin, V., Vršková, M., & Tvarožková, K. (2019). Prevalence and antimicrobial susceptibility of udder pathogens isolated from dairy cows in Slovakia. *The Journal of Dairy Research*, *86*(4), 436–439. doi:10.1017/S0022029919000694 PMID:31727191

Kaczorowski, Ł., Powierska-Czarny, J., Wolko, Ł., Piotrowska-Cyplik, A., Cyplik, P., & Czarny, J. (2022). The Influence of Bacteria Causing Subclinical Mastitis on the Structure of the Cow's Milk Microbiome. *Molecules (Basel, Switzerland)*, *27*(6), 1–11. doi:10.3390/molecules27061829 PMID:35335192

Kala, S. R., & Rani, N. L. (2022). Efficacy of antioxidant in the therapy of subclinical mastitis in buffaloes. *Original Article Buffalo Bulletin (January-March 2022)*, *41*(1).

Kala, S. R., Rani, N. L., Rao, V. V., & Subramanyam, K. V. (2021). Comparison of Different Diagnostic Tests for the Detection of Subclinical Mastitis in Buffaloes. *Buffalo Bulletin*, *40*(4), 653–659.

Kandeel, S. A., Morin, D. E., Calloway, C. D., & Constable, P. D. (2018). Association of California Mastitis Test Scores with Intramammary Infection Status in Lactating Dairy Cows Admitted to a Veterinary Teaching Hospital. *Journal of Veterinary Internal Medicine*, *32*(1), 497–505. doi:10.1111/jvim.14876 PMID:29222843

Kayitsinga, J., Schewe, R. L., Contreras, G. A., & Erskine, R. J. (2017). Antimicrobial treatment of clinical mastitis in the eastern United States: The influence of dairy farmers' mastitis management and treatment behavior and attitudes. *Journal of Dairy Science*, *100*(2), 1388–1407. doi:10.3168/jds.2016-11708 PMID:27939551

Kim, H., Min, Y., & Choi, B. (2019). Real-time temperature monitoring for the early detection of mastitis in dairy cattle: Methods and case research. *Computers and Electronics in Agriculture*, *162*(June 2018), 119–125. doi:10.1016/j.compag.2019.04.004

Krishnamoorthy, P., Goudar, A. L., Suresh, K. P., & Roy, P. (2021). Global and countrywide prevalence of subclinical and clinical mastitis in dairy cattle and buffaloes by systematic review and meta-analysis. *Research in Veterinary Science*, *136*(6450), 561–586. doi:10.1016/j.rvsc.2021.04.021 PMID:33892366

Kuchtík, J., Šustová, K., Sýkora, V., Kalhotka, L., Pavlata, L., & Konečná, L. (2021). Changes in the somatic cells counts and total bacterial counts in raw goat milk during lactation and their relationships to selected milk traits. *Italian Journal of Animal Science*, *20*(1), 911–917. doi:10.1080/1828051X.2021.1913077

Kumar, S., Mukherjee, M. M., & Varela, M. F. (2013). Modulation of Bacterial Multidrug Resistance Efflux Pumps of the Major Facilitator Superfamily. *International Journal of Bacteriology*, *2013*, 1–15. doi:10.1155/2013/204141 PMID:25750934

Lange, A., Grzenia, A., Wierzbicki, M., Strojny-Cieslak, B., Kalińska, A., Gołębiewski, M., Radzikowski, D., Sawosz, E., & Jaworski, S. (2021). Silver and copper nanoparticles inhibit biofilm formation by mastitis pathogens. *Animals (Basel)*, *11*(7), 1–11. doi:10.3390/ani11071884 PMID:34202806

Langoni, H., Da, D., Penachio, S., Borin, D., Felipe De Freitas Guimarães, N. I., Baldini, S., & Ii, L. (2012). 095 Somaticell ® as a screening method for somatic cell count from bovine milk. Somaticell ® as a screening method for somatic cell count from bovine milk *Avaliação do Somaticell ® como método de triagem para contagem de células somáticas do leite de bovi*. *42*(6), 1095–1101.

LeitnerG.LavonY.MerinU.JacobyS.BlumS. E.KrifucksO.SilanikoveN. (2019). Increasing the value of raw bulk milk quality based on mammary glands as production units vs. the udder in dairy cows with mastitis. bioRxiv. doi:10.1101/567271

Machado, N. A. F., Da Costa, L. B. S., Barbosa-Filho, J. A. D., De Oliveira, K. P. L., De Sampaio, L. C., Peixoto, M. S. M., & Damasceno, F. A. (2021). Using infrared thermography to detect subclinical mastitis in dairy cows in compost barn systems. *Journal of Thermal Biology*, *97*(February), 102881. doi:10.1016/j.jtherbio.2021.102881 PMID:33863445

Martí-De Olives, A., Peris, C., & Molina, M. P. (2020). Effect of subclinical mastitis on the yield and cheese-making properties of ewe's milk. *Small Ruminant Research*, *184*(April 2019), 106044. doi:10.1016/j.smallrumres.2019.106044

Massote, V. P., Zanateli, B. M., Alvez, G. V., Gonçalves, E. S., & Guedes, E. (2019). *Diagnóstico e controle de mastite bovina: uma revisão de literatura*. Revista Agroveterinária do Sul de Minas.

Medeiros, E. S., França, C. A., da Krewer, C. C., de Peixoto, R. M., de Souza Júnior, A. F., Cavalcante, M. B., da Costa, M. M., & Mota, R. A. (2011). Antimicrobial resistance of staphylococcus spp. isolates from cases of mastitis in buffalo in brazil. *Journal of Veterinary Diagnostic Investigation*, *23*(4), 793–796. doi:10.1177/1040638711407882 PMID:21908326

Merz, A., Stephan, R., & Johler, S. (2016). Staphylococcus aureus isolates from goat and sheep milk seem to be closely related and differ from isolates detected from bovine milk. *Frontiers in Microbiology*, *7*(MAR), 1–7. doi:10.3389/fmicb.2016.00319 PMID:27014240

Mimoune, N., Saidi, R., Benadjel, O., Khelef, D., & Kaidi, R. (2021). Alternative treatment of bovine mastitis. *Veterinarska Stanica*, *52*(6), 639–649. doi:10.46419/vs.52.6.9

Muhammad, G., Naureen, A., Asi, M. N., Saqib, M., & Fazal-ur-Rehman. (2010). Evaluation of a 3% surf solution (surf field mastitis test) for the diagnosis of subclinical bovine and bubaline mastitis. *Tropical Animal Health and Production*, *42*(3), 457–464. doi:10.100711250-009-9443-3 PMID:19731065

Oliveira, R. P., Bezerra, B. A., de Melo, R. P. B., da Silva, D. M. S., de Carvalho, R. G., Juliano, M. A., Farias, M. P. O., de Lira, N. S. C., & Mota, R. A. (2022). Bovine mastitis in northeastern Brazil: Occurrence of emergent bacteria and their phenotypic and genotypic profile of antimicrobial resistance. *Comparative Immunology, Microbiology, and Infectious Diseases*, *85*(November 2021). doi:10.1016/j.cimid.2022.101802

Oliveira, R. P., Silva, J. G., Aragão, B. B., Carvalho, R. G., Juliano, M. A., Frazzon, J., Farias, M. P. O., & Mota, R. A. (2022). Diversity and emergence of multi-resistant Staphylococcus spp. isolated from subclinical mastitis in cows in of the state of Piauí, Brazil. *Brazilian Journal of Microbiology*, *0123456789*(4), 2215–2222. doi:10.100742770-022-00822-1 PMID:36074251

Oliveira, S. C. C., Nishi, L., Mantovani, D., Pisano Mateus, G. A., Tonial dos Santos, T. R., Takaoka Alves Baptista, A., Guttierres Gomes, R., & Bergamasco, R. (2019). EXTRATOS DE PLANTAS BRASILEIRAS NO CONTROLE DA BACTÉRIA Staphylococcus aureus CAUSADORA DA MASTITE CONTAGIOSA EM BOVINOS LEITEIROS. *Revista Tecnológica*, *27*(1), 48–58. doi:10.4025/revtecnol.v27i1.43745

Pakhtunkhwa, K., Khan, I., Wazir, A. A., Khan, S., & Hassan, S. I. (2022). Prevalence of Sub-Clinical Mastitis in Buffaloes at Livestock Research and iMedPub Journals Prevalence of Sub-Clinical Mastitis in Buffaloes at Livestock Research and Development Station. *Paharpur and Surroundings of Tehsil Paharpur District D., 1*(June). doi:10.5281/zenodo.6626290

Panahi, M., & Saei, H. D. (2019). Genetic diversity and methicillin resistance of Staphylococcus aureus originating from buffaloes with mastitis in Iran. *Comparative Immunology, Microbiology, and Infectious Diseases, 62*(November 2018), 19–24. doi:10.1016/j.cimid.2018.11.014

Patel, R., Kunjadia, P., Koringa, P., Joshi, C., & Kunjadiya, A. (2019). Microbiological profiles in clinical and subclinical cases of mastitis in milking Jafarabadi buffalo. *Research in Veterinary Science, 125*(August 2016), 94–99. doi:10.1016/j.rvsc.2019.05.012

Peixoto, R. D. M., Aparecido, R., Mateus, M., Mastite, B., Pesquisa, B., Brasileira, V., Agrárias, C. C., & Federal, U. (2010). Mastite Em Pequenos Ruminantes: *Etiologia E Epidemiologia. 30*(9), 754–762.

Peixoto, R. de M., Silva, W. E. L., Almeida, J. R. G. S., Branco, A., & Costa, M. M. (2016). ANTIBACTERIAL POTENTIAL OF NATIVE PLANTS FROM THE CAATINGA BIOME AGAINST Staphylococcus spp. ISOLATES FROM SMALL RUMINANTS WITH MASTITIS. *Revista Caatinga, 29*(3), 758–763. doi:10.1590/1983-21252016v29n328rc

Pisanu, S., Cacciotto, C., Pagnozzi, D., Uzzau, S., Pollera, C., Penati, M., Bronzo, V., & Addis, M. F. (2020). Impact of Staphylococcus aureus infection on the late lactation goat milk proteome: New perspectives for monitoring and understanding mastitis in dairy goats. *Journal of Proteomics, 221*(January), 103763. doi:10.1016/j.jprot.2020.103763 PMID:32275959

Puppel, K., Kalińska, A., Kot, M., Slósarz, J., Kunowska-Slósarz, M., Grodkowski, G., Kuczyńska, B., Solarczyk, P., Przysucha, T., & Gołębiewski, M. (2020). The effect of staphylococcus spp., streptococcus spp. and enterobacteriaceae on the development of whey protein levels and oxidative stress markers in cows with diagnosed mastitis. *Animals (Basel), 10*(9), 1–11. doi:10.3390/ani10091591 PMID:32906645

Ribeiro, M. G., Costa, E. O., Leite, D. S., Langoni, H., Garino, F., Victória, C., & Listoni, F. J. P. (2006). Fatores de virulência em linhagens de Escherichia coli isoladas de mastite bovina. *Arquivo Brasileiro de Medicina Veterinária e Zootecnia, 58*(5), 724–731. doi:10.1590/S0102-09352006000500004

Rossi, R. S., Amarante, A. F., Correia, L. B. N., Guerra, S. T., Nobrega, D. B., Latosinski, G. S., Rossi, B. F., Rall, V. L. M., & Pantoja, J. C. F. (2018). Diagnostic accuracy of Somaticell, California Mastitis Test, and microbiological examination of composite milk to detect Streptococcus agalactiae intramammary infections. *Journal of Dairy Science, 101*(11), 10220–10229. doi:10.3168/jds.2018-14753 PMID:30146291

Saenz-de-Juano, M. D., Silvestrelli, G., Bauersachs, S., & Ulbrich, S. E. (2022). Determining extracellular vesicles properties and miRNA cargo variability in bovine milk from healthy cows and cows undergoing subclinical mastitis. *BMC Genomics, 23*(1), 1–15. doi:10.118612864-022-08377-z PMID:35255807

Sales, D. C., & Rangel, A. H. do N., Urbano, S. A., Tonhati, H., Galvão Júnior, J. G. B., Guilhermino, M. M., Aguiar, E. M., & Bezerra, M. de F. (2018). Buffalo milk composition, processing factors, whey constituents' recovery and yield in manufacturing Mozzarella cheese. *Food Science and Technology (Campinas), 38*(2), 328–334. doi:10.1590/1678-457X.04317

Da Silva, J. G., De Barros, M., Santos, N. D. D. L., Paiva, P. M. G., Napoleão, T. H., De Sena, M. J., Da Costa, M. M., De Oliveira, H. P., Moreira, M. A. S., & Mota, R. A. (2020). Antimicrobial activity of polypyrrole nanoparticles and aqueous extract of Moringa oleifera against Staphylococcus spp. carriers of multi-drug efflux system genes isolated from dairy farms. *The Journal of Dairy Research, 87*(3), 309–314. doi:10.1017/S0022029920000874 PMID:32958093

Silva, M. V. M., & Nogueira, J. L. (2010). Mastite: Controle E Profilaxia No Rebanho Bovino. *Revista Científica Eletrônica de Medicina Veterinária, 15*(VIII), 13.

Soares, J. C., Soares, A. C., Popolin-Neto, M., Paulovich, F. V., Oliveira, O. N. Jr, & Mattoso, L. H. C. (2022). Detection of Staphylococcus aureus in milk samples using impedance spectroscopy and data processing with information visualization techniques and multidimensional calibration space. *Sensors and Actuators Reports, 4*, 100083. doi:10.1016/j.snr.2022.100083

Souza, G. N., Brito, J. R. F., Moreira, E. C., Brito, M. A. V. P., & Silva, M. V. G. B. (2009). Variação da contagem de células somáticas em vacas leiteiras de acordo com patógenos da mastite. *Arquivo Brasileiro de Medicina Veterinária e Zootecnia, 61*(5), 1015–1020. doi:10.1590/S0102-09352009000500001

Spuria, L., Biasibetti, E., Bisanzio, D., Biasato, I., De Meneghi, D., Nebbia, P., Robino, P., Bianco, P., Lamberti, M., Caruso, C., Di Blasio, A., Peletto, S., Masoero, L., Dondo, A., & Capucchio, M. T. (2017). Microbial agents in macroscopically healthy mammary gland tissues of small ruminants. *PeerJ*, *2017*(11), 1–18. doi:10.7717/peerj.3994 PMID:29152416

Srikok, S., Patchanee, P., Boonyayatra, S., & Chuammitri, P. (2020). Potential role of MicroRNA as a diagnostic tool in the detection of bovine mastitis. *Preventive Veterinary Medicine*, *182*(July), 105101. doi:10.1016/j.prevetmed.2020.105101 PMID:32823253

Vakkamäki, J., Taponen, S., Heikkilä, A. M., & Pyörälä, S. (2017). Bacteriological etiology and treatment of mastitis in Finnish dairy herds. *Acta Veterinaria Scandinavica*, *59*(1), 1–9. doi:10.118613028-017-0301-4 PMID:28545485

Vásquez-García, A., Silva, T. dos S., de Almeida-Queiroz, S. R., Godoy, S. H. S., Fernandes, A. M., Sousa, R. L. M., & Franzolin, R. (2017). Species identification and antimicrobial susceptibility profile of bacteria causing subclinical mastitis in buffalo. *Pesquisa Veterinaria Brasileira*, *37*(5), 447–452. doi:10.15900100-736x2017000500004

Xudong, Z., Xi, K., Ningning, F., & Gang, L. (2020). Automatic recognition of dairy cow mastitis from thermal images by a deep learning detector. *Computers and Electronics in Agriculture*, *178*(May), 105754. Advance online publication. doi:10.1016/j.compag.2020.105754

Youssif, N. H., Hafiz, N. M., Halawa, M. A., & Saad, M. F. (2020). Influence of some hygienic measures on the prevalence of subclinical mastitis in a dairy farm. *International Journal of Dairy Science*, *15*(1), 38–47. doi:10.3923/ijds.2020.38.47

Zafalon, L. F., Filho, a N., Oliveira, J. V. De, Resende, F. D. De, Penteado, H., & Odessa, N. (. (2005). Alterações da Composição e da produção de leite oriundo de QUARTOS MAMÁRIOS DE VACAS COM E SEM MASTITE SUBCLÍNICA. *Arquivos do Instituto Biológico*, *72*(4), 419–426. doi:10.1590/1808-1657v72p4192005

Zalafon, L. F., Langoni, H., Benvenutto, F., Castelani, L., & Broccolo, C. R. (2008). 56 Aspectos Epidemiológicos Da Mastite Bovina Causada Por. *Veterinária e Zootecnia*, *15*(16), 56–65.

Chapter 7
The Use of Antibiotics in Dairy Farming:
Main Challenges and Alternative Approaches

Wendell Queiroz Leite
School of Agricultural and Veterinarian Sciences, São Paulo State University, Jaboticabal, Brazil

Mariane de Freitas Genari Severino
School of Agricultural and Veterinarian Sciences, São Paulo State University, Jaboticabal, Brazil

Eduardo Festozo Vicente
School of Sciences and Engineering, São Paulo State University, Tupã, Brazil

EXECUTIVE SUMMARY

The dairy chain is remarkably important in mammals' nutrition, highlighting the human food through the production of milk and its derivatives. Thus, antibiotics have become essential for life and health maintenance of animals, being widely used for treatment of bacterial infectious diseases. However, antimicrobial resistance resulted from uncontrolled usage of antibiotics for prophylactic purposes and animal growth promoter. The challenge of antimicrobial resistance and persistence of antibiotic residues in milk still occurs, which presents a potential risk to human and animal health. Therefore, it is crucial to develop alternative strategies to overcome the inappropriate use of antibiotics, especially for dairy farming. Thus, this chapter intends to raise awareness and clarify issues related to the use of antibiotics in dairy cattle, the major problems of antimicrobial resistance, residues in milk, and to report about the main alternative solutions to the use of conventional antibiotics in dairy production with scientific studies.

DOI: 10.4018/978-1-6684-5472-5.ch007

ANTIBIOTIC USE IN ANIMAL PRODUCTION

A Historical Overview

The late 1940s and early 1950s were considered the "era of wonder drugs" due to the introduction of antimicrobial drugs in the treatment of difficult to cure infections (Prescott, 2017; Prescott, 2019). After the World War II, between 1951 and 1955, antibiotics were introduced as feed additives in animal production as growth promoters, later antibiotics were used in the treatment of infections, mainly as intramammary drugs for the treatment of mastitis in dairy cows (Plastridge, 1958; Prescott, 2017; Prescott, 2019).

The main organisms associated with mastitis were Streptococcus agalactiae, Streptococcus uberis and Micrococcus pyogenes, which usually caused chronic mastitis, with or without clinical signs of the disease, but commonly causing significant losses in milk production (Plastridge, 1958). Plastridge (1958) highlights that several predisposing factors influence the susceptibility to mammary infection, such as exposure and duration of milking; exposure of the udder to cold; supply of protein-rich concentrate feed, milk flow rate, stage of lactation, age, and heredity. Among the main etiological forms, mastitis caused by *Streptococcus agalactiae* was the only one that could be eradicated (Plastridge, 1958).

Prior to the introduction of antibiotics, methods of controlling mastitis caused by *Streptococcus agalactiae* were based on performing on-farm sanitary practices, such as the use of quaternary ammonium molecules and chlorhexidine employed in the water for udder washing, teats immersion, and milking equipment washing (Plastridge, 1958). The author also reports that another method indicated to control contagious mastitis was the segregation and replacement of clinically affected animals by healthy animals, with the elimination of infection by *Streptococcus agalactiae* from the herd being reported between two to five years. However, this process is long and tedious, making it impractical in some herds. In this way, between 1946 and 1950, penicillin and streptomycin were allowed to be used in animals, and their use was evaluated empirically (Prescott, 2017).

Initially, the mastitis control program against *Streptococcus agalactiae* was established in USA (Connecticut, New York, and Massachusetts), England and Denmark. Specifically, in Connecticut, the control program began in 1940, with statewide laboratory sample testing. Penicillin treatments started in 1946, when the established intervals of mastitis retests were between two to three weeks and considered free of Streptococcus agalactiae at three to six months. In New York, the research and control program started in 1946. In England, the control measure started in 1955, which consisted of cultural examination of milk samples, treatment of all quarters of infected and dried cows using two doses of 100,000 units of procaine and penicillin in an oil base, with an interval of three days between treatments. The retest was six weeks after treatment. Two negative tests, followed by a re-test once a year, was followed by the removal of the animals that resisted the treatments. In Denmark, tests were performed on individual cows and all infected animals were treated in all quarters, with three doses of 25,000 units of sodium penicillin in an aluminum monostearate base, administered at 48-hour intervals (Plastridge, 1958).

Penicillin was the most effective antimicrobial drug in the treatment of mastitis caused by Streptococcus agalactiae (Plastridge, 1958). The author reports that penicillin-based treatment eliminated 80 to 90% of the microorganism from of infected quarters, maintaining the effective level in the mammary gland for three to five days. Treatment consisted of daily infusions of 25,000 to 100,000 units of penicillin in saline solution, or one or two infusions at 72-hour intervals of 100,000 units in a water-in-oil emulsion. Intramuscular injections of penicillin were also successful in the treatment of mastitis, with administrations of a total of 42,000,000 units, divided into 6,000,000 units in the first injection, and subsequently 3,000,000 units per injection every 12 hours during six days. There were cases of mastitis cure by Streptococcus agalactiae with two or three injections of 10,000,000 units.

Antimicrobial Resistance: An Alarming Problem

From 1958 to 1960, Smith and Crabb began to use the phage typing method and demonstrated the use of antimicrobial drugs as growth promoters was a causative factor of antibiotic resistance in Escherichia coli isolated from pig and chicken feces and alerted of the potential danger for humans (Lees et al., 2020). Moreover, when antibiotic susceptibility tests were performed on Staphylococcus aureus isolates from the nasal passages and skin of pigs and chickens kept in commercial conditions, it was found a higher prevalence of antibiotic resistant strains in animals kept on diets containing antibiotics than diets without it, and the same was found on the animals' handlers (Smith and Crabb, 1960a).

Currently, antimicrobial resistance is one of the greatest global public health problems (Nobrega et al., 2020). Zhang et al. (2022), in a meta-analysis bringing together works from 1992 and 2021, reported that penicillin showed higher rates of antibiotic resistance than Staphylococcus aureus, while other antibiotics also demonstrated increases in resistance over time.

In calves, when several antibiotic drugs were administered with the aim of reducing the population of E. coli in animals with diarrhea, an emergence of resistant E. coli was observed during the treatment period with chloramphenicol, streptomycin, oxytetracycline, chlortetracycline, sulfadimidine and phthalylsulfathiazole, but never with furazolidone, neomycin, polymyxin and furamazone (Smith and Crabb, 1960b). However, the resistant population was maintained only during the treatment period, as it evenly became partially resistant or sensitive population, depending on the interval of drug use.

Thus, due to "the age of miracle drugs", antimicrobial drugs were introduced for animal use even only as small number of controlled experiments have been performed by that time. Initial dosages were based on small and poorly controlled trials, so there were confusing mixes about optimal dosages that were widely divergent for many livestock diseases (Prescott, 2017). At the time, the widespread use of antibiotics contributed to disease control and improvements in animal welfare (Albright et al., 1961).

However, when methods of administration of antimicrobial drugs are used inappropriately in animal production, they can contribute to antibiotic residues in milk (Albright et al., 1961). The author reports that research carried out for 9 years, prior to the year 1960, revealed that out of 7,201 milk samples tested, 5.8% showed antibiotic residues. Plastridge (1958) also reports that some researchers were successful in producing resistant strains, which resisted in milk samples with up to three penicillin units, although exhibiting a limited resistance.

In addition to problems with antimicrobial resistance since the 1950s, several studies have been conducted identifying the presence of antibiotics in milk (Sisodia and Stowe, 2006). Despite having contributed to disease control and animal welfare, the use of antibiotics caused problems for the milk producer and consumer (Albright et al., 1961). The presence of antibiotics in milk have been related to the indiscriminate use of antimicrobials via intramammary route to treat bovine mastitis; failure to discard milk up to two days after administration, and the passive diffusion capacity of the non-ionized portion of the molecule (Sisodia and Stowe, 2006). As a remedial measure to antimicrobial resistance, in the period from 1966 to 1970 new drug analogues were evaluated as a solution to the transmissible resistance problems and multidrug resistance (Prescott, 2017).

In their studies of antimicrobial resistance, Anderson and Datt (1965), reported increasing proportions of resistant strains of Salmonella typhimurium isolated from the year 1961 onwards. These antimicrobial resistant bacteria were first found in the second half of 1962, and in which, 8% were resistant to ampicillin. In 1962, 17 strains of Salmonella typhimurium were isolated, in which the ampicillin resistance was always accompanied by resistance of other drugs (Anderson and Datt, 1965). The authors affirmed that antimicrobial drug resistance was contagious, as it could be transferred between mixed cultures for Escherichia coli. For a sensitive strain of Salmonella typhimurium, ampicillin resistance was related to the production of a penicillin-destroying enzyme, and many of the resistant strains from Salmonella that cause human infection could originate in cattle.

In 1965, Anderson and Datt broke the paradigm of vertical transmission of resistance through the possible demonstration of horizontal transfer, by exchanging

DNA fragments known as plasmids. In addition, horizontal transfer of resistance can occur between bacteria present in animal populations and bacteria associated with humans, highlighting the possibility of global environmental risks caused by antibiotic resistance (Lees et al., 2020).

Thus, between the years 1976 to 1980, there was a focus on pharmacokinetic studies and drug metabolism in production animals (Prescott, 2017). Mercer et al. (1977) presented a pharmacokinetic approach to the design of a drug residue profile in animals. The authors believed that a complete pharmacokinetic characterization of a drug in the target species would provide an essential database for the development of treatment protocols (dosage levels and intervals), withdrawal period and drug efficacy, which would facilitate the production of new drugs (Mercer et al., 1977).

The "modern era" of veterinary antimicrobial treatments began with the regulatory approval of enrofloxacin in 1987, followed by ceftiofur in 1988 (Lees et al., 2020). Most of the drugs approved for veterinary use were discovered in human antibiotic development programs (Shryock, 2004). Initially, the approval of antibiotics was intended for respiratory diseases and were introduced sequentially: Tilmicosin and Florfenicol, 1990; Danofloxacin,1991; Cefquinome and Marbofloxacin, 1995; Tulathromycin, 2003; Gamithromycin, 2008; and Tildipirosin, 2011, with additions to their current uses, such as new pathogens or diseases (Lees et al., 2020; Shryock, 2004).

MAIN CHALLENGES OF INAPPROPRIATE ANTIBIOTIC USE

Antibiotic Use in Dairy Farming

In animal production, antibiotics are used as growth promoters, treatments for disease, metaphylactic and prophylactic use (US GAO, 2011). The use of antibiotics as growth promoters aims to promote an increase in animal performance (Innes et al., 2020). The use of antimicrobials in the treatment of diseases is recommended after confirmation through clinical tests of bacterial infection. Metaphylactic use is the administration of broad-spectrum antibiotics to asymptomatic animals that have been exposed to infected animals. Prophylactic use is intended to remedy the emergence of bacterial infections in animals.

It is estimated that 73% of the world consumption of antimicrobials are destined for food production animals (Treiber and Beranek-Knauer, 2021) and the United States accounts for 80% (Van Boeckel et al., 2015). As a result of population growth, the demand for animal products will increase. As a result, it is estimated that the use of antimicrobials in animal production will increase by more than 67% by 2030 (Van Boeckel et al., 2015).

Animal products of great importance in the world are milk and its derivatives, being produced in different types of species, systems and technology (De Albuquerque Fernandes, et al., 2014). In dairy production, there is an excessive use of antimicrobials as growth promoters and as prophylactic use, therefore is necessary to modify their use in a rational way. (Sharma et al., 2018).

In dairy cows, mastitis is the most frequent disease responsible for large economic losses in milk production, even with the implementation of control programs (Pengov Kirbis, 2009). In addition to mastitis, bacterial infections in the respiratory system, uterus and feet are very common in dairy cows. These infections are usually treated with sulfonamides, ceftiofur, and other β-lactams (Pol; Ruegg, 2007).

Staphylococcus aureus is one of the main causative agents of contagious mastitis in the world, known for poor responses to conventional drugs and causing chronic intramammary infections, resulting in significant economic losses (Molineri et al., 2021). In 2012, it was estimated that intramammary infections caused by Staphylococcus aureus caused an economic loss in the US of US$ 2 billion/year and in the EU an average cost of € 485/cow, (Heikkila et al., 2012).

Langoni et al. (2017) considered as contagious agents of mastitis the following microorganisms: Streptococcus agalactiae; Staphylococcus aureus; Coagulase-negative Staphylococcus; Mycoplasma spp., and Corynebacterium bovis. The most frequent environmental mastitis causing microorganisms are Escherichia coli; Klebsiella pneumoniae; Enterobacter aerogenes; Proteus spp.; Pseudomonas spp.; Streptococcus uberis; Streptococcus dysgalactiae; yeasts, algae, and fungi (Langoni et al., 2017).

In their study conducted in Brazil, Rall et al. (2019) reported that the most prevalent species in the milk of cows with clinical mastitis was Staphylococcus aureus., In animals with subclinical mastitis, the Staphylococcus species most frequently isolated was Staphylococcus warneri, with an incidence of 27.5%. In animals considered healthy, the authors found a higher prevalence of Staphylococcus saprophyticus (30.2%) (Rall et al. 2019).

In China, Wang et al. (2015) analyzed the susceptibility profile of Staphylococcus aureus to antibiotics and found out that out of 2019 isolated strains from bovine mastitis cases, 34 antibiotic-resistant strains were identified. Moreover, over 70% of the antibiotic-resistant strains presented resistance to several antibiotics. The authors also report that Streptococcus aureus resistance is becoming increasingly complicated due to changes in the multidrug resistance mechanisms of these microorganisms (Wang et al. 2015). Zigo et al. (2017) evaluated the effectiveness of different antibiotics for agents causing mastitis during the first lactation of Holstein cows in eastern Slovakia. The results showed that Streptococcus agalactiae, Streptococcus uberis, Coagulase negative Staphylococcus had 100% sensitivity against dihydrostreptomycin, neomycin, novobiocin, prednisolone, procaine, benzylpenicillin and Streptococcus aureus had 94.4% sensitivity. The authors also inferred that more than 22% of Streptococcus aureus showed resistance to penicillin, amoxicillin, and streptomycin, while Streptococcus uberis showed 10.5% resistance to amoxicillin and 52.7% to streptomycin.

Antibiotics Residues in Milk

In addition to antimicrobial resistance in dairy cattle, there is a concern about antibiotic residues in human foods. Pharmacologically active molecules or their metabolites, found in foods derived from animals to which drugs has been administered are defined as residues (Beyene, 2015).

For food safety reasons, regulatory authorities such as EFSA (European Food Safety Authority), FDA (Food and Drug Administration) and Codex Alimentarius ("Food Code" is a set of dietary guidelines created by the Food and Agriculture Organization – FAO - and the World Health Organization – WHO - for consumer safety and food trade) have established the maximum residual limit (MRL) of antimicrobials in products of animal origin to ensure consumer protection (Quintanilla et al., 2019). When the antibiotic is administered to the animal, most of it is metabolized and excreted in the urine and, barely, in the feces as a detoxification mechanism. Nevertheless, part of the drug may remain for an interval in the animal and be present in milk or meat (Sachi et al., 2019).

Sachi et al. (2019) evaluated the frequency of scientific research topic "presence of antibiotic residues in milk" in the scientific literature between 1960 and 2017. The authors found 224 articles, in which antibiotic residues in cow milk samples were analyzed, quantitatively and qualitatively. However, most of the studies (82.14%) were related to detection methods, in which few samples were analyzed, and, in the majority, milk was spiked with known concentrations of antibiotics with the aim to improve methods of detection. In parallel, Treiber and Beranek-Knauer (2021) performed a literature review from 1999 to 2019 for articles evaluating residues of antibiotics in animal source foods, focusing on commercial products. A total of 73 scientific articles documented antimicrobial residues in animal products - which 27 studies documented antimicrobial residues in milk.

The presence of antibiotic residues in milk negatively affects the production of fermented dairy products, which can provoke decreases in the final products' quality. Therefore, this situation could have economic consequences for the dairy sector. Among the issues that the existence of antibiotics can cause in the manufacture of dairy products and were reported as: failures in the growth of starter cultures; failure in cheese ripening; failure in the curdling of milk; and failure in acid production, and flavor problems (Marth & Ellickson, 1959). In this way, many Lactic Acid Bacteria (LAB) are employed as starter cultures to make fermented dairy products. Moreover, non-starter LAB, originated from environment and raw material, collaborate to the normal development of dairy product features. Thus, the majority of the problems promoted by antibiotic residues are due they can inhibit the LAB development, completely or slightly, and retard the acid production.

In addition, lowering the pH is fundamental, i.e., in the cheese-making process, as it increases the enzymes' activity and the coagulation speed, which is of paramount importance especially for long and hard matured cheeses (Chiesa et al., 2020). Cogan (1972) analyzed the susceptibility to tetracycline, cloxacillin, penicillin, and streptomycin of eight single-strain lactic Streptococci, which three commercial cheese starters, and six lactic acid bacteria isolated from yogurt. Also, the authors found that the antibiotic ranges that caused 50% inhibition of the bacteria were (μg/L): tetracycline, 90–600; cloxacillin, 240–2,500; penicillin, 9–200; and streptomycin, 350–13,000. In conclusion, the yogurt isolates were more resistant to streptomycin and more sensitive to penicillin than the cheese starters.

Antimicrobial Resistance and its problems for One Health

Undoubtedly, antibiotics used in livestock feed as additives has largely contributed to animal productivity and health over the years. However, misuse and overuse of antimicrobials are major factors in the expansion of drug-resistant pathogens in animals, humans and commercial crops. Due to public health concerns, the European Union is progressively banning the use of antibiotics as growth promoters in livestock (Brown et al., 2016). The main elements that contribute to antimicrobial resistance are the misuse and excessive use of antibiotics in animals and humans, agricultural use of antibiotics, increasing income of developing countries is positively correlated with antibiotic use, accessible travel routes that contribute to spread of resistant pathogens, biological elements and information gaps (Dadgostar, 2019). Antibiotics used in very low doses or in high doses but with short or prolonged treatment time, incorrect or unnecessary prescription for illness, all have resulted over time in stronger infections, increased illness, and deaths. In this way, the discovery of each new antibiotic has been strictly followed by emerging resistance (WHO, 2014).

Antimicrobial resistance (AMR) appears in bacteria, fungi, viruses, and parasites through evolutionarily changes across the years and eventually these microorganisms do not respond to antibiotics and other drugs, making infections harder to treat and consequently increasing the threat of sickness spreading out of control, extreme infections and dying (WHO, 2021). Microorganisms with AMR can avoid the adverse consequences of antibiotics to which they were initially sensitive, resulting in their survival despite exposure to standard doses of antibiotics (Sharma et al. 2018). Nowadays, AMR is a worldwide public health concern and a threat to the achievement of the Sustainable Development Goals (SDGs), which directly affects poverty (SDG 1), food security (SDG 2), health security (SDG 3), economic growth (SDG 8) and requiring urgent actions to mitigate, prevent and control it (WHO, 2017).

Antibiotic resistance will occur due to natural selection. There are three descriptions of resistance: microbiological, pharmacological, and clinical resistance. Microbiological in vitro resistance includes a decrease of bacteria susceptibility to antibiotics above a threshold defined by the maximum limit for ordinary species' susceptibility (i.e. susceptible microorganisms). In vitro resistance can be verified with molecular techniques (i.e. genotypification), determining the presence of genes that code for any proteins associated with any resistance mechanism. Pharmacological resistance is established on pharmacokinetic parameters and general susceptibility of bacterial by minimum inhibitory attention (MIC) of the antibiotic for the microorganism. Clinical in vivo resistance occurs when an infection cannot be tackled promptly, and drugs failures are noticeable (Verraes et al., 2013).

Antimicrobial resistant organisms can proliferate between animals and humans, from foods of animal origin, and can be found virtually in all living organisms (i.e., crop plants) and the environment (WHO, 2021). Natural and human-caused evolutionary factors have contributed to the enormous increase and spread of drug-resistant bacteria. Natural and man-made evolutionary factors have contributed to the enormous increase and spread of drug-resistant bacteria. Bacteria can develop resistance through random mutations or acquire resistance genes through horizontal transfer by exchanging plasmids between different bacteria, both resulting from strong selective pressure caused by the heavy reliance on antimicrobial use in farm animals (Sharma et al., 2018). Hence, bacteria could develop resistance-mediating mutations or obtain resistance genes promoted by a selective pressure, due to an inappropriate antimicrobial use in animals (Verraes et al., 2013).

Multidrug-resistant strains emergence are responsible for approximately 15.5% of hospital-acquired infections and 700,000 deaths per year globally (Dafale et al., 2020; Mulani et al., 2019). It is estimated that by 2050 10 million lives per year could be lost and that the world economy could lost up to $100 trillion annually from the expected increase in these antibiotic resistant infections if no proactive solutions happen to slow or stop drug resistance (O'Neill, 2016). The main bacterial pathogens with high virulence and antimicrobial resistance received the denomination of "ESKAPE", which contains the initial of their scientific names. These pathogens are: Enterococcus faecium and Staphylococcus aureus in the Gram-positive group and Klebsiella pneumoniae, Acinetobacter baumannii, Pseudomonas aeruginosa and Enterobacter spp. in the Gram-negative group (Mulani et al., 2019). ESKAPE pathogens are different from others because they acquired greater resistance to penicillin, vancomycin, carbapenems and other antibiotics, commonly used in humans and animals (Santajit; Indrawattana, 2016).

Moreover, in 2017 the World Health Organization (WHO) created a priority list of antibiotic-resistant pathogens, including 12 families of bacteria, in order to guide research priorities and their funding. In this list pathogens are grouped into 3 categories according to priority: critical, high, and medium. The critical category includes Acinetobacter baumannii; Pseudomonas aeruginosa; and Enterobacteriaceae spp. In high priority, Enterococcus faecium and Staphylococcus aureus stand out. Of importance in the intermediate class, we have Helicobacter pylori, Campylobacter spp., Salmonellae spp. and Neisseria gonorrhoeae. In the medium priority we have Streptococcus pneumoniae, Haemophilus influenzae and Shigella spp. (WHO, 2017). Due to global concerns about antimicrobial resistance, alternative measures to antibiotics have been studied in the therapeutic and prophylactic use for diseases in farm animals including probiotics, prebiotics, symbiotics, organic acids, enzymes, phytogens, antimicrobial peptides, bacteriophages, metals, vaccines and essential oils (Lees et al., 2020).

ALTERNATIVE APPROACHES TO THE USE OF CONVENTIONAL ANTIBIOTICS IN ANIMAL PRODUCTION

Antimicrobial resistance and antibiotic residues in milk are global problems to public health and food safety, causing a lot of social, economic, and environmental damage. Thus, the global consumer demand for antibiotic-free animal products has surged. Therefore, the study and discovery of alternatives to conventional antibiotics are extremely necessary. Some alternatives to antibiotics for growth promotion and disease prevention are probiotics, prebiotics, antimicrobial peptides, phototherapies, phytochemicals and vaccines.

Probiotics and Prebiotics

Probiotics are live microorganisms that are beneficial to the native gut microbiota and are classified into defined or undefined probiotics. Defined probiotics are individual strains or mixtures of described microorganisms (i.e., species; the precise quantitative composition of the culture and the sequenced genomes of individual organisms in the combination to ensure the absence of any antibiotic resistance genes). Indefinite probiotics are composed of microbial mixtures that are not fully described but are prone to be more efficient than defined probiotics. Both types of probiotics are promising in disease prevention and, in some cases, their use results in better performance and growth promotion (Colin Hill et al., 2014; FAO, 2016). Some undefined probiotics, called competitive exclusion products, are usually administered after birth or hatch, and help animals to establish a community of good bacteria in the gut before colonization by pathogens, which resulted in high efficiency in preventing disease in young animals (Oliver et al. al., 2022; Callaway et al., 2003).

Despite scientific studies evaluating the effect of probiotics on growth promotion, positive impacts on disease prevention have also been demonstrated in cattle (FAO, 2007; Chiquette, 1995; Yutaka Uyeno et al. 2015). Data from 110 articles with 157 experiments and 376 treatments were applied in a quantitative meta-analysis to evaluate the influence of Saccharomyces cerevisiae supplementation on ruminal parameters and milk production in ruminants (Yutaka Uyeno et al. 2015). This meta-analysis aimed to highlight the main quantitative effects of live yeast supplementation on ingestion, rumen fermentation and milk production. The results showed that yeast supplementation increased rumen volatile fatty acid concentration (+2.17 mM on average), rumen pH (+0.03 on average), and tended to decrease ruminal lactic acid concentration (-0.9 mM on average) with no influence on the acetate-propionate ratio. Dry matter intake (DMI; +0.44 g/kg body weight - BW), milk production (+1.2 g/kg BW) and total organic matter digestibility increased (+0.8% average). These results suggest an improvement in rumen fermentation by yeast supplementation, although this has not been measured (Yutaka Uyeno et al. 2015).

Prebiotics are oligosaccharides included in the diet as a source of substrate for beneficial bacteria in the intestine, helping their growth. Both prebiotics and probiotics help beneficial bacteria to increase the speed of their growth, outperforming their harmful competitors, as well as helping to modulate the immune system. A recent review by the European Medicines Agency (EMA) concluded that prebiotics contribute to disease reduction and act as growth promoters (Vyas; Ranganathan, 2001; EMA, 2022).

The benefits of using prebiotics in cattle are limited to young animals, as they do not have a developed rumen. In older animals, the microbiota present in the rumen helps in the processing of complex carbohydrates of plant origin, such as cellulose, preventing the passage of these prebiotics to the intestine of these animals (Gaggìa et al., 2022). When prebiotics are added to milk replacers and fed to calves, they aid in the development of these animals, promoting growth and helping to prevent disease (Quigley et al., 1997). Calves fed with replacer with the addition of the prebiotic galactosyl-lactose, showed greater body weight gains when compared to animals fed with replacer without the inclusion of prebiotics. Regarding the health of calves, few studies have evaluated the effect of prebiotics on disease prevention, but significant improvements in the intestinal health of calves are cited (Heinrichs et al., 2003).

Among the functional oligosaccharides, manno-oligosaccharides (MOS) are prebiotics on the rise with potential bioactive properties. Currently, commercially available MOS comes from mannan from the yeast cell wall and is widely used in supplements as a prebiotic. Its main function is to stimulate the growth of beneficial microbiota, and it also confers anticancer and immunomodulatory effects by inducing different genetic markers in colon cells (Jana et al., 2021).

Ghosh & Mehla (2012) evaluated the effect of 4 g prebiotic MOS supplementation in 5-day-old calves. The parameters evaluated were body weight gain, feed intake, feed conversion, fecal score, fecal coliform count and feed cost. They found an increase in weight gain, feed intake and feed conversion efficiency. In contrast, there was a significant reduction in diarrhea severity in the MOS compared to the control treatment ($P < 0.01$). Feed cost was lower in MOS compared to control ($P < 0.01$), suggesting that MOS supplementation leads to better performance.

In addition to mannooligosaccharides, other different prebiotics used in calf supplementation are fructooligosaccharides (FOS) and cello-oligosaccharides (COS). Hasunuma et al. (2011) investigated the effect of cello-oligosaccharide or symbiotic combination of dextran and Lactobacillus casei in the feeding of fifty-two Holstein heifers on growth performance, fecal condition and hormone concentrations. At 90 days of age, heifers that received COS showed greater body weight gain and feed efficiency ($P < 0.05$) compared to control animals. EC treatment had less effect on the growth of Escherichia coli fecal counts in heifers compared to symbiotic treatment.

Grand et al. (2013) evaluated the effects of two different daily doses of short chain fructooligosaccharides (scFOS), a prebiotic ingredient added to a calf milk substitute. Growth performance, carcass characteristics and fecal concentrations of short-chain fatty acids from pre-ruminants were evaluated. The animals were 8 to 10 days old and reared up to 168 days of age. The calves received daily doses of 3 or 6 grams of scFOS. They found a dose-dependent improvement in feed conversion, tending to increase carcass weight. The authors suggest that the inclusion of scFOS allowed the performance of pre-ruminant calves to be improved, possibly because of the modification of ruminal fermentation.

Antimicrobial Peptides

Antimicrobial peptides (AMPs) have been considered promising substitutes for conventional antibiotics, given their potential for broad-spectrum biological activity, rapid bactericidal activity, and low propensity to develop antibiotic resistance (Magana et al., 2020). AMPs are low molecular weight molecules with broad inhibitory biological activity against bacteria, viruses, and fungi (Izadphanah; Gallo, 2005). Furthermore, when compared to conventional antibiotics, AMPs do not require interaction with a specific cellular target, which hampers the microorganisms 'ability to develop resistance mechanisms efficiently (Rydlo et al., 2006). AMPs belong to a diverse and abundant group of molecules that are produced by diverse organisms, both in plants and animals, and are grouped according to their intrinsic antimicrobial activity (Gallo et al., 2002; Brogden, 2005).

Research about AMPs is under continuous development and a considerable amount of data has been recorded and stored in specific databases (Huan et al., 2020). Many AMPs exhibit direct and rapid antimicrobial activity, causing disruption of physical integrity in the microbial membrane and/or translocation through the membrane into the bacterial cytoplasm to act on intracellular targets (Hancock; Sahl, 2006). In addition, their amino acid composition, amphipathicity, cationic charge and size make these peptides able to easily insert into lipid membranes, thus enabling the death of the target microorganism (Brogden, 2005; Izadphanah; Gallo, 2005).

Researchers have focused attention on AMPs in recent decades for the development of new natural antibiotic candidates because of its wide functional and structural variety. Currently, infections caused by the dangerous multidrug-resistant (MDR) bacteria have reached a level of pan-drug resistance, which means resistance to all conventional antibiotics available, and AMPs are an interesting alternative for the treatment of these infections due the lower propensity to generate bacterial resistance and potential activity against bacteria that are multi/extensively resistant to antibiotics (Roque-Borda et al., 2021, 2022). AMPs are promising candidates for replacing traditional antibiotics against multidrug-resistant microorganisms due to their broad biological activity against different types of pathology related to animal production and public health (Silveira et al., 2021).

Diez-González (2010) analyzed the efficiency of antimicrobial peptides in dairy cattle with potential use as growth promoters, and or as for the prevention and treatment of udder infections. Indeed, nisin, the first antimicrobial peptide studied, has been widely investigated for the prevention and treatment of udder infections during the dry period, and as sanitizing product for the udder before milking and has demonstrated significant reductions in udder pathogens in experimental studies (Pieterse; Todorov, 2022). Another study with Bovicin HC5, a lantibiotic identified and isolated from the ruminal bacterium Streptococcus bovis HC5, showed great antibacterial activity and the bacteria showed no adaptation to the lantibiotic (Mantovani et al., 2001, 2002).

Some probiotic strains produced bacteriocins, a type of antimicrobial peptide (Fernández et al., 2013). Probiotic strains can act against harmful intestinal bacteria in two ways: competing with harmful bacteria and at the same time killing the remaining harmful bacteria through the production of bacteriocins (PEW, 2017).

Phototherapies and Phytochemicals

The immune system in cows is constantly challenged and boosted by the marked physiological changes that occur between late pregnancy, parturition, and early lactation. This is due to the change in metabolic and nutritional status that has possible implications for the plasticity and flexibility of the immune system. The manipulation of nutritional status through different feeding and management strategies can be used to understand how the balance between metabolism, nutritional status and immune system is maintained in challenging conditions, such as an infection. Such models aim to explore the immune response to a challenge under widely varied nutritional and metabolic circumstances (Dänicke et al., 2018).

In addition to the dietary manipulation of cows for greater control of the metabolic state and, consequently, the immune state, alternatives based on phytotherapics can be used to help control diseases. Cui et al. (2015) evaluated the effectiveness of the herbal medicine, Sheng Hua Tang, as a prophylactic strategy to reduce the risks of puerperal metritis in dairy cows. The author argues that Sheng Hua Tang can induce anti-inflammatory and analgesic responses and modulate the immune response. In that study, 158 cows received once daily for three days 0.36 g of raw Sheng Hua Tang grass/kg body weight between 2 to 4 hours after calving, while 153 cows were the control group. The authors found a significant reduction in the incidence of puerperal metritis (12.1% vs 33.3%) for animals that received herbal treatment. The authors justify the use of Sheng Hua Tang in dairy cows due to the reduction in the use of antibiotics, since, in the study, cows affected with puerperal metritis are treated with intrauterine infusion of oxytetracycline to control bacterial growth at the site.

Moringa oleifera leaf meal supplementation has been shown to improve blood biochemistry, antioxidant status and udder health. Kekana et al., (2020) evaluated the effect of a supplementation with Moringa oleifera leaf meal over the hematological profile, milk production and antioxidant enzymes in Holstein cows. In the study, forty-eight cows received increasing levels of Moringa oleifera leaf meal. having a Group M30 was supplemented with 8.33 g/100 kg body weight and group M60 with 16.66 g/100 kg body weight while the control group did not receive any supplementation. The authors reported that the M60 group had higher red and white blood cell counts, higher serum superoxide dismutase, higher: total antioxidant capacity, milk fat, milk protein, and total milk antioxidant capacity.

Ma et al. (2020) evaluated the effects of Lonicera japonica extract on dairy performance, antioxidant status, endocrine status and immune function in mid-lactation heat stressed dairy cows. Twenty-four cows were allocated into four groups, the control and those that received 14, 28 and 56 g/d of Lonicera japonica extract for ten weeks. The addition of the extract to the diet increased the thyroxine concentrations quadratically, the glutathione peroxidase activity and the total antioxidant capacity in the serum and reduced the malondialdehyde concentration. Lonicera japonica extract supplementation had no negative effects on lactation, without alterations in milk production and composition, but helped with heat stress by reducing rectal temperature, improved endocrine, antioxidant status and immune functions.

In general, nutritional factors play a key role in modulating the immune response of animals and to maintain the metabolic state, due to the supplementation of micronutrients with antioxidant and immunostimulant properties. Also, phytotherapics have been used in ruminant nutrition due to their antimicrobial, antioxidant and anti-inflammatory properties that can provided increased resistance against infections, (Hashemzadeh-Cigari et al., 2014), thus helping to improve the immunological profile of animals and, consequently, reducing the use of antibiotics.

Phytochemicals are molecules with antimicrobial properties, and their use as food supplements provide a natural an alternative to the use of antibiotics (Tekippe et al., 2013). Phytochemicals are bioactive compounds derived from plants and have shown a wide range of antimicrobial properties in ruminants, with great potential to change rumen dynamics (Grazziotin et al., 2020). It is important to highlight the essential oils and tannins. Tannins are natural plant polyphenols classified as hydrolysable or condensed. Tannins can bind to dietary proteins through hydrogen bonds, forming a tannin-protein complex Because this complex is stable at pH 5.0 to 7.0, it only gets dissociated in the abomasum and consequently can reduce the rate of microbial protein degradation at the rumen (Frutos et al., 2004). Tannins has the potential to increase the non-degradable protein in the rumen and when combined with limiting amino acids it can increase the improvement in nitrogen use efficiency and consequently milk production (Grazziotin et al., 2020).

Essential oils are secondary plant metabolites responsible for the odor and color of plants and spices, and their antibacterial, antifungal and antioxidant properties make them useful as natural additives in animal feeds (Castillejos et al., 2006). In addition, essential oils can inhibit amino acid deamination and methanogenesis, and stimulate the production of propionate and butyrate in the rumen. In in vivo evaluations, essential oils have shown that they can alter microbial fermentation, and increases the digestibility of true organic matter, nitrogen, acid detergent fiber, and starch (Cite), Moreover, essential oils can also alter dry matter intake, milk yield, and milk fat. However, the results in the literature are shown to be variable, as several factors can influence the results, such as source of essential oils composition of the active ingredient, experimental conditions, among others (Tekippe et al., 2013).

Vaccines

Vaccines have been used to prevent diseases caused by certain bacteria, viruses and are good substitutes for some antibiotics (Oliver et al., 2007). Vaccine-induced protection is complex and depends on the interaction between several compartments of the host's immune system, effectiveness being measured by detection of the specific antibody, with higher levels being related to better protection. (Khardori, 2022). In dairy farming, mastitis is the most common reason for using antimicrobials as prophylactic or therapeutic methods, but widespread use is not sustainable. Therefore, responsible use of antibiotics and the inclusion of additional or replacement mastitis treatments procedures are recommended (Rainard et al., 2021; Ruegg, 2017). Several preventive strategies have been applied to minimize the incidence of bovine mastitis, including milking hygiene, optimization of milking procedures, antibiotic therapy, culling of persistently infected cows and vaccination, (Gurjar et al., 2013).

Rainard et al. (2021) showed that effective vaccines reduced the incidence of new infections and the occurrence of clinical cases. However, vaccine development has obstacles. Return on investment is one of them, which is conditioned by the size of the market and the acceptability of the vaccine, which is largely due to its proven effectiveness. Faced with these challenges, developers mitigate commitment to vaccine development, which subsequently reduces investment in research and progress.

Côté-Gravel & Malouin (2019) reported two main lines of vaccine creation strategies: 1. through targeting bacterial antigens expressed during infection; 2. inactivated and live attenuated vaccines. Antigen targeting vaccines can be constructed using a transcriptomic or immunoproteomic approach, which evaluates the expression of genes of microorganisms in response to infection, leading to the identification of candidate genes and proteins for vaccine development. Live attenuated vaccines are among the most used technologies today. They consist of weakened bacterial or viral strains considered harmless or less virulent, allowing them to transiently colonize the target host.

Staphylococcus aureus, one of the main pathogens of mastitis worldwide, is characterized by causing chronic intramammary infections that respond poorly to antimicrobial therapy, spreading in the herd and leading to great economic losses (Molineri et al., 2021). The high frequency of bovine mastitis necessitates the development of effective and safe vaccines (Campra et al., 2022). Streptococcus aureus is one of the main bacteria causing clinical and subclinical types of mastitis in cows (Hambali et al., 2018). An effective vaccine to prevent Staphylococcus aureus mastitis would improve animal health and reduce dependence on antibiotics (Misra et al., 2018). The fact of Staphylococcus aureus presents a large amount of virulence factors, makes it difficult to select them as vaccines targets (Scali et al., 2015). Hambali et al. (2018) sought to develop a monovalent mastitis vaccine using isolated Streptococcus aureus strains. They evaluated four bacterial concentrations (106, 107, 108 and 109 CFU/mL of Staphylococcus aureus) in potassium aluminum sulfate adjuvant. They found differences in body temperature in the first three hours compared to the control. The heart rates of the vaccinated animals were higher compared to the control group. The study highlights the importance of assessing vital signs in animals undergoing mastitis vaccine testing. The vaccine was not 100% effective, as 1 of the eight mammary quarters evaluated in the vaccinated group showed signs of swelling in the first week, but returned to normality 2 weeks after the challenge (Hambali, 2018).

Lubashevsky et al. (2003) developed a mastitis vaccine that is derived from three strains of Streptococcus aureus, having insoluble bacterial fragments from two of the strains and a supernatant fraction containing secreted antigens from the third strain. In their first study, the vaccines were administered to nine healthy cows and ten animals kept as controls. All animals were challenged with Streptococcus aureus strains, administered in two mammary quarters. Vaccinated cows had 70% protection against infection compared to 10% of control animals. In addition, vaccinated animals exhibited very mild inflammatory reactions and somatic cell counts (SCC) below 100,000. Yadlin, et al., (2003) in their second study, administered the vaccine to 228 animals and kept 224 in the control group. Vaccinated animals exhibited an antibody response five weeks after primary immunization and was maintained throughout the experimental period (300 days). As for the detection of Streptococcus aureus infection, there was no difference between the groups (1.3% vaccinated and 2.3% control). However, first and second lactation animals had 42 and 54%, respectively, lower SCC and higher milk production (+0.5 kg per day) than unvaccinated cows.

All commercially available vaccine formulations did not demonstrate sufficient protection and cost-effectiveness potential (Côté-Gravel & Malouin, 2019). No common protective antigen has yet been found against multiple strains of Streptococcus aureus, which is because selected antigens are not expressed by all strains, in addition to the in vivo environment being able to modulate gene expression (Rainard et al., 2018) Multiple factors account for the lack of protection, including inadequate vaccine targets, high diversity among strains that cause mastitis, cow-to-cow variation in the immune response, and a failure to induce an appropriate immune response to protect against a highly complex pathogen (Côté-Gravel & Malouin, 2019).

FINAL CONSIDERATIONS

Antibiotics have contributed over the years to the treatment and prevention of diseases, and also to improve animal performance. Despite the benefits, the inappropriate use of antibiotics has caused the development of antibiotic resistance in pathogenic microorganisms and antibiotic residues in products of animal origin, being a problem of public health and global food safety. In this way, the major global challenge is to solve these problems with scientific studies and the creation and discovery of natural and alternative products to the conventional antibiotics used in animal production.

In this sense, probiotics, prebiotics, antimicrobial peptides, phototherapies, phytochemicals, vaccines and other products have been studied as discussed in this chapter. Among these, antimicrobial peptides (AMPs) are an extremely promising alternative because they are natural molecules that do not leave residues in animal products, have a low propensity to develop microbial resistance, have a broad spectrum of action, and have a high biological activity against pathogenic microorganisms. Thus, AMPs can effectively contribute as growth promoters and treatment of animal diseases with specially in dairy cattle.

REFERENCES

Akkermans, J. P., & Vecht, U. (1994). Streptococcal infections as cause of death in pigs brought in for necropsy. *Tijdschrift voor Diergeneeskunde*, *119*, 123–128. PMID:8134911

Albright, J. L., Tuckey, S. L., & Woods, G. T. (1961). Antibióticos no Leite—Uma Revisão. *Journal of Dairy Science*, *44*(5), 779–807. doi:10.3168/jds.S0022-0302(61)89819-6

Anderson, E. S., & Datta, N. (1965). Resistance To Penicillins And Its Transfer In Enterobacteriaceae. *Lancet*, *285*(7382), 407–409. doi:10.1016/S0140-6736(65)90004-8 PMID:14238093

Beltrán, M. C., Althaus, R. L., Molina, A., Berruga, M. I., & Molina, M. P. (2015). Analytical strategy for the detection of antibiotic residues in sheep and goat's milk. *Spanish Journal of Agricultural Research*, *13*(1), e0501. https://revistas.inia.es/index.php/sjar/article/ view/6522. doi:10.5424jar/2015131-6522

Beyene, T. (2015). Veterinary Drug Residues in Food-animal Products: Its Risk Factors and Potential Effects on Public Health. *Journal of Veterinary Science & Technology*, *7*(01), 1–7. doi:10.4172/2157-7579.1000285

Brogden, K. A. (2005). Antimicrobial peptides: Pore formers or metabolic inhibitors in bacteria? *National Review*, *3*(3), 238–250. doi:10.1038/nrmicro1098 PMID:15703760

Brown, K., Uwiera, R. R. E., Kalmokoff, M. L., Brooks, S. P. J., & Inglis, G. D. (2016). Antimicrobial growth promoter use in livestock: A requirement to understand their modes of action to develop effective alternatives. *International Journal of Antimicrobial Agents*, *49*(1), 12–24. doi:10.1016/j.ijantimicag.2016.08.006 PMID:27717740

Callaway, T. R., & … . (2003). "Probiotics, Prebiotics and Competitive Exclusion for Prophylaxis Against Bacterial Disease," Animal Health Research Reviews 9, no. 02 (2008): 217-25; Roberto M. La Ragione and Martin J. Woodward, "Competitive Exclusion by Bacillus subtilis Spores of Salmonella enterica Serotype Enteritidis and Clostridium perfringens in Young Chickens. *Veterinary Microbiology*, *94*(3), 245–256. PMID:12814892

Campra, N. A., Reinoso, E. B., Montironi, I. D., Moliva, M. V., Raviolo, J., Ruiz Moreno, F., Marin, C., Camacho, N. M., Paredes, A. J., Morán, M. C., Estein, S. M., Maletto, B. A., Palma, S. D., & Cariddi, L. N. (2022). Spray-drying-microencapsulated *Minthostachys verticillata* essential oil and limonene as innovative adjuvant strategy to bovine mastitis vaccines. *Research in Veterinary Science*, *149*, 136–150. doi:10.1016/j.rvsc.2022.04.014 PMID:35792419

Castillejos, L., Calsamiglia, S., & Ferret, A. (2006). Effect of Essential Oil Active Compounds on Rumen Microbial Fermentation and Nutrient Flow in In Vitro Systems. *Journal of Dairy Science*, *89*(7), 2649–2658. doi:10.3168/jds.S0022-0302(06)72341-4 PMID:16772584

Chiesa, L. M., DeCastelli, L., Nobile, M., Martucci, F., Mosconi, G., Fontana, M., Castrica, M., Arioli, F., & Panseri, S. (2020). Analysis of antibiotic residues in raw bovine milk and their impact toward food safety and on milk starter cultures in cheese-making process. *Lebensmittel-Wissenschaft + Technologie*, *131*, 109783. doi:10.1016/j.lwt.2020.109783

Chiquette, J. (1995). Saccharomyces cerevisiae and Aspergillus oryzae, Used Alone or In Combination, as a Feed Supplement for Beef and Dairy Cattle. *Canadian Journal of Animal Science*, *75*(3), 405–415. doi:10.4141/cjas95-060

Cogan, T. M. (1972). Susceptibility of Cheese and Yogurt Starter Bacteria to Antibiotics. *Applied Microbiology*, *23*(5), 960–965. doi:10.1128/am.23.5.960-965.1972 PMID:4624215

Cui, D., Wang, S., Wang, L., Wang, H., Li, X., & Liu, Y. (2015). Prophylactic strategy with herbal remedy to reduce puerperal metritis risk in dairy cows: A randomized clinical trial. *Livestock Science*, *181*, 231–235. doi:10.1016/j.livsci.2015.09.005

Dadgostar, P. (2019). Antimicrobial Resistance: Implications and Costs. *Infection and Drug Resistance*, *2012*, 3903–3910. doi:10.2147/IDR.S234610 PMID:31908502

Dafale N.A., Srivastava, S., & Purohit, H.J. (2020). Zoonose: um link emergente para a resistência aos antibióticos sob "uma abordagem de saúde. *Indian Journal of Microbiology*, *60* (2), 139-152. . PMC 7105526. PMID 32255846 doi:10.1007/s12088-020-00860-z

Dänicke, S., Meyer, U., Kersten, S., & Frahm, J. (2018). Animal models to study the impact of nutrition on the immune system of the transition cow. *Research in Veterinary Science, 116*, 15–27. doi:10.1016/j.rvsc.2018.01.023 PMID:29428254

De Albuquerque Fernandes, S. A., Magnavita, A. P., Ferrao, S. P., Gualberto, S. A., Faleiro, A. S., Figueiredo, A. J., & Matarazzo, S. V. (2014). Daily ingestion of tetracycline residue present in pasteurized milk: A public health problem. *Environmental Science and Pollution Research International, 21*(5), 3427–3434. https://link.springer.com/article/10.1007/s11356-013-2286-5. doi:10.100711356-013-2286-5 PMID:24243094

Diez-Gonzalez, F. (2010). Applications of Bacteriocins in Livestock. *Current Issues in Intestinal Microbiology 8*. PMID:24031528

Pieterse, R., & Todorov, S. (2017). Bacteriocins: Exploring Alternatives to Antibiotics in Mastitis Treatment. *Brazilian Journal of Microbiology, 41*(3), 542–562.

EMA Committee for Medicinal Products for Veterinary Use (CVMP) and EFSA Panel on Biological Hazards (BIOHAZ), Murphy, D., Ricci, A., Auce, Z., Beechinor, J. G., Bergendahl, H., & Jukes, H. (2017). EMA and EFSA Joint Scientific Opinion on measures to reduce the need to use antimicrobial agents in animal husbandry in the European Union, and the resulting impacts on food safety (RONAFA). EFSA Journal, 15(1), e04666. doi:10.1111/jam.12081 PMID:23167818

Fernandez., ... (2013). Growth, Acid production and Bacteriocin Production by Probiotic Candidates under Simulated Colonic Conditions. Journal of Applied Microbiology114no. 3201387785 doi:10.1111/jam.12081

Frutos, P., Hervás, G., Giráldez, F. J., & Mantecón, A. R. (2004). Review. Tannins and ruminant nutrition. *Spanish Journal of Agricultural Research, 2*(2), 191–202. doi:10.5424jar/2004022-73

Gaggìa, F., Mattarelli, P., & Biavati, B. (2010, July 31). Probiotics and prebiotics in animal feeding for safe food production. *International Journal of Food Microbiology, 141*(Suppl 1), S15–S28. doi:10.1016/j.ijfoodmicro.2010.02.031 PMID:20382438

Gallo, R. L., Murakami, M., Ohtake, T., & Zaiou, M. (2002). Biology and clinical relevance of naturally occurring antimicrobial peptides. *The Journal of Allergy and Clinical Immunology, 110*(6), 823–831. doi:10.1067/mai.2002.129801 PMID:12464945

Ghosh, S., & Mehla, R. K. (2012). Influence of dietary supplementation of prebiotics (mannanoligosaccharide) on the performance of crossbred calves. *Tropical Animal Health and Production, 44*(3), 617–622. doi:10.100711250-011-9944-8 PMID:21805307

Grand, E., Respondek, F., Martineau, C., Detilleux, J., & Bertrand, G. (2013). Effects of short-chain fructooligosaccharides on growth performance of preruminant veal calves. *Journal of Dairy Science, 96*(2), 1094–1101. doi:10.3168/jds.2011-4949 PMID:23200477

Grazziotin, R. C. B., Halfen, J., Rosa, F., Schmitt, E., Anderson, J. L., Ballard, V., & Osorio, J. S. (2020). Altered rumen fermentation patterns in lactating dairy cows supplemented with phytochemicals improve milk production and efficiency. *Journal of Dairy Science, 103*(1), 301–312. doi:10.3168/jds.2019-16996 PMID:31733851

Gurjar, A. A., Klaessig, S., Salmon, S. A., Schukken, R. J., & Yancey, Y. H. (2013). Evaluation of an alternative dosing regimen of a J-5 mastitis vaccine against intramammary Escherichia coli challenge in nonlactating late-gestation dairy cows. *Journal of Dairy Science, 96*(8), 5053–5063. doi:10.3168/jds.2012-6456 PMID:23726420

Hambali, I. U. (2018). *Development of prototype killed vaccine against Staphylococcus aureus mastitis in dairy cows.* [Doctoral thesis, Universiti Putra Malaysia].

Hambali, I. U., Bhutto, K. R., Jesse, F. F. A., Lawan, A., Odhah, M. N., Wahid, A. H., Azmi, M. L. M., Zakaria, Z., Arsalan, M., Muhammad, N. A., & Jefri, M. N. (2018). Clinical responses in cows vaccinated with a developed prototype killed Staphylococcus aureus mastitis vaccine. *Microbial Pathogenesis, 124*, 101–105. doi:10.1016/j.micpath.2018.08.017 PMID:30114463

Hancock, R. E., & Sahl, H. G. (2006). Antimicrobial and host-defense peptides as new anti-infective therapeutic strategies. *Nature Biotechnology, 24*(12), 1551–1557. doi:10.1038/nbt1267 PMID:17160061

Hashemzadeh-Cigari, F., Khorvash, M., Ghorbani, G. R., Kadivar, M., Riasi, A., & Zebeli, Q. (2014). Effects of supplementation with a phytobiotics-rich herbal mixture on performance, udder health, and metabolic status of Holstein cows with various levels of milk somatic cell counts. *Journal of Dairy Science, 97*(12), 7487–7497. doi:10.3168/jds.2014-7989 PMID:25306268

Hasunuma, T., Kawashima, K., Nakayama, H., Murakami, T., Kanagawa, H., Ishii, T., Akiyama, K., Yasuda, K., Terada, F., & Kushibiki, S. (2011). Effect of cellooligosaccharide or symbiotic feeding on growth performance, fecal condition, and hormone concentrations in Holstein calves. *Animal Science Journal = Nihon Chikusan Gakkaiho, 82*(4), 543–548. doi:10.1111/j.1740-0929.2010.00861.x

Heikkila, A. M., Nousiainen, J. I., & Pyorala, S. (2012). Costs of clinical mastitis with special reference to premature culling. *Journal of Dairy Science, 95*(1), 139–150. doi:10.3168/jds.2011-4321 PMID:22192193

Heinrichs, A. J., Jones, C. M., & Heinrichs, B. S. (2003). Effects of Mannan Oligosaccharide or Antibiotics in Neonatal Diets on Health and Growth of Dairy Calves [Uyeno et al., "Effect of Probiotics/Prebiotics.]. *Journal of Dairy Science, 86*(12), 4064–4069. doi:10.3168/jds.S0022-0302(03)74018-1 PMID:14740845

Hill, C., Guarner, F., Reid, G., Gibson, G. R., Merenstein, D. J., Pot, B., Morelli, L., Canani, R. B., Flint, H. J., Salminen, S., Calder, P. C., & Sanders, M. E. (2014, August). Expert consensus document. The International Scientific Association for Probiotics and Prebiotics consensus statement on the scope and appropriate use of the term probiotic. *Nature Reviews. Gastroenterology & Hepatology, 11*(8), 506–514. doi:10.1038/nrgastro.2014.66 PMID:24912386

Côté-Gravel, J., & Malouin, F. (2019). Symposium review: Features of Staphylococcus aureus mastitis pathogenesis that guide vaccine development strategies. *Journal of Dairy Science, 102*(5), 4727–4740. doi:10.3168/jds.2018-15272 PMID:30580940

Huan Y., Kong Q., Mou H., Yi H. (2020). Antimicrobial peptides: Classification, design, application, and research progress in multiple fields. *Frontiers in Microbiology, 11*, 2559. PMID:33178164

Innes, G. K., Randad, P. R., Korinek, A., Davis, M. F., Price, L. B., So, A. D., & Heaney, C. D., (2020). External societal costs of antimicrobial resistance in humans attributable to antimicrobial use in livestock. *Annual review of public health, 41*, 141-157.

Izadpanah, A. E., & Gallo, R. L. (2005). Antimicrobial peptides. *Journal of the American Academy of Dermatology, 52*, 381–390.

Jana, U. K., Suryawanshi, R. K., Prajapati, B. P., & Kango, N. (2021). Prebiotic mannooligosaccharides: Synthesis, characterization, and bioactive properties. *Food Chemistry, 342*, 128328. doi:10.1016/j.foodchem.2020.128328 PMID:33257024

Kekana, T. W., Marume, U., Muya, M. C., & Nherera-Chokuda, F. V. (2020). Periparturient antioxidant enzymes, hematological profile and milk production of dairy cows supplemented with Moringa oleifera leaf meal. *Animal Feed Science and Technology*, *268*, 114606. doi:10.1016/j.anifeedsci.2020.114606

Khardori, N. (2022). Vaccines and vaccine resistance: Past, present, and future. *Indian Journal of Medical Microbiology*, *40*(2), 187–192. doi:10.1016/j.ijmmb.2021.12.008 PMID:34961641

Langoni, H., Salina, A., Oliveira, G. C., Junqueira, N. B., Menozzi, B. D., & Joaquim, S. F. (2017). Considerações sobre o tratamento das mastites. *Pesquisa Veterinaria Brasileira*, *37*(11), 1261–1269. doi:10.15900100-736x2017001100011

Lees, P., Pelligand, L., Giraud, E., & Toutain, P. (2020). A history of antimicrobial drugs in animals: Evolution and revolution. *Journal of Veterinary Pharmacology and Therapeutics*. doi:10.1111/jvp.12895 PMID:32725687

Leitner, G., Lubashevsky, E., Glickman, A., Winkler, M., Saran, A., & Trainin, Z. (2003). Development of a Staphylococcus aureus vaccine against mastitis in dairy cows: I. Challenge trials. *Veterinary Immunology and Immunopathology*, *93*(1–2), 31–38. doi:10.1016/S0165-2427(03)00051-5 PMID:12753773

Leitner, G., Yadlin, N., Lubashevsy, E., Ezra, E., Glickman, A., Chaffer, M., Winkler, M., Saran, A., & Trainin, Z. (2003). Development of a Staphylococcus aureus vaccine against mastitis in dairy cows. II. Field trial. *Veterinary Immunology and Immunopathology*, *93*(3–4), 153–158. doi:10.1016/S0165-2427(03)00062-X PMID:12814700

Ma, F. T., Shan, Q., Jin, Y. H., Gao, D., Li, H. Y., Chang, M. N., & Sun, P. (2020). Effect of Lonicera japonica extract on lactation performance, antioxidant status, and endocrine and immune function in heat-stressed mid-lactation dairy cows. *Journal of Dairy Science*, *103*(11), 10074–10082. doi:10.3168/jds.2020-18504 PMID:32896406

Magana, M., Pushpanathan, M., Santos, A. L., Leanse, L., Fernandez, M., Ioannidis, A., Giulianotti, M. A., Apidianakis, Y., Bradfute, S., Ferguson, A. L., Cherkasov, A., Seleem, M. N., Pinilla, C., de la Fuente-Nunez, C., Lazaridis, T., Dai, T., Houghten, R. A., Hancock, R. E. W., & Tegos, G. P. (2020). The value of antimicrobial peptides in the age of resistance. *The Lancet. Infectious Diseases*, *20*(9), e216–e230. doi:10.1016/S1473-3099(20)30327-3 PMID:32653070

Mantovani, H. C., Hu, H., Warobo, R. W., & Russell, J. B. (2002). Bovicin HC5, a bacteriocin from Streptococcus bovis HC5. *Microbiology*, *148*(11), 3347–3352. doi:10.1099/00221287-148-11-3347 PMID:12427926

Mantovani, H. C., Kam, D. K., Ha, J. K., & Russell, J. B. (2001). The antibacterial activity and sensitivity of Streptococcus bovis strains isolated from the rumen of cattle. *FEMS Microbiology Ecology, 37*(3), 223–229. doi:10.1016/S0168-6496(01)00166-0 PMID:11311432

Marth, E. H., & Ellickson, B. E. (1959). Problems Created by the Presence of Antibiotics in Milk—A Review. *Journal of Milk and Food Technology, 22*(9), 266–272. doi:10.4315/0022-2747-22.9.266

Mercer, H. D., Baggot, J. D., & Sams, R. A. (1977). Application of pharmacokinetic methods to the drug residue profile. *Journal of Toxicology and Environmental Health, 2*(4), 787–801. doi:10.1080/15287397709529479 PMID:857041

Misra, N., Wines, T. F., Knopp, C. L., Hermann, R., Bond, L., Mitchell, B., McGuire, M. A., & Tinker, J. K. (2018). Immunogenicity of a Staphylococcus aureus-cholera toxin A2/B vaccine for bovine mastitis. *Vaccine, 36*(24), 3513–3521. doi:10.1016/j.vaccine.2018.04.067 PMID:29739718

Molineri, A. I., Camussone, C., Zbrun, M. V., Suárez Archilla, G., Cristiani, M., Neder, V., Calvinho, L., & Signorini, M. (2021). Antimicrobial resistance of Staphylococcus aureus isolated from bovine mastitis: Systematic review and meta-analysis. *Preventive Veterinary Medicine, 188*, 105261. doi:10.1016/j.prevetmed.2021.105261 PMID:33508662

Mulani MS, Kamble EE, Kumkar SN, Tawre MS, Pardesi KR., (2019). Emerging Strategies to Combat ESKAPE Pathogens in the Era of Antimicrobial Resistance: A Review. *Front Microbiol.* . doi:10.3389/fmicb.2019.00539

Nobrega, D. B., Naqvi, S. A., Dufour, S., Deardon, R., Kastelic, J. P., De Buck, J., & Barkema, H. W. (2020). Critically important antimicrobials are generally not needed to treat non-severe clinical mastitis in lactating dairy cows: Results from a network meta-analysis. *Journal of Dairy Science, 103*(11), 10585–10603. doi:10.3168/jds.2020-18365 PMID:32896405

Oliver, & .... (2007). "Asas Centennial Paper: Developments and Future Outlook"; Els N.T. Meeusen et al., "Current Status of Veterinary Vaccines. *Clinical Microbiology Reviews, 20*(3), 489–510. PMID:17630337

OMS. Centro de Mídia, (2017). Comunicado de Notícias. OMS publica lista de bactérias para as quais novos antibióticos são urgentemente necessários. 2017 [citado em março]. *Disponível* em: https://www.who.int/mediacentre/news/releases/2017/bacteria-antibiotics-needed/en/

O'NEIL J. (2016). In A. Ro (Ed.), *Tackling drug-resistant infections globally: final report and recommendations* (Vol. 1, p. 84). Resistance.

Pengov, A., & Kirbis, A. (2009). Risks of antibiotic residues in milk following intramammary and intramuscular treatments in dairy sheep. *Analytica Chimica Acta, 637*(1-2), 13–17. doi:10.1016/j.aca.2008.09.021 PMID:19286006

Report, P. E. W. (2017). *Alternatives to Antibiotics in Animal Agriculture*. PEW trusts. https://www.pewtrusts.org/~/media/assets/2017/07/alternatives_to_antibiotics_in_animal_agriculture.pdf. 2017.

Pieterse, R., & Todorov, S. D. (2010). Bacteriocins: exploring alternatives to antibiotics in mastitis treatment. *Brazilian Journal of Microbiology, 41*, 542-562. Plastridge, WN., (1958). Bovine Mastitis: A Review. *Journal of Dairy Science, 41*(9), 1141–1181.

Pol, M., & Ruegg, P. L. (2007). Treatment Practices and Quantification of Antimicrobial Drug Usage in Conventional and Organic Dairy Farms in Wisconsin. *Journal of Dairy Science, 90*(1), 249–261. doi:10.3168/jds.S0022-0302(07)72626-7 PMID:17183093

Prescott, J. F. (2017). History and Current Use of Antimicrobial Drugs in Veterinary Medicine. *Microbiology Spectrum, 5*(6), 5.6.05. doi:10.1128/microbiolspec.ARBA-0002-2017 PMID:29271338

Prescott, JF., (2019). History of antimicrobial use in agriculture: an overview. *Antimicrobial resistance in bacteria of animal origin,* 19-27.

Quigley, J. D. III, Drewry, J. J., Murray, L. M., & Ivey, S. J. (1997). Body Weight Gain, Feed Efficiency, and Fecal Scores of Dairy Calves in Response to Galactosyl-Lactose or Antibiotics in Milk Replacers. *Journal of Dairy Science, 80*(8), 1751–1754. doi:10.3168/jds.S0022-0302(97)76108-3 PMID:9276816

Quintanilla, P., Doménech, E., Escriche, I., Beltrán, M. C., & Molina, M. P. (2019). Food Safety Margin Assessment of Antibiotics: Pasteurized Goat's Milk and Fresh Cheese. *Journal of Food Protection, 82*(9), 1553–1559. doi:10.4315/0362-028X.JFP-18-434 PMID:31424292

Rainard, P., Foucras, G., Fitzgerald, J. R., Watts, J. L., Koop, G., & Middleton, J. R. (2018). Knowledge gaps and research priorities in Staphylococcus aureus mastitis control. *Transboundary and Emerging Diseases, 65*, 149–165. doi:10.1111/tbed.12698 PMID:28984427

Rainard, P., Gilbert, F. B., Germon, P., & Foucras, G. (2021). Invited review: A critical appraisal of mastitis vaccines for dairy cows. *Journal of Dairy Science, 104*(10), 10427–10448. doi:10.3168/jds.2021-20434 PMID:34218921

Rall, V. L. M., Miranda, E. S., Castilho, I. G., Camargo, C. H., Langoni, H., Guimarães, F. F., Araújo Júnior, J. P., & Fernandes Júnior, A. (2014). Diversity of Staphylococcus species and prevalence of enterotoxin genes isolated from milk of healthy cows and cows with subclinical mastitis. *Journal of Dairy Science, 97*(2), 829–837. doi:10.3168/jds.2013-7226 PMID:24359821

Roque-Borda, C. A., Bento da Silva, P., Rodrigues, M. C., di Filippo, L. D., Duarte, J. L., Chorilli, M., Vicente, E. F., Garrido, S. S., & Rogério Pavan, F. (2022). Pharmaceutical nanotechnology: Antimicrobial peptides as potential new drugs against WHO list of critical, high, and medium priority bacteria. *European Journal of Medicinal Chemistry, 241*, 114640. doi:10.1016/j.ejmech.2022.114640 PMID:35970075

Roque-borda, C. A., da Silva, P. B., Rodrigues, M. C., Azevedo, R. B., di Filippo, L., Duarte, J. L., Chorilli, M., Vicente, E. F., & Pavan, F. R., (2021). Challenge in the Discovery of New Drugs: Antimicrobial Peptides against WHO-List of Critical and High-Priority Bacteria. *Pharmaceutics 2021, 13*(6), 773. doi:10.3390/pharmaceutics13060773

Ruegg, P. L. (2017). A 100-Year Review: Mastitis detection, management, and prevention. *Journal of Dairy Science, 100*(12), 10381–10397. doi:10.3168/jds.2017-13023 PMID:29153171

Rydlo, T., Rotem, S., & Mor, A. (2006). Antibacterial Properties of Dermaseptin S4 Derivatives under Extreme Incubation Conditions. *Antimicrobial Agents and Chemotherapy, 50*(2), 490–497. doi:10.1128/AAC.50.2.490-497.2006 PMID:16436701

Sachi, S., Ferdous, J., & Sikder, M. H. Hussani, (2019). S.M.A.K. Antibiotic residues in milk: Past, present, and future. *J. Adv. Vet. Anim. Res., 6*, 315–332. https://www.ncbi.nlm.nih.gov/pmc/articles/PMC6760505/

Santajit, S., & Indrawattana, N. (2016). Mechanisms of Antimicrobial Resistance in ESKAPE Pathogens. *BioMed Research International, 2475067*, 1–8. doi:10.1155/2016/2475067 PMID:27274985

Scali, F., Camussone, C., Calvinho, L. F., Cipolla, M., & Zecconi, A. (2015). Which are important targets in development of S. aureus mastitis vaccine? *Research in Veterinary Science, 100*, 88–99. doi:10.1016/j.rvsc.2015.03.019 PMID:25975626

Sharma, Chetan; Rokana, Namita; Chandra, Mudit; Singh, Brij Pal; Gulhane, Rohini Devidas; Gill, Jatinder Paul Singh; Ray, Pallab; Puniya, Anil Kumar; Panwar, Harsh, (2018). Antimicrobial Resistance: Its Surveillance, Impact, and Alternative Management Strategies in Dairy Animals. *Frontiers in Veterinary Science*, 4(), 237–. doi:10.3389/fvets.2017.00237

Shryock, T. R. (2004). O futuro dos produtos anti-infecciosos na saúde animal. *Nature Reviews. Microbiology*, *2*(5), 425–430. doi:10.1038/nrmicro887 PMID:15100695

Silveira, R. F., Roque-Borda, C. A., & Vicente, E. F. (2021). Antimicrobial peptides as a feed additive alternative to animal production, food safety and public health implications: An overview. *Animal Nutrition*, *7*(3), 896–904. doi:10.1016/j.aninu.2021.01.004 PMID:34632120

Sisodia, C. S., & Stowe, C. M., (2006). THE MECHANISM OF DRUG SECRETION INTO BOVINE MILK. *Annals of the New York Academy of Sciences*, *111*(2), 650–661. doi:. 1964.tb doi:10.1111/j.1749-6632

Smith, H. W., & Crabb, W. E. (1960a). The effect of diets containing tetracyclines and penicillin on the Staphylococcus aureus flora of the nose and skin of pigs and chickens and their human attendants. *The Journal of Pathology and Bacteriology*, *79*(2), 243–249. doi:10.1002/path.1700790204 PMID:13831990

Smith, H. W., & Crabb, W. E. (1960b). O efeito da quimioterapia sobre o número de colis bacterianas nas fezes de bezerros. *Jornal de Patologia Comparada e Terapêutica*, *70*, 126–134. doi:10.10160368-1742(60)80011

Tekippe, J. A., Tacoma, R., Hristov, A. N., Lee, C., Oh, J., Heyler, K. S., Cassidy, T. W., Varga, G. A., & Bravo, D. (2013). Effect of essential oils on ruminal fermentation and lactation performance of dairy cows. *Journal of Dairy Science*, *96*(12), 7892–7903. doi:10.3168/jds.2013-7128 PMID:24119814

Treiber, F. M., & Beranek-Knauer, H. (2021). Antimicrobial residues in food from animal origin—A review of the literature focusing on products collected in stores and markets worldwide. *Antibiotics (Basel, Switzerland)*, *10*(5), 534. doi:10.3390/antibiotics10050534 PMID:34066335

U.N. Food and Agriculture Organization. (2016). Probiotics in Animal Nutrition. *FAO Animal Production and Health Paper No. 179*, https://www.fao.org/3/a-i5933e.pdfJoan, S. Jeffrey. (1999). Use of Competitive Exclusion Products in Poultry. *Poultry Fact Sheet No. 30*. Cooperative Extension, University of California http://animalsciencey.ucdavis.edu/avian/pfs30.htm

Food, U.N. and Agriculture Organization. (2012). "Probiotics in Animal Nutrition. Research in Veterinary Science, 93(1), 250-58. doi:10.1111/j.1863-2378.2007.01059.x PMID:17803515

Sargeant, J. M. (2020). Pre-Harvest Interventions to Reduce the Shedding of E. coli O157 in the Faeces of Weaned Domestic Ruminants: A Systematic Review. Zoonoses and Public Health, 54. doi:10.1111/j.1863-2378.2007.01059.x

US GAO. (Gov. Account. Off.). (2011). Antibiotic resistance: agencies have made limited progress addressing antibiotic use in animals GAO-11-801, *Comm. Rules*, House Represent, Washington, DC: https://www.gao.gov/assets/330/323090.pdf

Usha Vyas and Natarajan Ranganathan, (2001). Probiotics, Prebiotics, and Symbiotic: Gut and Beyond. *Gastroenterology Research and Practice.*

U.N. Food and Agriculture Organization and World Health Organization. (2012). *Report of a Joint FAO/WHO Expert Consultation on Evaluation of Health and Nutritional Properties of Probiotics in Food Including Powder Milk with Live Lactic Acid Bacteria*. UN.

Van Boeckel, T. P., Brower, C., Gilbert, M., Grenfell, B. T., Levin, S. A., Robinson, T. P., Teillant, A., & Laxminarayan, R. (2015). Global trends in antimicrobial use in food animals. *Proceedings of the National Academy of Sciences of the United States of America, 112*(18), 5649–5654. doi:10.1073/pnas.1503141112 PMID:25792457

Verraes, C., Van Boxstael, S., Van Meervenne, E., Van Coillie, E., Butaye, P., Catry, B., de Schaetzen, M.-A., Van Huffel, X., Imberechts, H., Dierick, K., Daube, G., Saegerman, C., De Block, J., Dewulf, J., & Herman, L. (2013). Antimicrobial Resistance in the Food Chain: A Review. *International Journal of Environmental Research and Public Health, 10*(7), 2643–2669. doi:10.3390/ijerph10072643 PMID:23812024

Wang, D., Wang, Z., Yan, Z., Wu, J., Ali, T., Li, J., Lv, Y., & Han, B. (2015). Bovine mastitis Staphylococcus aureus: Antibiotic susceptibility profile, resistance genes and molecular typing of methicillin-resistant and methicillin-sensitive strains in China. *Infection, Genetics and Evolution, 31*, 9–16. doi:10.1016/j.meegid.2014.12.039 PMID:25582604

World Health Organization. (2021). *Antimicrobial resistance.* WHO. https://www.who.int/news-room/fact-sheets/detail/antimicrobi al-resistance

World Health Organization. (2014). *WHO's First Global Report on Antibiotic Resistance Reveals Serious, Worldwide Threat to Public Health.* WHO. https://www.who.int/mediacentre/news/releases/2014/amr-report/en/

Uyeno, Y., Shigemori, S., & Shimosato, T. (2015). Effect of Probiotics/Prebiotics on Cattle Health and Productivity. *Microbes and Environments*, *30*(126), 126–132. doi:10.1264/jsme2.ME14176 PMID:26004794

Zhang, J., Wang, J., Jin, J., Li, X., Zhang, H., & Zhao, C. (2022). Prevalence, antibiotic resistance, and enterotoxin genes of Staphylococcus aureus isolated from milk and dairy products worldwide: A systematic review and meta-analysis. *Food Research International*, *162*, 111969. doi:10.1016/j.foodres.2022.111969 PMID:36461219

Zigo, F., Yasiľ, M., Elečkov, J., Zigová, M., & Farkašová, Z. (2017). Mastitis Pathogens Isolated from Samples of Milk in Dairy Cows and Their Resistance against Antimicrobial Agents. *Journal of Food Science and Engineering*, *7*(2). doi:10.17265/2159-5828/2017.02.007

Chapter 8
Effect of Evaporative Cooling System on Behavior, Milk Yield, and Milk Quality of Holstein Cows in a Tropical Climate

Cristiane Gonçalves Titto
São Paulo University, Brazil

Evaldo Antonio Lencioni Titto
São Paulo University, Brazil

Rafael Martins Titto
São Paulo University, Brazil

Alfredo Manuel Franco Pereira
https://orcid.org/0000-0001-9430-9399
Universidade de Évora, Portugal

Messy Hannear de Andrade Pantoja
São Paulo University, Brazil

João Alberto Negrão
São Paulo University, Brazil

EXECUTIVE SUMMARY

In a tropical climate, air temperature can be stressful for Holstein cows throughout the year. This chapter's objective was to determine how the use of ventilation and fogging on a freestall barn can positively change the behavior, milk production, and milk quality of Holstein cows in all seasons. Twenty-eight lactating cows were divided into two groups: cooling group (CG) and non-cooled group (NCG). For different weather patterns throughout the year, behaviors (position, posture, and activity) were observed every 30 min during the day. Milk production was measured and milk samples were collected for analysis of protein, fat, lactose, total solids, and somatic cell count. All animals spent most of the day standing in the shade eating,

DOI: 10.4018/978-1-6684-5472-5.ch008

ruminating, and idling regardless of the season (P<0.05). Also, animals under cooling system showed greater milk production, fat content, and higher somatic cell count (P<0.01). Provision of cooling system during the warmer months of the year proved to be efficiently increasing the feeding time, milk production, and milk quality of lactating cows.

BACKGROUND

Milk production can be negatively affected by heat, especially in animals of high genetic merit (Ahmed et al., 2022), and heat stress can have negative effects on the welfare (Lacetera, 2019), which can be rated by physiological and behavioral responses. Climatic factors, such as temperature, relative humidity, and solar radiation, may cause increase in rectal temperature and respiratory rate (Li et al., 2020) and reduction in daily activities (Hut et al 2022) and dry matter intake (Chang-Fung-Martel et al., 2021).

Decrease in lying and eating behavior are the first changes on daily activities when the air temperature rises (; Chang-Fung-Martel et al., 2021; Tullo et al., 2019; DelCurto-Wyffels et al., 2021). Besides the decline in milk production, heat stress may also reduce milk quality: somatic cells increase (Zeinhom et al., 2016) and protein and fat contents become smaller (Bernabucci et al., 2015). Animals under heat stress reduced milk yield from 34.3 kg/day to 22.5 kg/day when compared to thermoneutral conditions (Fontoura et al., 2022), in addition, findings by Bertocchi et al. (2014) showed higher somatic cell scores in heat (4.613) compared to cold (4.287).

However, there is evidence that providing shade or other cooling methods is beneficial for cattle, based on changes in respiratory rate and body temperature (Schütz et al., 2011; Brown-Brand, 2018), diurnal feeding (Portugal et al., 2000), and plasmatic cortisol and IGF-I (Chaiyabutr et al., 2008; Brown-Brandl et al 2017).In freestall barns, cows stay inside during the hottest hours of the day to obtain shelter from intense solar radiation; however, the barn itself can cause heat stress if not well ventilated or has high roofing. Environmental modification on dairy barn has demonstrated improvement in microclimate and a positive effect on cow's performance with forced ventilation (Marumo et al., 2022), ventilation with misting (Titto et al., 2013), and ventilation with sprinkling (Román et al., 2019). Studies that investigate animal behavior to show animal preferences is important to increase welfare (Arnott et al., 2017).

Within this context, this study aimed to evaluate the behavior in relation to the environmental condition system (fans with fogging), milk yield, and milk quality under different climatic conditions throughout the year.

MATERIAL AND METHODS DESCRIPTION

Animals and Facilities

The experiment was conducted at the FZEA-USP Division of Dairy Cattle located in Pirassununga, SP, Brazil (21°80’00[2]’’ S, 47°25’42’’[2] W, altitude 634 m) during summer, and was approved by the Ethics Committee of FZEA/USP. The study utilized 28 Holstein cows between the first and third lactations, with an average milk yield of 20 kg cow^{-1}. day^{-1}. The animals were divided into two groups of 14 homogeneous cows each, housed postpartum in two pens: one cooled group (CG) with an environment conditioning system (fans and fogging), and the other non-cooled group (NCG) with natural ventilation and shade. Cow selection was performed postpartum based on the calving number and milk yield to create homogeneous groups. Two side-by-side freestall barns were placed northwest-southeast, with a cement floor, covered with fiber cement roof panels, 3.5 m high with sand bedding. The feeding and water troughs were positioned under the roof. The cows had unlimited access to a 0.3 ha paddock of coast-cross grass without shade.

The cooled area was equipped with five axial flow fans (125 cm diameter; 300 m^3. min^{-1} maximum airflow rate) installed on the east side of the feed passage (Figure 1). The fans were spaced 3 m apart on either side of the drive-through feed alley and positioned to provide airflow in the direction of prevailing winds. The fans were mounted at a height of approximately 2.5 m and angled downward approximately 10° from the vertical. A set of two fogging lines was mounted above the feed alley and bedding. The evaporative cooling system was thermostatically controlled and was switched on at 26°C. A plastic wall separated the two experimental pens to avoid interference.

The evaporative cooling system was triggered automatically during all seasons when the air temperature exceeded 26°C, a value determined for the system call.

Figure 1. Sketch of free stall and coast-cross area (not scaled)
Dotted lines indicate each region. White area covered by fiber cement tiles; grey area not covered. Numbers 1 to 5 indicate the region of installation occupied by the animals. The circles represent the axial flow fans. The rectangle indicates the water trough. Double Solid lines are the fogging lines: 1) trough; 2) corridors; 3) unshaded cemented area inside freestall; 4) bed; 5) coast-cross area.

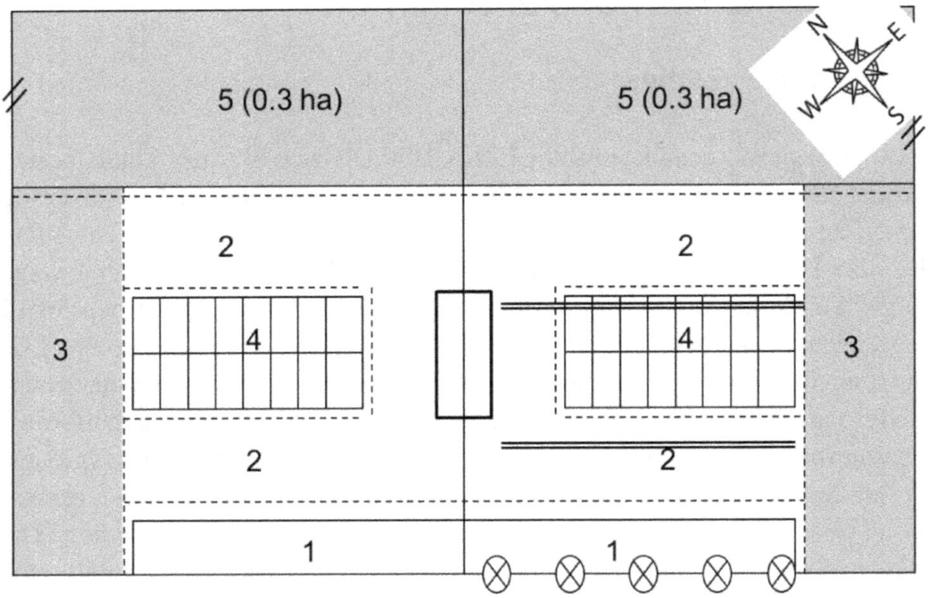

Macroclimate and Microclimate Parameters

Weather variables including air temperature, relative humidity, and solar radiation were recorded by an electronic weather station (Table 1). Inside each barn, the microclimate was measured with a black globe thermometer with data logger located at cow height in the center area and Temperature-Humidity Index (THI) – a measure of the degree of discomfort or stress placed on living animals was calculated (West, 2003). Each season of the year was characterized by temperature, relative humidity and solar radiation measured at the meteorological station of the Construction and Ambience Laboratory at FZEA/USP (see Table 1). These data were used to divide the seasons: Autumn (April, May, June), Winter (June, July, August), Spring (September, October, November), Dry Summer (December to January), Rainy Summer (January to February).

Table 1. Climatic data for seasons were recorded by weather station electronics and a black globe thermometer

	Autumn	Winter	Spring	Dry summer	Rainy summer
Air temperature min (°C)	8.9	3.4	11.9	14.4	15.4
Air temperature max (°C)	29.6	27.7	33.3	34.8	35.3
THI min	42	21	50	57	57
THI mean	69	59	75	78	78
THI max	92	87	99	99	99
Relative humidity min (%)	42	30	26	24	35
Relative humidity max (%)	100	100	99	99	98
Solar radiation mean (W/m^2)	376.0	299.7	403.1	510.5	511.8
Accumulated rainfall	82.0	46.4	156.6	32.4	272.8
BGT cooled pen (°C)	24.0-32.0	20.0-28.5	24.0-29.0	25.3-28.3	25.3-32.3
BGT non-cooled pen (°C)	25.0-32.0	20.0-28.5	24.0-33.0	25.3-29.3	25.7-33.3

Behavior

Cow behaviors were recorded using instantaneous scan sampling (Martin & Bateson, 1993) every 30 min during the hottest hours of the day (9:30 am-6:00 pm) on three consecutive days in each different weather pattern (Autumn: April, May, June; Winter: June, July, August; Spring: September, October, November; Dry summer: December until January 15; Rainy summer: January 16 until February 28) in a total of 15 days. The observations were interrupted by milking between 2:00 pm and 3:00 pm (milking lasted 68 ± 10 min, mean ± S.E.). It was recorded the location of the individual animal at the facility within the freestall or picket defined by the numbers from 1 to 5 delimited by dotted lines: 1) trough; 2) corridors; 3) cemented area inside the freestall; 4) bed; 5) coast-cross area (Figure 1), time spent standing in the shade, and their daily activities such as eating, drinking, ruminating, interactions (positive, negative, sexual, and external), and idleness (Table 2).

Production and Quality of Milk

Samples were collected during autumn, winter, spring, dry summer, and rainy summer seasons in bottles containing bronopol tablets for analysis of somatic cell count (SCC) and other main components of milk: protein, fat, lactose, total solids, and nonfat dry extract. SCC was determined by flow cytometry (Somacount, 300, Bentley Instruments Inc., Chaska, MN) and the other main components by infrared techniques (Bentley 2000, Bentley Instruments Inc., Chaska, MN). Total milk production was also measured for each collection.

Table 2. Description of behavior observed in experimental cows

Behavior	Description
Position	
In the shade	At least with the head in the shade
In the barn	Trough; corridors; discovery area (cemented inside the freestall); bed; coast-cross area (see Figure 1)
Posture	
Standing	4 with ground anchors
Lying down	ñank in contact with the ground
Activity	
Eating	Ingesting silage and/or feed and/or fodder and/or salt
Ruminating	Chewing movements without food intake
Drinking	Ingesting the drinking water, with the mouth in water or water dripping from the mouth shortly after the stay in the water trough
Idling	No apparent activity
Interaction	Licking/scratching, fighting, mounts, external interface

Statistical Analysis

To analyze the behavioral data, a nonlinear, but linearizable model was adjusted by applying the Generalized Linear Model (Nelder & Wenderburn, 1972), using the GENMOD procedure of SAS software (version:9, Cary, NC, EUA). The model for behavioral variables, such as position in the shade, standing posture, and activities (eating, ruminating, drinking water, idling, and interaction), included the animal effect as the replicate measure to evaluate the fixed effect of treatment (CG or NCG) and time along the daylight and air temperature as covariates. The effect of the seasons on the average animal behavior with and without the availability of a cooling system was compared using ANOVA and means comparison by Tukey-Kramer at a significance level of 0.05. To compare variables between animals of the experimental groups (CG or NCG), a t-test was performed with Satterthwaite's approximation for degrees of freedom at the 0.05 significance level.

Milk production was analyzed in the seasons with a random effect of days in milk by variance analysis and mean comparison by Tukey-Kramer at a significance level of 0.05. A similar procedure was used for variable milk quality (fat, protein, lactose, non-fat dry extract, total solids, and SCC). Comparisons between treatments were performed by t-test with Satterthwaite's approximation for degrees of freedom at a significance level of 0.05. All values are presented as mean ± standard error of the mean

RESULTS AND DISCUSSION

Behavior

The animals remained standing in the shade for most of the day (84.2% of the observed time), regardless of the season and the availability of the evaporative cooling system ($P = 0.18$). In the non-cooled animals, the standing time decreased from 11 h and increased after 14 h (Figure 2), and it was linked with reduced feed and increased rumination. When lying down, the cows preferred to lie in bed (86% winter and an average of 64% in the other seasons), and in the remaining time, the animals preferred to lie on the grass and avoid lying on the concrete floor. Although rumination activity is associated with the lying position, a decrease in this behavior was observed when the days were hottest ($P < 0.05$), and there was a difference in the availability of cooling systems, especially in dry summer, when there was a greater number of animals that preferred ruminating while standing ($P < 0.01$) or remained in leisure at bay without cooling.

Rafiee et al. (2022) also noted a predominance of the standing position in cows, and longer time for eating was mainly with high milk producing animals. Cows in free stall with availability of fans and sprinkling over the trough spent more time standing rather than lie on the bed to promote heat loss (Calegari et al., 2014). With the increase in the temperature inside the shed, there was a decrease in the number of animals lying (Allen et al., 2015; Vieira, 2021). The standing posture maximizes the surface area exposed to the body environment and increases the airflow around the animal, facilitating heat loss by convection (Igono et al., 1987; Wang et al., 2018). After milking, food being available animals preferred eating and standing for the next 40-60 min (DeVries et al., 2011). Immediately after this period, most of the cows were observed in idling. In the period before milking, more than half of the animals were no longer lying. Longer standing may occur due to increased intramammary pressure and early start of milk ejection (DeVries et al., 2011; Melvin et al., 2019). It was found that in animals without the availability of a conditioned environment, there was a greater decrease in milk production between periods that are warm.

Figure 2. Mean frequency of standing behavior observed from 9:30 to 18:00 h of Holstein cows without (NCG) the availability of evaporative cooling system during the year

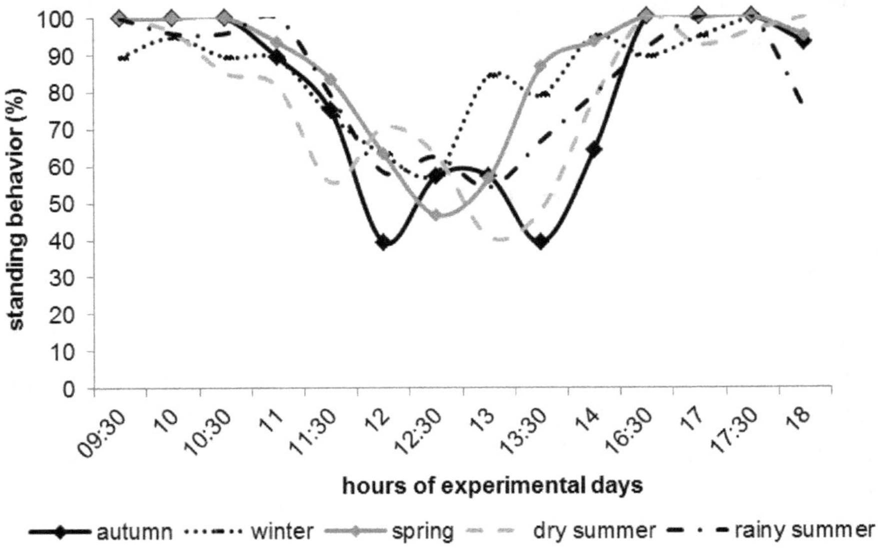

Eating activity was most frequent during daylight hours (P < 0.05). There was no difference between the experimental seasons, except during the dry summer and rainy summer seasons, when the animals in the non-cooled group spent less time eating (P = 0.03, Figure 3). Throughout the year, animals in the non-cooled group had lower ingestion during summer (P<0.05); however, the cooled group did not reduce their ingestion in summer (P = 0.26, Table 3). The frequency of water consumption was higher during winter in the non-air-conditioned barn (P = 0.03) despite the low values observed throughout the day. There were no differences between seasons for the cooled group, and there was no effect on this activity (P > 0.05).

Higher feeding frequencies were observed shortly after the start of milking activity (9:30 am and 4:30 pm) when a new diet was offered (Figure 3). One hour after these peaks, there was a decrease in the number of animals engaged in eating activity. The lowest frequencies occurred during the hottest hours of the day, between 11:30 am and 2:00 pm, when the frequency of rumination increased (P < 0.05). The same pattern was observed in animals from the two experimental groups (P > 0.05).

The animals remained in the covered corridors (area 2) near the water trough for most of the observed hours, with higher frequencies found during winter (61.3% and 58.5% for CG and NCG, respectively, P > 0.05) and lower frequencies during summer in the non-cooled barn (45.6%, P < 0.01). The animals oscillated between the bed (Area 4) and feeder (Area 1) during the remaining period. The uncovered area that was cemented was used less (less than 2% of the time in both groups at all times of the year), following the picket, which was most used at the end of the day, when the sun was setting.

In summer, cows change their eating patterns during daytime and nighttime to avoid the heat increment caused by dry matter intake during the hottest hours of the day (Portugal et al., 2000). The shorter feeding time of the non-cooled animals during the summer may have been influenced by higher exposure of animals to high air temperature and THI values, and this may have affected thermoregulation, dry matter intake, and probably the induction of metabolic changes as part of the physiological adaptations (Ferrazza et al., 2017). Hillman et al. (2005) found a greater decrease in the body temperature of cows in free stalls with fans, misting, and standing, when compared with animals lying. Even with access to artificial shade and cooling systems, cows in the present study spent more time standing under the structure, probably as a way of losing more heat.

Table 3. Mean frequency of daily activities observed from 9:30 am to 6:00 pm of Holstein cows with (CG) and without (NCG) the availability of evaporative cooling system during the year

	Autumn	Winter	Spring	Dry Summer	Rainy Summer	P season
Eating						
CG	49.75^{A}	53.51^{A}	50.98^{A}	45.09^{A}	43.15^{A}	0.26
NCG	51.23^{Aa}	52.56^{Aa}	44.12^{Aa}	37.20^{Bb}	36.11^{Bb}	0.05
Ruminating						
CG	13.80^{Ac}	15.81^{Ac}	18.40^{Abc}	26.49^{Aab}	27.97^{Aa}	0.017
NCG	11.25^{Ac}	14.57^{Abc}	18.78^{Aabc}	20.99^{Bab}	24.70^{Aa}	0.05
Drinking water						
CG	3.99	5.56	3.33	5.38	3.17	0.53
NCG	4.50^{ab}	6.88^{a}	2.68^{bc}	2.91^{bc}	1.49^{c}	0.03
Interaction						
CG	6.13	5.03	7.23	4.17	3.56	0.31
NCG	8.50	2.83	7.32	8.97	8.55	0.38
Idling						
CG	26.33	20.09	20.06	28.87	22.19	0.47
NCG	24.52	23.14	27.10	29.93	29.16	0.42

Means within a row with different lower-case letters and means within a column with different upper-case letters differs significantly by Tukey-Kramer test <0.05.

Figure 3. Eating behavior of Holstein cows without (NCG) availability of an evaporative cooling system during the year
** (P=0.03)*

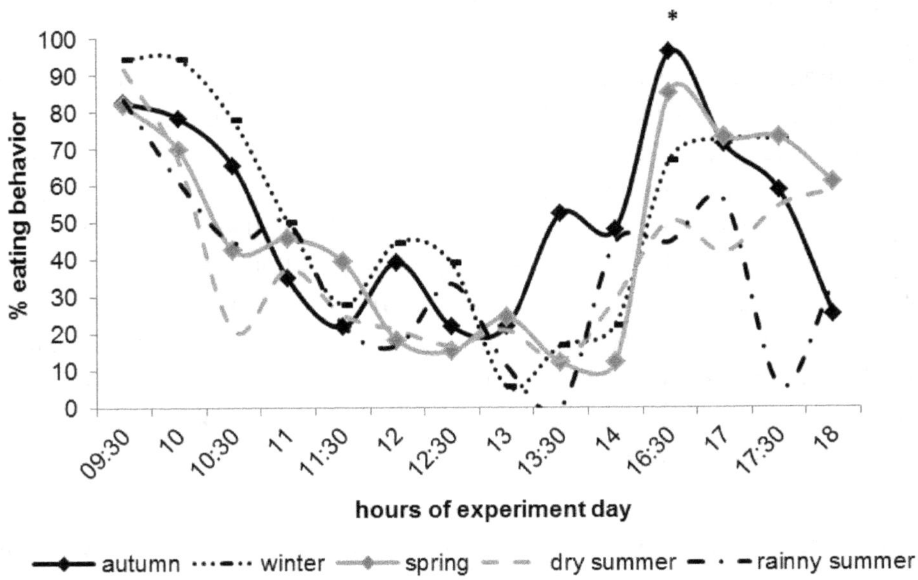

Animals seek shade in response to an increase in the absorbed thermal load (Van Laer et al., 2015). In the present study, the roof of the shed provided shade, and there was even absorption of heat from the radiation facility. An evaporative cooling system would be a way to reduce the increase in radiant heat load of the structure by cooling the ambient air, providing greater comfort for cows. And that greatest comfort does not seem to have been evident when observing frequencies in standing; however, despite eating activity during the dry summer and rainy summer being decreased from 1 to 1.5 hours when compared to winter, where there was the greatest time of eating, animals kept in a cooled environment spent more time in this activity during summer. Therefore, the frequency of rumination during the two phases of summer was higher in the cooled group. This result shows that thermal comfort should have been higher to allow the animal to ruminate, an activity preferably carried out when the animal is under comfort.

In contrast, animal under heat stress may exhibit reduced of pH, passage rates and consequently affects the production of volatile fatty acids (Nonaka et al., 2008; King et al., 2011). Which are important modifiers of rumen fermentation and influence on microbial synthesis (Wang et al., 2021; Wang et al., 2022). This impairs the digestion process that needs microbial population decreasing the dry matter intake and rumination, and it can result in later metabolic disorders and reduce in milk yield (Beauchemin et al., 2018).

Milk Quality and Production

Season had no effect on milk quality (P > 0.05). The lack of a cooling system resulted in a lower content of dry extract, with no significant difference (P > 0.05), and there was no difference between protein and lactose content (P > 0.05). The mean values of fat were lower (2.56%) than those of the animals kept in the cooling barns (3.05%, P < 0.05, Table 4).

Table 4. Milk composition in Holstein cows with and without the availability of evaporative cooling system during the year (mean ± S.E.)

	Fat (%)	Protein (%)	Lactose (%)	Total solids (%)	Dry extract (%)
Non-cooled animals	2.56 ± 0.12b	2.97 ± 0.03	4.42 ± 0.02	11.27 ± 0.14	8.32 ± 0.03
Cooled animals	3.05 ± 0.14a	3.08 ± 0.07	4.30 ± 0.06	11.35 ± 0.14	8.30 ± 0.08

Source: Prepared by the authors.
Means within a column with different upper-case letters differs significantly by Tukey-Kramer test <0.05

On evaluating the milk composition of Holstein subjected to heat stress, during summer a lower concentration of milk protein and lactose were found, but milk fat content was not changed (Garcia et al., 2015). In the present study, we did not find changes for protein and lactose in non-cooled group animals, but found lower percentage of fat. In non-cooled group animals a decline in fat percentage of 15.2% was observed. The effect of increasing temperature was also seen by Toghdory et al. (2022) found that with increasing temperature of 6.2 to 31.3 °C, the fat and protein decreased 4.09% and 5.75%, respectively.

The fat reduction in NCG occur because animals were not able to overcome the thermal challenge and probably suffered thermal stress due to high air temperature and THI recorded in spring and summer, according to Gorniak et al. (2014) elevation THI may contribute with decreased amount of fat in milk. The increase of the temperature can cause lower dry matter intake (Pereira et al., 2019), besides alterations in the ruminal fermentation parameters, reducing pH values and consequently acetate and butyrate concentration (Zhao et al., 2019). Therefore, directly influencing amount of milk fat.

Difference was found between the experimental groups in relation to the SCC, with lowest mean for the cooled animals in autumn and winter ($P < 0.01$, Figure 4) and higher values during the warmer season ($P < 0.01$). As the SCC data had no normal distribution, it was transformed to log base 10. There was effect of the seasons, with higher SCC found in winter for non-cooled group, in the summer dry and rainy for the cooled group ($P < 0.01$). Milk production decreased during warmer months in non-cooled animals ($P < 0.01$; Figure 5).

Figure 4. Somatic cell count (x1000 cel.ml^{-1}) in milk of Holstein cows with and without availability of evaporative cooling system during the year

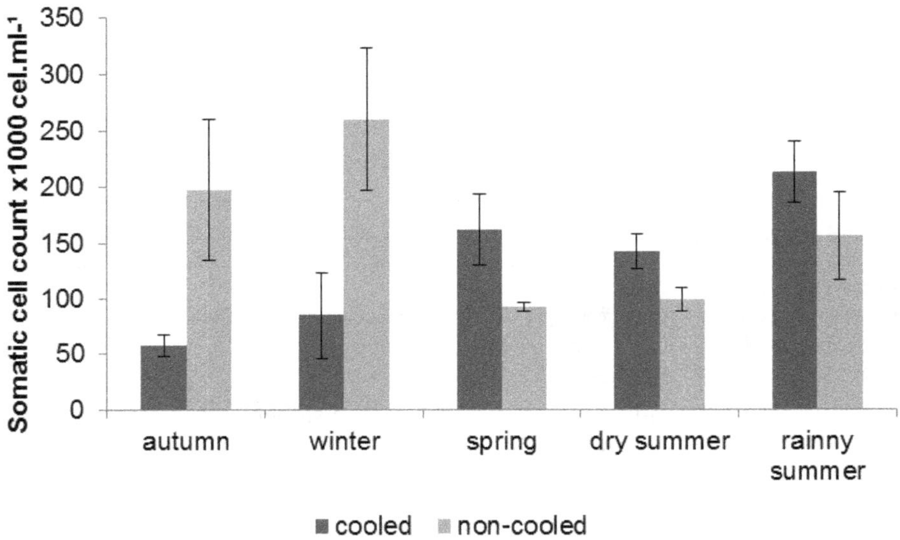

The lower CCS in non-cooled group during winter and autumn can be explained by the lower milk production since milk production has a negative correlation with CCS, so the occurrence of infections and the presence of pathogenic microorganisms in dairy cows are generally associated with a reduction in the volume of milk produced (Picinin et al. 2019). On the other hand, the greater SCC in animals cooled group in the summer and spring seasons may be a result of warmer and wetter periods that present the greatest challenges in relation to the hygiene and sanitary quality of milk, due to increased accumulation of mud on dairy facilities and a higher incidence of dust on the ceilings and roofs during milking. Such factors associated with failures in the milking routine may be responsible for high initial contamination and increased prevalence of mastitis in the herd evidenced by the greater tendency toward higher SCC and cause major risk for the decline in the quality and volume of milk produced during unfavorable weather conditions, such as rainfall and higher temperatures (Sant'Anna & Paranhos da Costa, 2011). However, despite the fluctuation in the values throughout the year, in most samples the SCC remained below 200.000 cells/mL, a threshold value for the absence of infection (Alves et al. 2014; Moroni et al., 2018).

*Figure 5. Milk production (kg.cow^{-1}. day^{-1}) in cooled and non-cooled Holstein cows with and without the availability of an evaporative cooling system during the year * (P<0.01)*

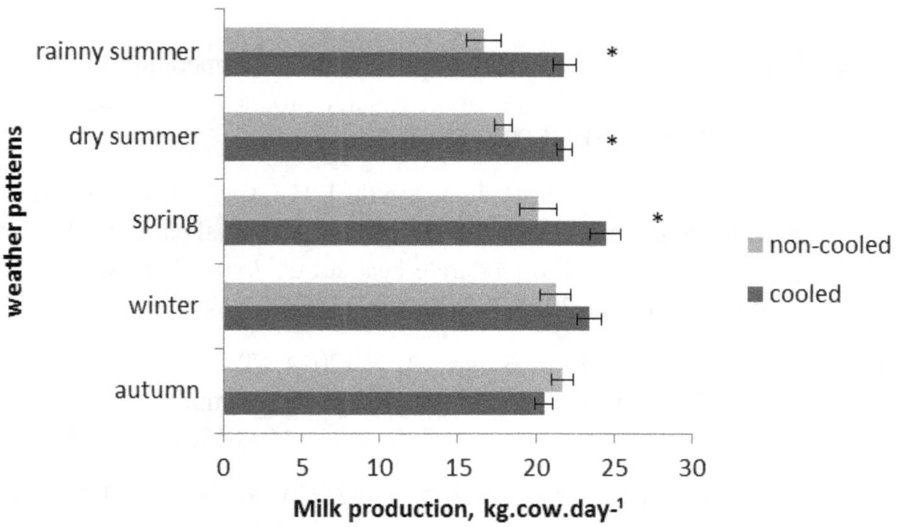

The thermal environment is one of the main factors that can affect milk production in dairy cattle (Nardone et al., 2010). During spring and summer, were observed temperature peaks reached 35.3 °C. The THI reached mean values of 75 and 78, respectively, indicating an environmental warning situation for dairy cows, under these conditions, the productivity of the animal can be seriously impaired (M'Hamdi et al., 2021). The NCG animals presented lower milk yield in warmer periods, this may have occurred due of stressful environmental conditions within the experimental area that represented the greatest thermal challenge for this group, this may have caused changes in the biological functions of these animals, which include decrease in food intake (Marai et al., 2008). In a heat stress situation, the animal increases its effort to dissipate the excess of accumulated thermal energy, in an attempt to avoid detrimental changes to the body (Rashamol et al., 2018). However, if the thermolytic capacity is compromised, the energy deviation occurs that would be used in other metabolic and productive processes, to maintain homeothermy resulting in less milk synthesis (Das et al., 2016).

The use of an evaporative cooling system is not necessary in the cooler months of the year (autumn, winter, and spring); thus, the spending on energy and water can be reduced. The evaporative cooling environment for lactating cows proved effective during the hottest months of the year in increasing the daytime feeding time and milk yield, and better milk quality was observed with a higher percentage of fat.

REFERENCES

Ahmed, H., Tamminen, L. M., & Emanuelson, U. (2022). Temperature, productivity, and heat tolerance: Evidence from Swedish dairy production. *Climatic Change, 175*(1-2), 10. doi:10.100710584-022-03461-5

Allen, J. D., Hall, L. W., Collier, R. J., & Smith, J. F. (2015). Effect of core body temperature, time of day, and climate conditions on behavioral patterns of lactating dairy cows experiencing mild to moderate heat stress. *Journal of Dairy Science, 98*(1), 1–10. doi:10.3168/jds.2013-7704 PMID:25468707

Alves, E. C., Dahmer, A. M., & Borges, A. F. (2014). Total bacterial count and somatic cell count in refrigerated raw milk stored in communal tanks. *Brazilian Journal of Food Technology, 17*(3), 221–225. doi:10.1590/1981-6723.1114

Arnott, G., Ferris, C. P., & O'Connell, N. E. (2017). Review: Welfare of dairy cows in continuously housed and pasture-based production systems. *Animal, 11*(2), 261–273. doi:10.1017/S1751731116001336 PMID:27364762

Beauchemin, K. A. (2018). Invited review: Current perspectives on eating and rumination activity in dairy cows. *Journal of Dairy Science, 101*(6), 4762–4784. doi:10.3168/jds.2017-13706 PMID:29627250

Bernabucci, U., Basiricò, L., Morera, P., Dipasquale, D., Vitali, A., Cappelli, F. P., & Calamari, L. (2015). Effect of summer season on milk protein fractions in Holstein cows. *Journal of Dairy Science, 98*(3), 1–13. doi:10.3168/jds.2014-8788 PMID:25547301

Brown-Brandl, T. M. (2018). Understanding heat stress in beef cattle. *Revista Brasileira de Zootecnia, 47*(0), e20160414. doi:10.1590/rbz4720160414

Brown-Brandl, T. M., Chitko-McKown, C. G., Eigenberg, R. A., Mayer, J. J., Welsh, T. H. Jr, Davis, J. D., & Purswell, J. L. (2017). Physiological responses of feedlot heifers provided access to different levels of shade. *Animal, 11*(8), 1344–1353. doi:10.1017/S1751731116002664 PMID:28007043

Calegari, F., Calamari, L., & Frazzi, E. (2014). Fan cooling of the resting area in a free stalls dairy barn. *International Journal of Biometeorology, 58*(6), 1225–1236. doi:10.100700484-013-0716-1 PMID:24122338

Chaiyabutr, N., Chanpongsang, S., & Suadsong, S. (2008). Effects of evaporative cooling on the regulation of body water and milk production in crossbred Holstein cattle in a tropical environment. *International Journal of Biometeorology, 52*(7), 575–585. doi:10.100700484-008-0151-x PMID:18427839

Chang-Fung-Martel, J., Harrison, M. T., Brown, J. N., Rawnsley, R., Smith, A. P., & Meinke, H. (2021). Negative relationship between dry matter intake and the temperature-humidity index with increasing heat stress in cattle: A global meta-analysis. *International Journal of Biometeorology, 65*(12), 2099–2109. doi:10.100700484-021-02167-0 PMID:34283273

Das, R., Sailo, L., Verma, N., Bharti, P., Saikia, J., Imtiwati, & Kumar, R. (2016). Impact of heat stress on health and performance of dairy animals: A review. *Veterinary World, 9*(3), 260–268. doi:10.14202/vetworld.2016.260-268 PMID:27057109

DelCurto-Wyffels, H. M., Dafoe, J. M., Parsons, C. T., Boss, D. L., DelCurto, T., Wyffels, S. A., Van Emon, M. L., & Bowman, J. G. P. (2021). The Influence of Environmental Conditions on Intake Behavior and Activity by Feedlot Steers Fed Corn or Barley-Based Diets. *Animals (Basel), 11*(5), 1261. doi:10.3390/ani11051261 PMID:33925628

DeVries, T. J., Deming, J. A., Rodenburg, J., Seguin, G., Leslie, K. E., & Barkema, H. W. (2011). Association of standing and lying behavior patterns and incidence of intramammary infection in dairy cows milked with an automatic milking system. *Journal of Dairy Science, 94*(8), 3845–3855. doi:10.3168/jds.2010-4032 PMID:21787921

Ferrazza, R. A., Garcia, H. D. M., Aristizábal, V. H. V., Nogueira, C. H., Veríssimo, C. J., Sartori, J. R., Sartori, R., & Ferreira, J. C. P. (2017). Thermoregulatory responses of Holstein cows exposed to experimentally induced heat stress. *Journal of Thermal Biology, 66*, 68–80. doi:10.1016/j.jtherbio.2017.03.014 PMID:28477912

Fontoura, A. B. P., Javaid, A., Sáinz de la Maza-Escolà, V., Salandy, N. S., Fubini, S. L., Grilli, E., & McFadden, J. W. (2022). Heat stress develops with increased total-tract gut permeability, and dietary organic acid and pure botanical supplementation partly restores lactation performance in Holstein dairy cows. *Journal of Dairy Science, 105*(9), 7842–7860. doi:10.3168/jds.2022-21820 PMID:35931486

Garcia, A. B., Angeli, N., Machado, L., Cardoso, F. C., & Gonzalez, F. (2015). Relationships between heat stress and metabolic and milk parameter in dairy cows in southern Brazil. *Tropical Animal Health and Production, 47*(5), 889–894. doi:10.100711250-015-0804-9 PMID:25851928

Gorniak, T., Meyer, U., Südekum, K., & Dänicke, S. (2014). Impact of mild heat stress on dry matter intake, milk yield and milk composition in mid-lactation Holstein dairy cows in a temperate climate. *Archives of Animal Nutrition, 68*(5), 358–369. doi:10.1080/1745039X.2014.950451 PMID:25176045

Hillman, P.E., Lee, C.N., & Willard, S.T. (2005). Thermoregulatory responses associated with lying and standing in heat-stressed dairy cows. *American Society of agricultural and Biological Engineers, 48*, 795-801.

Hut, P. R., Scheurwater, J., Nielen, M., van den Broek, J., & Hostens, M. M. (2022). Heat stress in a temperate climate led to adapted sensor-based behavioral patterns of dairy cows. *Journal of Dairy Science, 105*(8), 6909–6922. doi:10.3168/jds.2021-21756 PMID:35787319

Igono, M. O., Johnson, H. D., Steevens, B. J., Krause, G. F., & Shanklin, M. D. (1987). Physiological, productive, and economic benefits of shade, spray, and fan systems versus shade for Holstein cows during summer heat. *Journal of Dairy Science, 70*(5), 1069–1079. doi:10.3168/jds.S0022-0302(87)80113-3 PMID:3597932

King, C.C., Dschaak, C.M., Eun, J.-S., Fellner, V., Young, A.J. (2011). Quantitative analysis of microbial fermentation under normal or high ruminal temperature in continuous cultures. *The Professional Animal Scientist, 27*(4), 319-327.. doi:10.15232/S1080-7446(15)30495-2

Lacetera, N. (2019). Impact of climate change on animal health and welfare. *Animal Frontiers, 9*(1), 26–31. doi:10.1093/af/vfy030 PMID:32002236

Li, G., Chen, S., Chen, J., Peng, D., & Gu, X. (2020). Predicting rectal temperature and respiration rate responses in lactating dairy cows exposed to heat stress. *Journal of Dairy Science, 103*(6), 5466–5484. doi:10.3168/jds.2019-16411 PMID:32278558

M'Hamdi, N., Darej, C., Attia, K., Znaidi, I. L., Khattab, R., Djelailia, H., Bouraoui, R., Taboubi, R., Marzouki, L., & Ayadi, M. (2021). Modelling THI effects on milk production and lactation curve parameters of Holstein dairy cows. *Journal of Thermal Biology*, *99*, 102917. doi:10.1016/j.jtherbio.2021.102917 PMID:34420599

Marai, I. F. M., El-Darawanya, A. A., Fadielc, A., & Abdel-Hafezb, M. A. M. (2008). Reproductive performance traits as affected by heat stress and its alleviation in sheep. *Tropical and Subtropical Agroecosystems*, *8*, 209–234.

Martin, P., & Bateson, P. (1993). *Measuring behavior: an introductory guide*. Cambridge University Press. doi:10.1017/CBO9781139168342

Marumo, J. L., Lusseau, D., Speakman, J. R., Mackie, M., & Hambly, C. (2022). Influence of environmental factors and parity on milk yield dynamics in barn-housed dairy cattle. *Journal of Dairy Science*, *105*(2), 1225–1241. doi:10.3168/jds.2021-20698 PMID:34802739

Melvin, J. M., Heuwieser, W., Virkler, P. D., Nydam, D. V., & Wieland, M. (2019). Machine milking–induced changes in teat canal dimensions as assessed by ultrasonography. *Journal of Dairy Science*, *102*(3), 2657–2669. doi:10.3168/jds.2018-14968 PMID:30639026

Moroni, P., Nydam, D. V., Ospina, P. A., Scillieri-Smith, J. C., Virkler, P. D., Watters, R. D., Welcome, F. L., Zurakowski, J., Ducharme, N. G., & Yeager, A. E. (2018). 8 diseases of the Teats and Udder. In Peek, S.F., & Thomas J. Divers (Eds), Rebhun's Diseases of Dairy Cattle (389-465). Elsevier. doi:10.1016/B978-0-323-39055-2.00008-5

Nardone, A., Ronchi, B., Lacetera, N., Ranieri, M. S., & Bernabucci, U. (2010). Effects of climate changes on animal production and sustainability of livestock systems. *Livestock Science*, *130*(1-3), 57–69. doi:10.1016/j.livsci.2010.02.011

Nelder, J. A., & Wedderburn, R. W. M. (1972). Generalized Linear Model. *Journal of the Royal Statistical Society. Series A (General)*, *135*(3), 370–384. doi:10.2307/2344614

Nonaka, I., Takusari, N., Tajima, K., Suzuki, T., Higuchi, K., & Kurihara, M. (2008). Effects of high environmental temperatures on physiological and nutritional status of prepubertal Holstein heifers. *Livestock Science*, *113*(1), 14–23. doi:10.1016/j.livsci.2007.02.010

Pereira, A. M. F., Titto, E. A. L., & Almeida, J. A. A. (2019). *Adaptação dos ruminantes aos climas quentes*. Appris Editora.

Picinin, L. C. A., Bordignon-Luiz, M. T., Cerqueira, M. M. O. P., Toaldo, I. M., Souza, F. N., Leite, M. O., Fonseca, L. M., & Lana, A. M. Q. (2019). Effect of seasonal conditions and milk management practices on bulk milk quality in Minas Gerais State – Brazil. *Arquivo Brasileiro de Medicina Veterinária e Zootecnia, 71*(4), 1355–1363. doi:10.1590/1678-4162-10063

Portugal, J. A. B., Pires, M. F. A., & Durães, M. C. (2000). Effect of air temperature and humidity on frequency of feeding, water ingestion and rumination in Holstein cows. *Arquivo Brasileiro de Medicina Veterinária e Zootecnia, 52*, 154–159. doi:10.1590/S0102-09352000000200013

Rafiee, H., Alikhani, M., & Ghorbani, G. R. (2022). Effect of dietary protein level and corn processing on behavior activity of high producing dairy cows. *Acta Scientiarum. Animal Sciences, 44*, e54603. doi:10.4025/actascianimsci.v44i1.54603

Rashamol, V. P., Sejian, V., Bagath, M., Krishnan, G., Archana, P. R., & Bhatta, R. (2018). Physiological adaptability of livestock to heat stress: An updated review. *Journal of Animal Behaviour and Biometeorology, 6*(3), 62–71. doi:10.31893/2318-1265jabb.v6n3p62-71

Román, L., Saravia, C., Astigarraga, L., Bentancur, O., & La Manna, A. (2019). Shade access in combination with sprinkling and ventilation effects performance of Holstein cows in early and late lactation. *Animal Production Science, 59*(2), 347. doi:10.1071/AN16571

Sant'Anna, A. C., & Paranhos da Costa, M. J. (2011). The relationship between dairy cow hygiene and somatic cell count in milk. *Journal of Dairy Science, 94*(8), 3835–3844. doi:10.3168/jds.2010-3951 PMID:21787920

Schütz, K. E., Rogers, A. R., Cox, N. R., Webster, J. R., & Tucker, C. B. (2011). Dairy cattle prefer shade over sprinklers: Effects on behavior and physiology. *Journal of Dairy Science, 94*(1), 273–283. doi:10.3168/jds.2010-3608 PMID:21183037

Silva, M. V., Almeida, G. L. P., Pandorfi, H., Moraes, A. S., Macêdo, G. A. P. A., Batista, P. H. D., Silva, R. A. B., & Guiselini, C. (2021). Influence of meteorological elements on behavioral responses of gir cows and effects on milk quality. *Acta Scientiarum. Animal Sciences, 43*, e52604. doi:10.4025/actascianimsci.v43i1.52604

Titto, C. G., Negrão, J. A., Titto, E. A., Canaes, T. S., Titto, R. M., & Pereira, A. M. (2013). Effects of an evaporative cooling system on plasma cortisol, IGF-I, and milk production in dairy cows in a tropical environment. *International Journal of Biometeorology*, *57*(2), 299–306. doi:10.100700484-012-0554-6 PMID:22580965

Toghdory, A., Ghoorchi, T., Asadi, M., Bokharaeian, M., Najafi, M., & Ghassemi Nejad, J. (2022). Effects of environmental temperature and humidity on milk composition, microbial load, and somatic cells in milk of Holstein dairy cows in northeast Iran. *Animals (Basel)*, *12*(18), 2484. doi:10.3390/ani12182484 PMID:36139344

Tullo, E., Mattachini, G., Riva, E., Finzi, A., Provolo, G., & Guarino, M. (2019). Effects of Climatic Conditions on the Lying Behavior of a Group of Primiparous Dairy Cows. *Animals (Basel)*, *9*(11), 869. doi:10.3390/ani9110869 PMID:31717823

van Laer, E., Moons, C. P. H., Ampe, B., Sonck, B., Vandaele, L., De Campeneere, S., & Tuyttens, F. A. M. (2015). Effect of summer conditions and shade on behavioural indicators of thermal discomfort in Holstein dairy and Belgian Blue beef cattle on pasture. *Animal*, *9*(9), 1536–1546. doi:10.1017/S1751731115000804 PMID:25994098

Vieira, F. M. C., Soares, A. A., Herbut, P., Vismara, E. S., Godyń, D., Dos Santos, A. C. Z., Lambertes, T. D. S., & Caetano, W. F. (2021). Spatio-thermal variability and behavior as bio-thermal indicators of heat stress in dairy cows in a compost barn: A case study. *Animals (Basel)*, *11*(5), 1197. doi:10.3390/ani11051197 PMID:33919438

Wang, X., Zhang, G., & Choi, C. Y. (2018). Effect of airflow speed and direction on convective heat transfer of standing and reclining cows. *Biosystems Engineering*, *167*, 87–98. doi:10.1016/j.biosystemseng.2017.12.011

Wang, Z., Niu, K., Rushdi, H. E., Zhang, M., Fu, T., Gao, T., Yang, L., Liu, S., & Lin, F. (2022). Heat Stress Induces Shifts in the Rumen Bacteria and Metabolome of Buffalo. *Animals (Basel)*, *12*(10), 1300. doi:10.3390/ani12101300 PMID:35625146

Wang, Z., Yang, D. S., Li, X. Y., Yu, Y. N., Yong, L. Y., Zhang, P. H., He, J. H., Shen, W. J., Wan, F. C., Feng, B. L., Tan, Z. L., & Tang, S. X. (2021). Modulation of rumen fermentation and microbial community through increasing dietary cation-anion difference in Chinese Holstein dairy cows under heat stress conditions. *Journal of Applied Microbiology*, *130*(3), 722–735. doi:10.1111/jam.14812 PMID:32757409

West, J., Mullinix, B. G., & Bernard, J. K. (2003). Effects of hot, humid weather on milk temperature, dry matter intake, and milk yield of lactating dairy cows. *Journal of Dairy Science*, *86*(1), 232–242. doi:10.3168/jds.S0022-0302(03)73602-9 PMID:12613867

Zeinhom, M. M., Aziz, R. L. A., Mohammed, A. N., & Bernabucci, U. (2016). Impact of seasonal conditions on quality and pathogens content of milk in Friesian cows. *Asian-Australasian Journal of Animal Sciences*, *29*(8), 1207–1213. doi:10.5713/ajas.16.0143 PMID:27165021

Zhao, S., Min, L., Zheng, N., & Wang, J. (2019). Effect of Heat Stress on Bacterial Composition and Metabolism in the Rumen of Lactating Dairy Cows. *Animals (Basel)*, *9*(11), 925. doi:10.3390/ani9110925 PMID:31694321

Chapter 9
Criteria for Production and Quality Analyses in Milk Produced in Brazil

Priscilla Ayleen Bustos Mac-Lean
School of Sciences and Engineering, São Paulo State University, Brazil

Eduardo Guilherme Satolo
https://orcid.org/0000-0002-8176-2423
School of Sciences and Engineering, São Paulo State University, Brazil

Ligia Cardoso dos Santos
School of Sciences and Engineering, São Paulo State University, Brazil

Silvana Gomes Gonzalez
https://orcid.org/0000-0002-4054-7620
School of Sciences and Engineering, São Paulo State University, Brazil

Leonardo Alexander de Oliveira Campos
School of Sciences and Engineering, São Paulo State University, Brazil

Isabela Garcia Mendes de Araujo Santos
School of Sciences and Engineering, São Paulo State University, Brazil

Julia Fleuri Barbosa
School of Sciences and Engineering, São Paulo State University, Brazil

EXECUTIVE SUMMARY

Milk and its derivatives are among the foods most consumed by the population, capable of meeting important nutritional needs of humans. As the consumer market is increasingly demanding in terms of quality of the milk produced, the difficulty

DOI: 10.4018/978-1-6684-5472-5.ch009

in maintaining and controlling these points is an issue present in several studies. Quality can be influenced from the type of production system in which the milk is produced, passing through nutritional and sanitary management, to the environment in which the animals remain during their productive life. This chapter presents a brief characterization of the dairy production chain in Brazil and the federal legislation that governs the production and quality of the milk produced. Some results of research on milk quality in shared tanks carried out by the Kamby Research Group with small rural producers in the Tupã/SP region are also presented. Finally, factors that affect the quality of milk and future actions to improve the dairy production chain are pointed out.

INTRODUCTION

Brazil is the third largest milk producer in the world, with an annual production of 34 billion liters produced, mostly by small and medium-sized properties in 98% of Brazilian municipalities (IBGE, 2021). As the consumer is increasingly demanding in terms of the quality of the milk produced, it is difficult to maintain and control these issues and this is an aspect present in several studies.

This quality can be influenced by several factors, among which stand out the different milking management practices, sanitary practices, the type of food supplied to the animals and the climatic variables to which they are submitted, causing changes both in the composition of the main nutrients and in the to the microbiological characteristics of milk (Gomes, 1999).

Maintaining the quality and natural composition of the milk that is produced is one of the great responsibilities of producers and authorities directly linked to the health area. To maintain the standardization of dairy production in the country, the Ministry of Agriculture, Livestock and Supply (MAPA) presented, in 2018, the update of the Normative Instructions (NI) that are currently in the country, relates to refrigerated raw milk and pasteurized milk, known as NI 76/2018 and NI 77/2018. The NI 76/2018 seeks to address the identification and the minimum quality levels established for raw and pasteurized milk, while NI 77/2018 brings the standards that must be established to obtain quality milk, taking into account factors such as hygiene, maintenance of equipment and property installations, the precautions that must be taken from collection to delivery to dairy products and issues that must be raised to ensure the qualification of professionals involved in all collection and transport of milk, which includes concern for the animal management to avoid diseases such as mastitis (MAPA, 2018).

The main analyzes carried out to verify the quality parameters, required by current legislation, are those that indicate the nutritional composition and some microbiological parameters of the milk, usually carried out at strategic points in the process, making it possible to find possible points of contamination or alteration of the product. In addition to this, there are analyzes that make it possible to track whether there has been adulteration in the milk, which is considered fraud during the production process.

According to Spink and Moyer (2011) food fraud can be defined as "intentional deliberation and substitution, addition, adulteration or misrepresentation of food, food ingredients od false or misleading statement made about a product, for economic gain". In addition, there is still great difficulty regarding the traceability of milk and the changes that can occur, mainly in shared tanks of cooperatives, associations or even, although in a smaller proportion, in individual tanks.

Thus, it is possible to note the difficulties of obtaining a quality product due to the numerous factors that may occur throughout the process. By making use of zootechnical practices and raising awareness of the players in the dairy production chain, it is possible to guarantee the delivery of better-quality milk and thus provide a bonus from dairy products, increasing the economic and financial results of the producer (Pinheiro et al., 2015).

The chapter aims to present the main criteria established in the legislation for the production of milk in Brazil, and to describe some results of laboratory analyzes of the milk quality. In addition, it is intended to present analyses that detect possible adulterations carried out during the collection or delivery process to modify the composition of the product.

THE DAIRY PRODUCTION CHAIN IN BRAZIL

The dairy production chain can be described as simple and is divided into three segments: (i) dairy farming; (ii) processing and packaging; and (iii) retail (Kazancoglu et al., 2018). However, this chain, considered simple, can be affected when the dairy production scenario involves a high number of small-scale producers, a very recurrent situation in Brazil (Assis et al., 2018).

In dairy farming, the production chain is made up of several interrelated actors, starting with the production of agricultural inputs, the first link in the chain, formed by companies that manufacture raw materials, credit, equipment, services and research. The second link in the chain comprises the milk production itself, where the producers, animals and employees of the property are located. The third link, on the other hand, is represented by the industry, where the companies in charge of the logistics of milk collection and distribution of industrialized products are located. Finally, there is the final consumer, representing the fourth link in the chain (Sparemberger et al., 2009).

Brazil occupies a prominent role in the works when talking about milk production capacity, due to its high production yield and earning potential, which attracts investors from other countries to bet on milk and its derivatives. According to Brazilian Agricultural Research Corporation (EMBRAPA, 2020), a transformation has occurred in the dairy sector in recent years, leading to an 80% increased in milk production, with the same number of heads in its herd. What happened is justified by the implementation of different technologies in the production process, in the properties and even in the workforce.

BRAZILIAN MILK PRODUCTION SYSTEMS

In order to obtain quality milk production, it is necessary to choose, on a case by case, the production system that will be applied in the creation of the dairy herd. The ideal production method for each case takes into account the factors available on the property, namely: the area destined for breeding, animal genetics, technologies available in the region, food availability, size of production and property and, mainly, the producer profile. These properties can adhere to three types of production systems, namely intensive, semi-intensive and extensive (Viana, 2021).

In the intensive production system, the animals are confined, limiting their presence to open environments. The property's facilities must be designed for the physiological and behavioral characteristics of the animal that will be raised, preserving good ventilation, artificial or natural shading and control of the temperature and humidity of the place (Lopes, 2012). This system is widely used in places with little territorial space and dry, with higher construction and maintenance costs.

Within the intensive system, there are several types of structure, the most used in the dairy chain being the Free Stall and the Compost Barn (Marion and Segatti, 2010). According to Zanin et al. (2015), the Free Stall system is widely used in the confinement of dairy cows in several countries; this system offers individual beds, a care track, and halls access for the animals. This structure promotes greater control of environmental conditions, enabling adequate animal comfort. The Compost Barn structure gathers dairy cattle in a large common area, and the facilities form a collective bed with comfortable materials that can be transformed into organic compost that is reused later. When well-managed, in this system, animal comfort and herd welfare are evident, in addition to the reduction of production costs when related to the Free Stall. The composting of the waste produced by the animals, as well as the beds, promote an environmentally correct destination for the waste generated (Damasceno, 2012), allowing a financial return for the sale of the compost or an aid in reducing the cost of fertilization on the same property in pastures or other cultures or areas, when needed.

Unlike the intensive system, the semi-intensive production raises the animals on pasture, with the aid of a diet increased by forage, bulky supplements, and feed, which can be temporary or permanent, throughout the creation (Marion, 2007). The cows remain in the pens during the day, where they are fed and milked. This system provides good sanitary conditions for the animals, with the presence of a production control, enabling the use of technologies, such as breastfeeding and artificial insemination of the animals. Therefore, this system is widely used in suburban areas, around large urban centers that do not have a very large area available for creations (Willers, 2014) and has a lower construction and maintenance cost when compared to the intensive system.

The extensive system, in contrast, has as its main feature the creation of dairy cattle exclusively in the field, using the natural resources available on the property, reducing the presence of equipment, labor and food control. The predominant meal is natural pasture, without or with a minimum of supplements and feed to the animals. This type of breeding can be applied to both single and mixed species and the property can choose to irrigate the pastures or not (Santos, Marion, Segatti, 2008). This system is little used in Brazil, since there is a need for large areas and investment in pasture quality technologies.

LEGISLATION IN DAIRY PRODUCTION CHAIN REGARDING PRODUCTION AND QUALITY ASSESSMENT

All care related to milk production in Brazil is guided by the normative instructions (IN) 76 and 77 published on November 26, 2018, by the Ministry of Agriculture, Livestock and Supply (MAPA). Some of the precautions are related to the composition of the milk, which generally consists of several solids, diluted in water, which give the product characteristics that indicate its quality (Prado et al., 2016). In addition to the concentration of solids, storage temperature, care in the milking process, animal care, hygienic-sanitary conditions, climate, water quality and feeding of lactating cows, are determining factors for the milk quality. Among the factors that alter its quality, stand out those related to nutrition, diseases, sanitary management, milking management, rational management, climate, and genetics (Brasil, 2018).

These norms serve as a reference for the quantitative and qualitative parameters of milk, whose main objective I to give the best possible quality to the product, in addition to guaranteeing the health and well-being of the consumer, through a series of steps that must be followed, in order to achieve that objective. Among the solids in force in NI 76/2018 and NI 77/2018, there are the percentages of protein, fat in addition to the characteristics that involve the sanitary aspect of the milk, somatic cell count (SCC) and standard plate count (SPC). These are of paramount importance in terms of quality, as they illustrate the animal's welfare, the milking environment, storage mode, in addition to the hygienic-sanitary conditions present before, during and after milking.

Furthermore, protein and fat levels, for example, are indicative of the amount of total solids present in milk, which will indicate the best destination given to the product, which can be fermentation, fresh consumption, production of cheese or other derivatives. Likewise, the concentration of SCC in milk indicates the presence of diseases such as subclinical mastitis, an infection in the teat that causes the displacement of somatic cells (defense cells; leukocytes) to fight the disease, and the SPC indicates the sanitary conditions to which the milk was exposed (colony-forming unit per milliliter of milk). In addition to these, aspects such as storage, sanitary conditions, transport, identification, herd, equipment, production, reception, collection, and analysis of milk were guided and detailed by these norms.

Therefore, regardless of the milking system used, the importance of good milking practices is highlighted so that the milk is within the quality parameters required by the regulations in force and can be consumed, generating added value and benefiting producer, herd and consumer.

Cases Studies on Milk Quality

In studies carried out by the Kamby Research and Extension Group, authors of this chapter, based at the São Paulo State University (Unesp), School of Sciences and Engineering, Tupa Campus, results are shown regarding the quality of milk according to the different methods of milking and the form of storage.

In these studies, carried out between 2019 and 2021, the quality and possible changes in the milk from shared tanks of small dairy farms in Tupã and region, São Paulo State, Brazil, were analyzed.

For the quality study, the fat (F), protein (P), lactose (LACT), and somatic cell count (SCC) levels of refrigerated raw milk stored in collective tanks and of the producers who feed them in five rural properties were evaluated. The analyzes were carried out by collecting samples from each producer and, subsequently, from the tank in which it was stored. Data were subjected to analysis of variance, correlation analysis and means comparison test (Tukey at 5%). Significant differences were found in the fat content of two tanks in relation to the results of the producers that delivered in this tank, while the SCC indexes remained within the limits allowed by national legislation in all producers and consequently in each tank (Table 1).

Table 1. Analysis of variance of composition parameters (fat, protein - PROT and lactose - LACT) and somatic cell count (SCC) of milk from collective tanks and the respective producers that supply them

Parameters	***Tank A***	***ProdA1***	***ProdA2***	***ProdA3***							
FAT (%)	3.04±0.89b	3.97±0.7ab	5.05±1.52a	3.32±0.68ab							
PROT(%)	2.71±0.55a	2.98±0.28a	3.04±0.25a	2.86±0.54a							
LACT (%)	3.92±0.81a	4.31±0.4a	4.37±0.4a	4.13±0.7a							
SCC (K SC/mL)	<90a	<90a	<90a	<90a							
Parameters	***Tank B***	***ProdB1***	***ProdB2***	***ProdB3***	***ProdB4***						
FAT (%)	3.7±0.11a	4.37±0.37a	3.65±0.42a	3.65±0.74a	3.01±1.17a						
PROT (%)	3.19±0.06a	3.10±0.06a	3.09±0.18a	2.56±0.36a	2.94±0.54a						
LACT (%)	4.62±0.09a	4.47±0.11a	4.48±0.26a	3.66±0.54a	4.27±0.79a						
SCC (K SC/mL)	107.3± 31.8a	<90a	107.3± 31.8a	107.3± 31.8a	<90a						
Parameters	***Tank C***	***ProdC1***	***ProdC2***	***ProdC3***	***ProdC4***						
FAT (%)	3.74±0.04a	4.30±0.21a	3.14±0.93a	3.95±0.52a	3.17±0.24a						
PROT (%)	3.28±0.01a	3.06±0.31a	2.90±0.61a	3.35±0.05a	3.23±0.03a						
LACT (%)	4.76±0.01a	4.42±0.46a	4.2±0.9a	4.86±0.08a	4.71±0.04a						
SCC (K SC/mL)	<90a	<90a	<90a	<90a	<90a						
Parameters	***Tank D***	***ProdD1***	***ProdD2***	***ProdD3***	***ProdD4***	***ProdD5***	***ProdD6***	***ProdD7***			
FAT (%)	3.43±0.076a	3.56±0.82a	3.01±1.56a	2.49±0.37a	3.07± 0.71a	3.07±0.12a	3.31±0.32a	3.46±0.11a			
PROT (%)	3.14±0.35a	3.15±0.19a	2.88±0.68a	3.52±0.01a	3.08±0.59a	3.25±0.04a	3.14±0.23a	3.28±0.02a			
LACT (%)	4.56±0.52a	4.57±0.32a	4.17±0.9a	5.15±0.01a	4.48±0.87a	4.71±0.03a	4.56±0.35a	4.76±0.02a			
SCC (K SC/mL)	<90a	<90a	148±83.4a	<90a	<90a	<90a	<90a	<90a			
Parameters	***Tank E***	***ProdE1***	***ProdE2***	***ProdE3***	***ProdE4***	***ProdE5***	***ProdE6***	***ProdE7***	***ProdE8***	***ProdE9***	***ProdE10***
FAT (%)	3.66±0.05ab	3.09±0.83b	3.01±0.2b	4.12±0.91ab	3.14±0.29b	3.59±0.52ab	3.97±0.69ab	5.2±0.27a	4.65±0.65ab	3.16±0.12b	3.58±1.11ab
PROT (%)	3.30±0.04a	2.79±0.54a	3.18±0.03a	3.01±0.57a	3.31±0.04a	3.34±0.04a	3.17±0.19a	3.29±0.06a	3.24±0.18a	3.25±0.06a	2.8±0.53a
LACT (%)	4.77±0.05a	3.99±0.89a	4.63±0.05a	4.35±0.85a	4.81±0.05a	4.71±0.21a	4.58±0.26a	4.74±0.1a	4.68±0.28a	4.03±0.78a	4.73±0.10a
SCC (K SC/mL)	<90a	<90a	<90a	<90a	<90a	<90a	<90a	110.3±37a	<90a	<90a	<90a

* Lowercase letters between columns differ by Tukey's test ($p < 0.05$)

According to Cunha et al. (2016), milk from collective tanks showed predominantly higher bacterial counts when compared to milk from individual tanks throughout all seasons. Mastitis is responsible for altering the physical, Chemical, and microbiological patterns of milk and the mammary gland, resulting in a reduction in fat levels (Langoni, 2000), and making it difficult to uniformize the quality of the milk of those who share the same tank.

Also in this study, a positive correlation was found between fat and protein, as well as the correlation that occurred between fat and lactose, reaching values of 0.438 and 0.432, respectively (Table 2). Protein is one of the components that suffers the most variability due to environmental and nutritional factors, second only to fat. As the protein content of milk increases, there is an increase in its total production, a behavior that is not repeated on the same scale with fat when production increases (Carvalho, 2002), thus accusing the low relationship between both.

Lactose and protein showed a positive and significant correlation of 0.988 (Table 2). Among the proteins present in cow's milk is α-lactalbumin, corresponding to 2.5% of the total proteins and acts as one of the subunits of the enzyme lactase-synthetase responsible for the synthesis of lactose (González and Noro, 2011), making it the high significance of this correlation is notable.

Table 2. Pearson's correlation between the parameters of fat, protein (PROT, %), lactose (LACT, %) and somatic cell count (SCC, K SC/mL of milk)

Parameters	FAT (%)	PROT (%)	LACT (%)
PROT (%)	0.438*		
LACT (%)	0.432*	0.988*	
SCC (K SC/mL)	0.333 [NS]	0.102 [NS]	0.091 [NS]

*There was correlation; [NS] There was no correlation

These changes in milk can occur through the addition of water, alkalis to mask the acidity, preservatives, density and cryoscopic index reconstitution, the presence of antibiotic residues and milk skimming (Kartheek et al., 2011). All these maneuvers are intended to increase milk volume and the addition of substances is carried out to cover changes in density, acidity, cryoscopy, and other indices, so that the milk is within the parameters required by legislation.

In a second study, the milk quality parameters were evaluated according to current legislation (NI 76/2018), identifying whether there are changes in the milk collected in shared tanks. Analyze of pH, Dornic acidity, density, alcohol test, alizarol test and density restorative test (chloride and starch) were carried out following the indications made by the Adolfo Lutz Institute (2008) and the Ministry of Agriculture, Livestock and Supply (Brazil, 2018).

Quantitative analyzes demonstrated a lack of uniformity in the milk that is delivered to collective tanks, with cases of producers reaching pH values of 4.95, with Dornic acidity of 61.33°D and situations of samples with a pH of 6.86, Dornic acidity 14°D and density of 14° Quenneve (ºQ), being completely outside the borderline values delimited by Brazilian legislation (Table 3). However, in Brazil, according to NI 76/2018, released by the Ministry of Agriculture, Livestock and Supply (BRASIL, 2018) and according to the methodology adopted by the Adolfo Lutz Institute (2008), refrigerated raw milk must meet certain values of reference limits for quantitative analyses, therefore, it should remain with a pH around 6.4 and 6.8, acidity by the Dornic method between 14°D and 18°D and the Quenneve scale for density with indices between 28 at 32°, when analyzing the quantitative parameters of milk quality.

In addition, for the qualitative variables, none of the tanks reached the ideal color in the alizarol test and the three tanks had producers with alcohol test instability, with coagulation, affecting the quality of the milk in the tank (Figure 1).

As for the different milking methods, the collection and analysis of 14 milk samples from rural producers who deliver milk to an association of small rural producers in the Tupã region was carried out and, using electronic equipment EKOMILK

SCAN Somatic Cells Analyzer® and EKOMILK M®, based on the results of the concentration of total proteins, fat and somatic cell count, it was verified whether the samples presented differences in quality according to the milking method performed by the producer (manual or mechanized). The values obtained were submitted to an ANOVA analysis of variance, and Tukey's test was applied, with 95% confidence, detecting that there was no significant difference for the levels of somatic cell count and protein between the types of milking, and that there was difference for fat levels, in which milk from manual milking had higher levels ($p<0.05$), being a viable option for the production of cheese and dairy products, due to the higher concentration of total solids (Table 4) .

Table 3. Analysis of variance of quantitative parameters (pH, Dornic acidity and density) of milk quality from collective tanks to verify adulterations in milk

Tank	pH		Dornic Acidity (°D)		Density (°Q)	
Producers Tank 1						
Tank1	6.65±0	A	16.67±0.58	G	31±0	A
T1PA	6.45±0.23	ABC	20.67±1.16	ABCDE	33±0	B
T1PB	6.62±0.07	AB	15.67±0.58	G	29±0	B
T1PC	6.57±0.03	AB	18±0	CDEFG	30±0	B
T1PD	6.57±0.01	AB	17.33±0.58	EFG	31±0	B
T1PE	6.42±0.01	ABC	20.67±3.06	ABCDE	30±0	B
T1PF	6.43±0.01	ABC	19±1	ABCDEFG	30±0	B
T1PG	6.55±0.05	AB	21±0	ABCD	33±0	B
T1PH	6.48±0.05	AB	16.33±1.53	G	28±0	B
T1PI	6.5±0.01	AB	18±0	CDEFG	29±0	B
T1PJ	6.34±0.01	BCD	20.33±0.58	ABCDEF	30±0	B
T1PK	6.17±0.02	CD	22.33±0.58	A	30±0	B
T1PL	6.64±0.02	A	16.67±1.16	G	31±0	B
T1PM	6.52±0.1	AB	17.67±1.16	DEFG	29±0	B
T1PN	6.45±0.09	ABC	17.67±0.58	DEFG	32±0	B
T1PO	6.58±0.01	AB	18.33±2.08	BCDEFG	31±0	B
T1PP	6.58±0.02	AB	17.67±0.58	DEFG	30±0	B
T1PQ	6.13±0.03	D	21.67±1.53	AB	31±0	B
T1PR	6.48±0.02	AB	21.33±1.16	ABC	29±0	B
T1PS	6.18±0.3	CD	18.67±0.58	BCDEFG	26±0	B
T1PT	6.68±0.01	A	17±0	FG	32±0	B
Producers Tank 2						
Tank2	6.45±0	D	16±0	CD	25±0	F

continues on following page

Table 3. Continued

Tank	pH		Dornic Acidity (°D)		Density (°Q)	
T2PA	6.62±0	B	17±1	CD	29±0	C
T2PB	6.52±0	C	17±1	CD	28±0	D
T2PC	6.66±0	C	17±2	CD	30±0	B
T2PD	6.62±0	C	16.33±0.58	CD	26.33±0.58	E
T2PE	6.65±0	C	16.67±0.58	CD	32±0	A
T2PF	6.68±0	C	9.33±0.58	E	17±0	H
T2PG	6.68±0	C	15.33±0.58	D	29±0	C
T2PH	6.86±0	A	5±0	F	14±0	I
T2PJ	6.78±0	A	10±0	E	18±0	G
T2PK	6.63±0	B	15±0	D	29±0	C
T2PL	6.54±0	D	17.33±0.58	CD	28±0	D
T2PM	6.67±0	B	18±0	BC	30±0	B
T2PN	6.29±0	D	23±1.73	A	29±0	C
T2PO	6.55±0	D	20±0	B	30±0	B
Producers Tank 3						
Tank3	6.49±0	I	19.67±0.58	CDE	29±0	F
T3PA	6.58±0.01	DE	14.33±1.16	H	31±0	D
T3PB	6.64±0.01	C	15±0	GH	33±0	B
T3PC	6.65±0.01	C	15.67±0.58	FGH	30±0	E
T3PD	6.54±0	G	18±1	DEFGH	34±0	A
T3PE	6.67±0.01	A	16.33±1.53	EFGH	30±0	E
T3PF	6.64±0.01	C	17.67±0.58	DEFGH	32±0	C
T3PG	6.58±0.01	DE	61.33±1.53	A	30±0	E
T3PH	6.68±0	A	14.67±1.53	GH	30±0	E
T3PI	4.95±0.01	N	61.33±1.53	A	33±0	B
T3PJ	6.64±0.01	C	16.33±0.58	EFGH	29±0	F
T3PK	6.52±0.01	H	19±0	CDEF	30±0	E
T3PL	6.56±0.01	EF	20.33±0.58	CD	28±0	F
T3PM	6.66±0.01	AC	16.33±0.58	EFGH	27±0	G
T3PN	6.32±0.01	L	22.33±1.53	BC	29±0	F
T3PO	6.38±0	K	21.33±0.58	CD	25±0	G
T3PP	6.59±0	D	19.33±0.58	CDEF	31±0	D
T3PQ	6.55±0.01	FG	19±1	CDEF	29±0	F
T3PR	6.56±0.01	EF	16.33±0.58	EFGH	30±0	E
T3PS	6.56±0.01	EF	18.33±0.58	DEFG	31±0	D
T3PT	6.06±0.01	M	25.33±3.79	B	32±0	C
T3PU	6.47±0.01	J	20.67±0.58	CD	31±0	D

Means that do not share the same letter in the columns are significantly different by Tukey's test of means (P<= 5%).
*Subtitle: TxPy, where Tx is the tank (x= 1 to 3), Py is the producer who delivers to the tank (y= A to U).

Figure 1. Frequency of occurrences for the qualitative analyzes (alizarol test and alcohol test) of milk from small producers who deliver in collective tanks

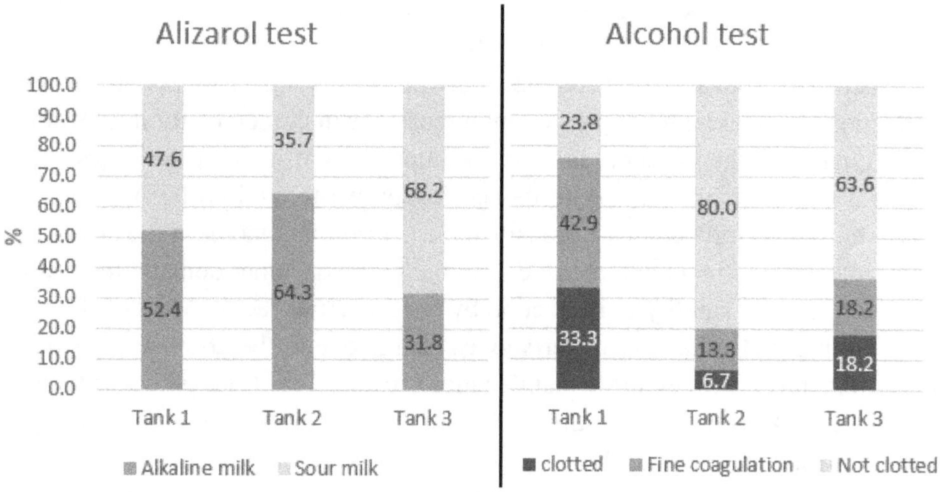

Table 4. Means and standard deviation (ANOVA) of protein, fat and somatic cell count (SCC) levels of milk samples from different milking systems

Variable analyzed	Milking System			
	Mechanical Milking		Manual Milking	
Protein (%)	3.35	±0.33a	3.48	±0.44a
Fat (%)	3.98	±0.72b	4.46	±0.54a
SCC (K SC/mL of milk)	344.68	±187.42a	355.81	±297.45a

*Means that do not share the same letter between columns differ from each other by the Tukey's test means at 95% confidence.

FACTORS AFFECTING MILK QUALITY

Data on milk production and consumption have been positive over time. On the other hand, there are still challenges to be overcome, such as the low quality of the milk and the low volume produced per property (Defante et al., 2019; Bánkuti et al., 2020), which indicates the demand for new projects, that provide technical assistance to milk producing families, in addition to diagnosing points that should be improved on the properties (Rosa et al., 2017).

With the high demand for milk, small actions, such as hygiene and handling, can induce an improvement in the quality of the final product (Santo et al., 2017). Dairy products often show interest in quality milk and are willing to pay a different price to purchase the product. However, it is necessary to take some precautions for this to happen, such as managing people on the properties, providing efficiency in the management processes, cleaning the milking environment and equipment, resulting in the desired quality (BELOTI et al., 2020).

Scalco and Souza (2006), pointed out the quality management in the milk production chain, as being of paramount importance in the search for improvement in the quality of the product, and for the improvement in the procedures of the activities of the chain, seeking to reduce losses, waste, and costs. According to Reis et al. (2007) milk quality is considered a determining factor in the production chain, intensifying more and more, due to the high consumer demand for quality, generating greater rigidity in inspection by the industrial sector. Most of the time, the low quality of milk and its derivatives occurs due to the absence or errors in animal handling practices, in the maintenance and cleaning of equipment, in hygiene during the milking process, in the refrigeration of the milk and in the training of the labor (Santana et al., 2001).

Consequently, the number of investments in the sector is low, which causes problems throughout the chain, highlighting the low training of employees, low use of technologies, lack of sanitary control and adequate hygienic conditions in the milking process, in storage and in transport. Factors, these, that compromise the commercialization of the raw material by the small producer, who is often forced to resort to informal sales of the product (Molina et al., 2015).

Good hygiene practices in milking are an alternative to improve milk quality, resulting in better prices paid to the producer (Defante et al., 2019). They are the initial step to obtain a good microbiological quality of the product, since the moment of milking is one of the main entrance doors of the microorganisms in the milk, thus compromising the physicochemical characteristics of the product, due to the high index of contamination (Beloti, 2015), in addition to ensuring food safety and meeting the requirements imposed by federal legislation, milk quality control also makes the product more competitive, helping it to remain in the market (Rosa et al., 2017). Even providing benefits, the success of good practices can be compromised by the lack of credibility in their short-term results, since producers may fail to implement them because they do not believe in their benefits (Scalco and Souza, 2006).

The main sources of contamination by bacteria in the milking process are the milking equipment, cooling tank, teats, and the udder. In addition, factors such as the health of the mammary gland and the sanitary conditions of the facilities where milking occurs, also cause microbial contamination (Molineri et al., 2012). According to Almeida et al. (2013), improving the quality of milk helps in several points such as: food safety, increasing the shelf life of processed products, complying with legislation, increasing industrial yield and profitability for producers and for the company. industry. Also noteworthy is the use of good milking practices, as an efficient aspect in the search for milk quality, which occurs mainly due to the simplicity of its implementation, which can also be used in small-scale properties.

FINAL CONSIDERATIONS

It is essential to provide guidance and technical information among agents of the means of production, owners and employees who are related to creation. This action brings important benefits to disseminate knowledge, confidence and leadership, important factors for successfully performing tasks (Rojo-Gimeno et al., 2016; Tolosa et al., 2016). Educational actions allow the professional to see the need for change, exposing which factors are important to fulfill these needs, such as management techniques and agricultural technologies in constant change (Zuin, 2008).

Stimulation and constant interaction between technicians and producers are positive guidelines that offer a mutation of information and experiences that develop greater knowledge, a necessary factor in modifying the milking process, where all parties are involved in the process (Kilpatrick, 2000; Valeeva, Lam, Hogeveen, 2007; Sembada et al., 2016). By the way, the guidelines make it possible to improve the productivity of the property and especially its income. Proper management and good practices prevent mastitis by avoiding contamination of milk by chemicals and antibiotics, factors that benefit cattle performance (Jansen et al., 2009; Matsubara et al., 2011; Zareie, Zamani, Karami, 2016).

The short period of guidance and training given to producers can influence reductions in the standard plate count (SPC) in milk in such a way that they are minimized, factors that impair production and demonstrate that this training, although important, needs to be longer lasting or better distributed (Tolosa et al., 2016; Ponsano et al., 2011). Even with the divergence of results in the reduction of SPC where, in properties, it was obtained and in others not, the reflection of the application of good practices has good effects, where the producer seeks to understand and apply correct methods in the production process.

Summarizing, it is very important to apply frequent guidelines to milk producers, as these influence quality production, offering adequate time for training, improving efficiency, and providing producers with content that will be more fixed, giving time for development, application and emergence of doubts that can be solved by studying case by case, and property by property.

REFERENCES

Almeida, T. V. Parâmetros de qualidade do leite cru bovino: contagem bacteriana total e contagem de células somáticas. *Dissertação* (Mestrado em Ciência Animal) – Universidade Federal de Goiás, Goiânia, 2013.

Assis, J., Ferreira, J. D., Martins, H. H., & Schneider, M. B. (2016). Cadeia produtiva do leite no Brasil no contexto do comércio internacional. *Revista de Ciências Empresariais*, *17*(1), 63–93.

Bánkuti, F. I., Prizon, R. C., Damasceno, J. C., De Brito, M. M., Pozza, M. S. S., & Lima, P. G. L. (2020). Farmers'actions toward sustainability: A typology of dairy farms according to sustainability indicators. *Animal*, *14*(2), 417–423. doi:10.1017/S1751731120000750 PMID:32290889

Beloti, V. (2015). *Leite: Obtenção, Inspeção e Qualidade*. Editora Planta.

Beloti, V., Dias, J. A., & Oliveira, A. M. Ordenha e boas práticas de produção. In: Pecuária Leiteira na Amazônia, Embrapa. Brasília, DF: Embrapa, p. 105-130, 2020.

Brasil, R. B., Silva, M. A. P., Carvalho, T. S., Cabral, J. F., Nicolau, E. S., & Neves, R. B. S. (2012, December). *Revista do Instituto de Laticínios Cândido Tostes*. Avaliação da qualidade do leite cru em função do tipo de ordenha e das condições de transporte e armazenamento. *Santa Terezinha*, *67*(389), 34–42.

Carvalho, G. F. (2002). Milk yield, somatic cell count and physicochemical characteristics of raw milk collected from dairy cows in Minas Gerais state. In: Congresso Panamericano de Qualidade do Leite e Controle da Mastite. Ribeirão Preto.

Cunha, A. F. (2010). Evolução anual da qualidade do leite cru de tanques individuais e comunitários do Vale do Rio Doce (MG). *Boletim do Centro de Pesquisa de Processamento de Alimentos*, [S.l.], *34*(1).

Damasceno, F. A. (2012). Compost bedded pack barns system and computational simulation of airflow through naturally ventilated reduced model.

Defante, L.; Damasceno, J. C.; Bánkuti, F. I.; Ramos, C. E. C. D. O. (2019). Typology of dairy production systems that meet Brazilian standards for milk quality. *Revista Brasileira de Zootecnia*.

EMBRAPA. Anuário Leite 2020. 1. Ed., Embrapa gado de leite, Elyon, 2020.

Gomes, S. T. (1999). Diagnóstico e perspectivas da produção de leite no Brasil. In: Restrições técnicas, econômicas e institucionais ao desenvolvimento da cadeia produtiva do leite no Brasil, Brasília, p. 19-35.

González, F. H. D., Noro, G. (2011). Variações na composição do leite no subtrópico brasileiro. In: González, F.H.D.; Pinto, A.T.; Zanella, M.B.; Fischer, V.; Bondan, C. Qualidade do leite bovino: variações no trópico e no subtrópico, Passo Fundo: UPF Editora, 2, 28-53.

Instituto Adolfo Lutz (São Paulo). (2008). *Métodos físico-químicos para análise de alimentos /coordenadores Odair Zenebon, Neus Sadocco Pascuet e Paulo Tiglea — São Paulo*. Instituto Adolfo Lutz.

Jansen, J. B. H. P., Van Den Borne, B. H. P., Renes, R. J., Van Schaik, G., Lam, T. J. G. M., & Leeuwis, C. (2009, November). Explaining mastitis incidence in Dutch dairy farming: The influence of farmers' attitudes and behavior. *Preventive Veterinary Medicine, Amsterdam, 92*(3), 210–223. doi:10.1016/j.prevetmed.2009.08.015 PMID:19800700

Kartheek, M., Smith, A. A., Muthu, A. K., & Manavalan, R. (2011). Determination of adulterants in food: A Review. *Journal of Chemical and Pharmaceutical Research, 3*(2), 629–636.

Kazancoglu, Y., Ozkan-Ozen, Y. D., & Ozbiltekin, M. (2018). Minimizing losses in milk supply chain with sustainability: An example from an emerging economy. *Resources, Conservation and Recycling, 139*, 270–279. doi:10.1016/j.resconrec.2018.08.020

Kilpatrick, S. (2000). Education, and training impacts on farm management practice. *Journal of Agricultural Education and Extension, Abingdon, 7*(2), 105–116. doi:10.1080/13892240008438811

Langoni, H. (2000). Tendências de modernização do setor lácteo: Monitoramento da qualidade do leite pela contagem de células somáticas. *Revista Educação Continuada. CMRV-SP., 3*, 57–64.

Lopes, M. A., & Santos, G. (2012). Comparativo econômico entre propriedades leiteiras em sistema intensivo de produção: Um estudo multicasos. *Revista Brasileira de Saúde e Produção Animal, 13*(3), 591–603. doi:10.1590/S1519-99402012000300001

Marion, J. C. (2007). *Contabilidade Rural: Contabilidade Agrícola, contabilidade da Pecuária, Imposto de Renda*. Pessoa Jurídica.

Marion, J. C., & Segatti, S. (2010). *Contabilidade da pecuária. 9*. Atlas.

Matsubara, M. T., Beloti, V., Tamanini, R., Fagnani, R., Da Silva, L. C. C., Monteiro, A. A., & Barros, M. D. A. F. (2011). Good practices in milking to reduce microbiological contamination of milk in agreste of Pernambuco. Semina. Ciencias Agrarias, Londrina.

Ministro Do Estado Da Agricultura Pecuaria E Abstecimento. (2018). *Mapa 76 e 77*. Disponível em: https://saudeinspecaoanimal.com.br/in76-77.

Molina, C. H. A., Centenaro, G. S., & Furlan, V. J. M. (2015). Qualidade do leite cru comercializado informalmente no município de Itaqui-RS. *Vigilância Sanitária em Debate: Sociedade. Ciência & Tecnologia*, *3*(4), 106–113.

Molineri, A. I., Signorini, M. L., Cuatrín, A. L., Canavesio, V. R., Neder, V. E., Russi, N. B., & Bonazza, J. C. (2012). Calvinho, L.F. Association between milking practices and psychrotrophicbacterial counts in bulk tank milk. *Revista Argentina de Microbiologia*, *44*, 187–194. PMID:23102468

Philpot, N. W., & Nickerson, S. C. (2002). *Vencendo a luta contra a mastite*. Westfalia Surge/Westfalia Landtechnik do Brasil.

Pinheiro, F. E.; Lopes, M. A.; Mendes, J. R. R. (2015). Efeito do pagamento por qualidade do leite na rentabilidade da atividade leiteira. *Revista do Instituto de Laticínios Cândido Tostes*, [S.l.], *70*(1), 24-34.

Ponsano, E. H. G., Pinto, M. F., Grassi, T. F., Avanço, S. V., & Lima, L. K. F. (2011). Capacitação de produtores rurais para a melhoria da qualidade do leite cru produzido na região de Araçatuba-SP. *Revista Ciência em Extensão*, *7*(1), 91–101.

Prado, R. M., Paulin, M. F., Prado, I. N., Santos, G. T., Benchaar, C., & Petiti, H. V. (2016). Milk yield, milk composition, and hepatic lipid metabolismo in transition dairy cows fed flaxseed or linola. DOI: . 11. ed., Sherbrook, American Dairy Science Association, 99, 8831-8846. doi:10.3168/jds.2016-11003

Reis, G. L., Alves, A. A., Lana, A. M. Q., Coelho, S. G., Souza, M. R., Cerqueira, M. M. P., Penna, C. F. A. M., & Mendes, E. D. M. (2007). Procedimentos de coleta de leite cru individual e sua relação com a composição físico-química e a contagem de células somáticas. Ciência Rural, Santa Maria.

Rojo-Gimeno, C., Postma, M., Dewulf, J., Hogeveen, H., Lauwers, L., & Wauters, E. (2016). Farm-economic analysis of reducing antimicrobial use whilst adopting improved management strategies on farrow-to-finish pig farms. *Preventive Veterinary Medicine*, *129*, 74–87. doi:10.1016/j.prevetmed.2016.05.001 PMID:27317325

Rosa, J. F., Domingues, L. D. A. S. P., Silva, R. H., De Lima, H. G., & Cereser, N. D. (2017, January-June). Pontos críticos de contaminação na produção leiteira. *Expressa Extensão*, *22*(1), 90–103. doi:10.15210/ee.v22i1.7808

Santana, E. H. W., Beloti, V., Barros, M., Moraes, L. B., Gusmão, V. V., & Pereira, M. S. (2001). de A. F.; Moraes, L. B.; Gusmão, V. V.; Pereira, M. S. Contaminação do leite em diferentes pontos do processo de produção: I. Microrganismos aeróbios mesófilos e psicrotróficos. *Semina. Ciências Agrárias*, *22*(2), 145–154. doi:10.5433/1679-0359.2001v22n2p145

Santo, R. S. E. et al. (2017). Fatores e técnicas de produção e sua influência na produtividade e qualidade do leite bovino.

Santos, J. G., Marion, J. C., & Segatti, S. (2008). *Administração de Custos na Agropecuária* (Vol. 3). Atlas.

Scalco, A. R.; Souza, R. C. (2006). Qualidade na cadeia de produção de leite: diagnóstico e proposição de melhorias. *Organizações Rurais & Agroindustriais*, *8*(3).

Sembada, P., Duteurtre, G., Purwanto, B. P., & Suryahadi. (2016). Improved milk production performance of smallholder farms in West Java (Indonesia). *Tropical Animal Health and Production*, *48*(4), 793–799. doi:10.100711250-016-1029-2 PMID:26969314

Sparemberger, A., Büttenbender, P. L., Zamberlan, L., & Hofler, C. E. (2009). *Inovações tecnologicas nas cadeias do agronegócio de alimentos da região fronteira noroeste do Rio Grande do Sul*. COINI - Congreso Argentino de Ingeniería Industria.

Spink, J., & Moyer, D. C. (2011). Defining the Public Health Threat of Food Fraud. *Journal of Food Science*, *76*(9), R157–R163. doi:10.1111/j.1750-3841.2011.02417.x PMID:22416717

Tolosa, T., Verbeke, J., Piepers, S., Tefera, M., Getachew, Y., Supré, K., & Devliegher, S. (2016). *Milk production, quality, and consumption in Jimma (Ethiopia): facts and producers', retailers', and consumers' perspectives* (Vol. 124). Preventive Veterinary Medicine.

Valeeva, N. I., Lam, T. J. G. M., & Hogeveen, H. (2007). Motivation of dairy farmers to improve mastitis management. *Journal of Dairy Science*, *90*(9), 4466–4477. doi:10.3168/jds.2007-0095 PMID:17699068

Viana, E. (2021). Sistemas de produção na pecuária leiteira. *Esteio Gestão Agropecuária*. . Disponível em: https://esteiogestao.com.br/sistemas-de-producao-na-pecuaria-leiteira/.

Willers, C. D. (2014). *Avaliação dos impactos ambientais da pecuária de corte semi-intensiva.*

Zanin, A. (2015). Apuração de custos e resultado econômico no manejo da produção leiteira: Uma análise comparativa entre o sistema tradicional e o sistema freestall. *Organizações Rurais & Agroindustdddddriais. Lavras, 17*(4), 431–444.

Zareie, R., Zamani, G. H., & Karami, E. (2016). The effect of dairy farmers' technical knowledge on milk bacterial count. *Academic Web Journal of Agricultural Research, Delta, 1*(1), 9–13.

Zuin, L. F. S., & Zuin, P. B. (2008). *Produção de Alimentos Tradicionais: Extensão Rural.* Idéias & Letras.

Chapter 10
Applications of Cheese Whey in Dairy Production Chains:
Waste to Raw Material of High Value-Added Products

Mayara Mari Murata
University Pitágoras Unopar Anhanguera, Brazil

Luiz Rodrigo Ito Morioka
University Pitágoras Unopar Anhanguera, Brazil

Josemeyre Bonifacio da Silva Marques
University Pitágoras Unopar Anhanguera, Brazil

Hélio Hiroshi Suguimoto
University Pitágoras Unopar Anhanguera, Brazil

EXECUTIVE SUMMARY

Cheese whey is considered the main by-product obtained by the dairy industry and it is derived from coagulation during cheese production. Half of cheese whey production is used in well-established processes, like production of fresh products, animal feed, and a source of high-added-value compounds, such as whey protein and β-galactosidase. Unfortunately, the use of cheese whey as raw material requires expensive processes. Therefore, a high proportion of this product is wasted and discarded into industrial effluent treatment stations. Different research groups have been studying potential strategies to valorize cheese whey. The goal of this chapter is to present the main characteristics of cheese whey and introduce the possibilities of using this by-product for new biotechnological purposes. The challenge is the arrangement of these processing systems with the players involved in the dairy industry. A joint action between them can turn cheese whey waste into high-added-value products producing zero dairy waste and honoring environmental legislation and public policies.

DOI: 10.4018/978-1-6684-5472-5.ch010

INTRODUCTION

Cheeses are traditional products widely consumed throughout the world and in 2021, their production overcame 500 million tons (USDA, 2022). This number has been growing, since consumption of dairy products is increasing steadily worldwide (Pires et al., 2021; USDA, 2022). During cheese production, casein is used as a basic cheese component, while soluble whey proteins are separated into liquid whey waste. Therefore, cheese whey is considered the main by-product obtained by the dairy industry (Panghal et al., 2017).

Cheese whey composition depends on numerous factors such as milk origin, cheese production process, amount of water and addition of sanitizing agents. Regardless of the process, this by-product has proven to contain potential ingredients for the development of food products with improved nutritional characteristics and other functionalities (Pires et al., 2021). Therefore, from the total cheese whey production, 50% is currently being used in well-established processes, for example: 1) inside the cheese factories to produce fresh products (such as ricotta and whey butter), 2) to feed farm animals and 3) a source of high-added-value compounds for the food or pharmaceutical industries, mainly proteins of high biological value (whey protein) and lactose (Murari et al., 2019; Carrillo-Lopez et al., 2021).

Unfortunately, the use of cheese whey as raw material requires expensive processes such as ultrafiltration, evaporation, and fermentation, which are rarely adopted by small and medium-sized enterprises that cannot afford these costs. Therefore, a high proportion of cheese whey is still wasted and discarded (Bolognesi et al., 2021).

Due to the increase in food consumption and to the severe environmental regulations, the management of food waste and by-products is a challenge for the agri-food industries (Pires et al., 2021). The high level of lactose makes cheese whey a highly polluting element (Carvalho et al., 2013). This characteristic adversely impacts aquatic life due to the decrease of oxygen availability in water. Overtime, it causes a rapid consumption of oxygen, eutrophication, formation of foam, salinization, and acidification. In the soil, this industrial effluent can alter the physical and chemical composition of soil, resulting in decreased crop yield (Ahmad et al., 2019).

As the limiting factor for whey physic-chemical and biological treatment is its high cost for the industry, especially small and medium-sized producers, turning waste use practically unfeasible (De Almeida et al., 2021). In this context, different research groups have been studying potential strategies to valorize cheese whey and attract industries to invest in other whey-based products. These initiatives have a potential to turn this environmental problem into high value-added ecological products or by-products in food, fuel, health, pharmaceuticals, and plastic industries (Murari et al., 2019; Chalermthai et al., 2021; Argenta et al., 2021; Pires et al., 2021). In addition, the commitment of the entire dairy production chain towards a sustainable use of cheese whey in compliance with current environmental policies can contribute to zero waste disposal (Romani et al., 2018).

Finding lucrative alternatives for cheese whey applications are opportunities to solve environmental waste pollution problems. The goal of this chapter is to present the main characteristics of this important by-product from cheese production and introduce the possibilities of using cheese whey for other biotechnological purposes, reducing the impact of its disposal on nature, and developing competitive bioproducts and bioprocesses. In the future, several products with greater added value can be developed and cheese whey waste will decrease in the cheese productive chain.

CHEESE INDUSTRY: OVERVIEW

In 2021, the total fluid milk produced by the major producers in the world reached 539,6 million tons. Milk production is dominated by two important regions, the European Union (EU) and the United States (US). Other important milk producers are India, China, Russia, Brazil, Mexico, and Ukraine. New Zealand, Argentina and Australia are considered major milk exporters (USDA, 2022). Milk is internationally traded mainly in the form of processed dairy products (OECD, 2021). China is the major importer of dairy products and Japan, Russia, Mexico, the Middle East and North Africa are also considered important net importers of dairy products (OECD, 2021). The EU produced 160 million tons of milk in 2020. From this total, 91% of raw milk was delivered to dairies to produce other products such as fresh drinking milk, cheese, butter, powder milk, whey, and others (EUROSTAT, 2020). Two thirds of all whole milk available to dairies in the EU is used to make cheese and butter. In average, 16.7 million tons of skimmed milk together with 57.6 million tons of whole milk were used to produce 10.3 million tons of cheese (EUROSTAT, 2020).

In 2020, Brazil produced 37.21 million tons of milk (IBGE, 2021) and is considered a traditional dairy importer. Due to the pandemic, imports have increased significantly and in 2020, Brazil imported 174.24 thousand tons of dairy products, including 31.02 thousand tons of cheese (Ximenes, 2021). Brazil has an internal cheese production that reached 760 thousand tons of cheese in 2021 (USDA, 2022).

Fresh milk is preferably consumed (42.9%) worldwide, and among dairy products, cheese is the most preferred (25.2%), followed by butter (23.1%), skimmed milk powder (5.1%), and whole milk powder (3.7%) (FAO, 2015; Santos et al., 2017). According to these data, cheese is the dairy product category accounting for the largest percentage of the U.S. milk-fat supply, and that percentage has been growing, since cheese consumption rose in the past two decades (USDA, 2022). This increasing demand for cheese production reflects worldwide and it has led to the surplus of cheese whey as the major by-product.

CHEESE WHEY: WASTE FROM CHEESE INDUSTRY

During the cheese-making process, milk is concentrated by the addition of acids or proteolytic enzymes, which promote the coagulation of casein, the main milk protein, and lead to the formation of curds. Part of the solid components, such as proteins and fats, are concentrated in the curd, while the whey proteins, lactose and soluble solids are removed in the cheese whey (Kindstedt, 2014).

The coagulation of casein by the addition of acids, promote the milk pH drops to 4.6, which is the isoelectric point of casein. The accumulation of hydrogen ions neutralizes the polarity of the surfaces of casein micelles, making them incapable of interacting with water molecules. The micelles are then forced to interact with each other, resulting in the formation of aggregates and micelle chains. As the coagulation progresses, the micellar chains increase in length and intertwine to form a three-dimensional network-like matrix that traps water and the solid components (mainly lactose, fat, whey proteins, and salts) of milk (Kindstedt, 2014) (Figure 1).

Figure 1. Schematic representation of the acidic coagulation process

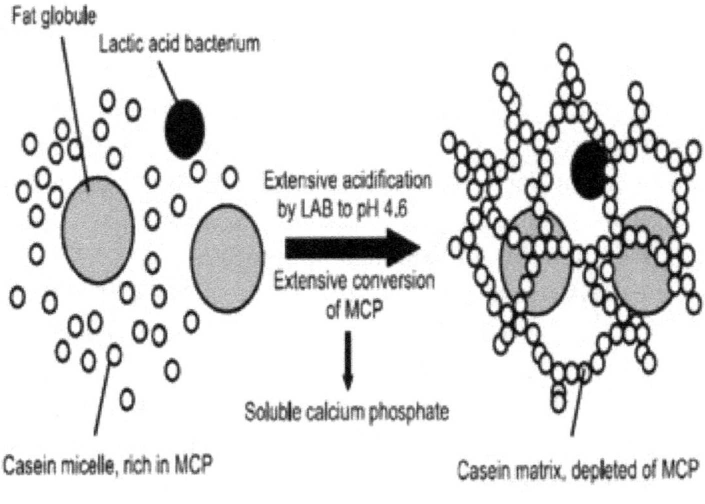

Milk is then transformed from a liquid state to a soft, fragile gel, or curd. The casein matrix that gives rise to the curd is highly demineralized because most of the micellar calcium phosphate (MCP) is dissolved by the high concentration of lactic acid and the low pH required to initiate curding (Kindstedt, 2014; Jelen, 2003).

In enzymatic coagulation, casein micelles are coagulated by the addition of rennet, whose main enzyme is chymosin. Coagulation occurs in two steps: hydrolysis of the casein micelle stabilizing protein called κ-casein, and aggregation and gelation of micelles modified by rennet (Fox et al., 2017). κ-casein is highly sensitive to chymosin and is therefore hydrolyzed when rennet is added to milk. This hydrolysis occurs specifically in the linkage between the amino acids phenylalanine at position 105 (Phe105) and methionine at position 106 (Met106). κ-casein molecules form the surface layer of the casein micelle (Figure 2), stabilizing this particle and releasing caseinomacropeptide (CMP). This process causes the micelles to aggregate in the form of a matrix rich in micellar calcium phosphate (MCP).

Figure 2. Schematic diagram of the enzymatic coagulation process

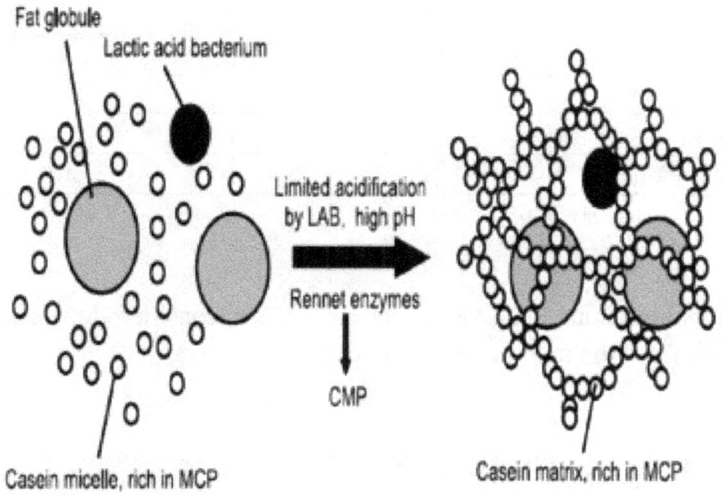

The yellowish-green whey is the watery portion that separates from the mass formed during the coagulation of milk for the cheese manufacture, and which retains about 55% of the milk's nutrients (Addai et al., 2020).

The type and composition of cheese whey basically depend on the techniques that are used to remove casein from liquid milk in addition to the source of milk (beef, goat, sheep, among others) (Panesar et al., 2007). Thus, when casein removal is performed by rennet at pH 6.5, this type of whey is referred as sweet whey. Acid whey is when the removal occurs by fermentation processes or addition of organic or mineral acids, as in the manufacture of fresh cheese or in most industrial caseins at pH < 5.

In general, cheese whey contains 93% water, and its protein fraction contains 65% β-lactoglobulin, 25% α-lactoalbumin, immunoglobulins, protease peptones and bovine serum albumin (BSA) (Fox et al., 2017). In smaller amounts, cheese whey contains lactoferrin, lysosome, glycomacropeptide, phospholipoprotein and lactoperoxidase. Its constitution also includes mineral salts, lactic and citric acid, and B-complex vitamins. The differences between these two types, sweet and acid, are the mineral content, acidity and composition of the whey protein fraction (Table 1).

Table 1. Chemical composition of sweet and acid cheese whey

Component	Sweet whey (g/L)	Acid whey (g/L)
Total solids	63-70	63-70
Lactose	46-52	44-46
Protein	6-10	6-8
Calcium	0.4-0.6	1.2-1.6
Phosphate	1-3	2-4.5
Lactato	2	6.4
Cloride	1.1	1.1

Source: Adapted from: Jelen (2003)

Despite its high nutritional value, cheese whey is considered the main residue or by-product from the manufacture of cheese or casein.

CHEESE WHEY GLOBAL MARKET

Depending on the cheese manufacturing, it is estimated that producing 1 kg of cheese generates 6 to 9 L of cheese whey (Pires et al., 2021). Worldwide cheese whey production is around 160 million tons per year (Okamoto et al., 2019) and it is estimated that each year, there is an increase of 1 to 2% in the cheese whey annual production. This residue is rich in proteins and lactose, containing about 1.5 million tons of proteins of high nutritional value and 8.6 million tons of lactose (Panghal et al., 2017).

In the past, 50% of the total cheese whey produced was transformed into various food products, in which 45% was used directly in the liquid form or as powdered whey, 15% as lactose and lactose-free by-products, and 40% as whey protein concentrates. Nowadays, cheese whey can also be reused within the cheese industry to produce other fresh products such as ricotta cheese and whey butter (Yadav et al., 2015).

As a value-added product, cheese whey can be transformed into different fractions, such as whey powder, whey protein concentrate, whey protein isolate, whey permeate and lactose (Barba, 2021). One third of cheese whey produced in the world is processed into whey powder that is used in various products in food industry. The other fractions are also used in the food and pharmaceutical industries (Çelik et al., 2020).

In animal feed, it provides high quality proteins and lactose as energy sources in addition to calcium, phosphorus, sulfur and water-soluble vitamins (Sienkiewicz & Riedel, 1990). The in natura use for soil irrigation is a frequent practice due to the ability of plant growth stimulation and nitrogen fixation through rhizobian activity, improving soil structure by increasing water retention efficiency (Erman et al., 2011). However, both in animal feed and soil application, there are problems with cheese whey use, since excess of lactose can cause digestion problems in animals and its dispersion in the soil leaves high saline deposits impairing fertility and vegetable growth.

The implementation of bioprocessing systems using cheese whey as a raw material for the production of functional or health-promoting foods, leading to the production of bioproteins, enzymes, carotenoids, biopreservatives, biological gums, exopolysaccharides, microbial oil and bioplastics contribute to adding value to this product (Murari et al., 2019). Due to its functional and technological properties, such as gelling capacity, foaming and emulsifying capacity, it also allows its application as an ingredient in various foods (Carrillo-Lopez et al., 2021).

Cheese whey has been gaining ground in the cosmetics industry, since technological processes are under development to encourage this industry to increase the use of natural resources in production and decrease environmental impacts (Çelik et al., 2020). Cheese whey can be used as a natural cosmetic ingredient primarily as a protein and it is considered safe. Hydrolyzed whey protein has potential of use as skin moisturizer, soap, lotion and shampoo, due to your specific properties such as water binding, foam formation, emulgator and gelling (Çelik et al., 2020).

The Figure 3 shows the main uses of cheese whey in the global market.

CONCERNS OF CHEESE WHEY WASTE: ENVIRONMENTAL IMPACTS

Cheese whey is considered a dairy industrial residue since approximately 47% of total worldwide cheese whey is discarded in bodies of water and into the soils without any process (Bolognesi et al., 2021). Cheese whey has considerable nutritional value and a strong organic matter (mainly lactose, proteins, lipids and carbohydrates) that can cause serious disposal problems when dumped untreated into the environment.

Figure 3. Overall uses of cheese whey
Legend: WP: whey powder, WPC: whey protein concentrate, WPI: whey protein isolate, WPr: whey permeate

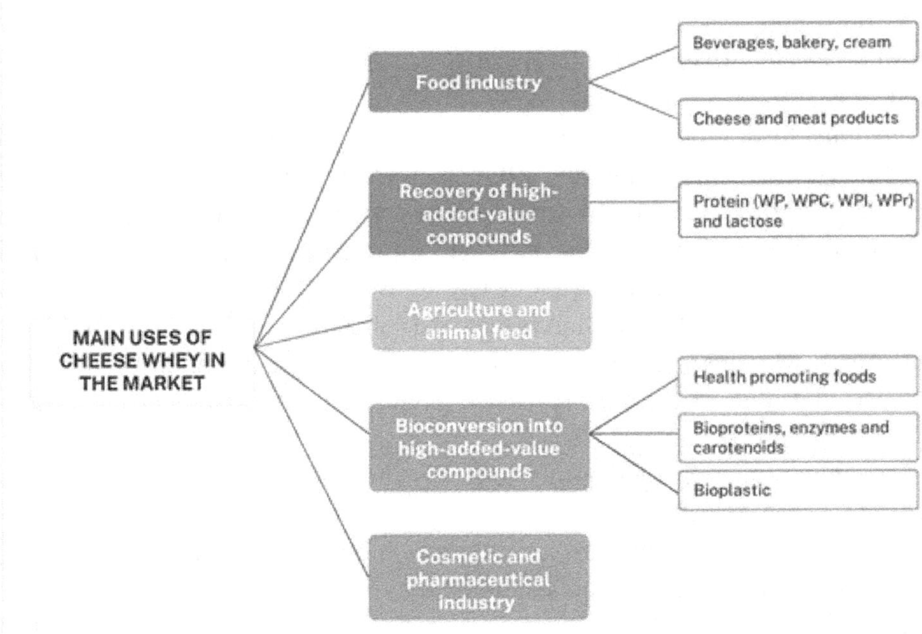

Therefore, in the dairy industry, cheese whey is the main pollutant waste, because of its high organic and volumetric load (Ahmad et al., 2019), containing lactose. The values of biological oxygen demand (BOD) (30–50 g L-1) and chemical oxygen demand (COD) (60-80 g L-1) present in cheese whey are above standards, along with large amount of phosphorus and nitrogen. These characteristics adversely impact aquatic life due to the decrease of oxygen availability in water. Its continuous discharge without treatment causes a rapid consumption of oxygen, eutrophication, formation of foam, salinization and acidification. If discarded without treatment in the soil, this industrial effluent could alter the physical and chemical composition of soil, due to its high organic and nutrient concentration resulting in decrease of soil quality and crop yield by acidification leading to environmental degradation (Ahmad et al., 2019, De Almeida et al. 2021).

Current ways of cheese whey management to decrease these environmental impacts include:

- biological treatment (e.g., anaerobic digestion and fermentation);
- physicochemical treatment;
- direct land application or its use as animal feed.

However, the limiting factor of these treatments is the high cost for the industry considered as an expensive procedure, size of the cheese producers, lack of legislation and the cheese whey biological instability. Therefore, different research groups have been studying potential strategies to valorize cheese whey and attract industries to invest in other whey-based products.

BIOTECHNOLOGY APPLICATIONS OF CHEESE WHEY

β-Galactosidase

β-galactosidase (β-D-galactohydrolase, EC 3.2.1.23) has a wide applicability, and is not intended solely for the manufacture of products without lactose. It can be used to control the crystallization of lactose in desserts, since lactose is hygroscopic, in the treatment of cheese whey, and to transform it into products such as ethanol and sweet syrup. It can be used in confectionery and bakery (Oliveira et al., 2011, Saquib, 2017) as well. Another feature of β-galactosidase enzyme is the synthesis of prebiotics such as galacto-oligosaccharides (GOS), by controlled kinetic reaction process of transgalactosylation of lactose, process of great interest to the dairy industry (Anisha, 2017).

The β-galactosidase enzyme can be obtained from different sources such as microorganisms (bacteria, fungi, yeasts, and microalgae), animals, plants, fruits, and tissues, but those of microbial species origin have higher yields, being more technologically attractive by the industry (Ahmed et al., 2016, Bentahar et al., 2019). Therefore, the production of the enzyme requires a development efficiency in the fermentation process in order to guarantee a high production of this enzyme (Anisha, 2017; Choonia & Lele, 2013). Due to the importance of providing an adequate destination for cheese whey and having this as a potential use to produce the β-galactosidase enzyme, cheese whey becomes an interesting raw material that can be used as a culture medium for microorganisms capable of producing β-galactosidase.

Cheese whey has been used for various purposes through biotechnological processes, for example, in the production of whey protein and has currently been used as a culture medium in order to obtain products with high added value, such as the β-galactosidase enzyme. β-Galactosidases catalyze different reactions of industrial interest (Jin et al., 2016, Xavier et al., 2018). β-Galactosidases from *Kluyveromyces lactis* (Kl-β-gal) are among the most reported in the literature (Pereira-Rodríguez et al., 2010, Becerra et al., 1998), owing to the numerous applications they have enabled in the environmental, food, and biotechnological industries.

In the food industry, this enzyme is essential in the manufacture of lactose-free dairy products since there is a vast and increasing number of people who cannot digest the carbohydrate (Argenta et al., 2021). β-Galactosidase possesses hydrolytic and transgalactosylation activities (Guerrero et al., 2015, Vera et al., 2020). Both activities can occur simultaneously, but the latter is prevalent when the lactose concentration is high. In the hydrolysis reactions catalyzed by β-galactosidase, the formation of an enzyme–galactosyl complex occurs upon a simultaneous liberation of glucose and a transfer of the enzyme–galactosyl complex to an acceptor that contains a hydroxyl group (Martins et al., 2019). Glucose and galactose are formed by a hydrolytic reaction from lactose, and galactose is obtained as a product if the acceptor in the reaction is the water (Saquib et al., 2017, Otieno, 2010). Kl-β-gal is also able to produce 6¢ galactobiose (Gal-(1,6)-β-D-Gal), allolactose (Gal-(1,6)-β-DGlc) and the trisaccharide 6¢ galactosyllactose (Gal-(1,6)-β-D-Gal-(1–4)-DGlc) in large quantities by transglycosylation, and it shows a remarkably high hydrolytic activity when using lactose as a substrate (Martínez-Villaluenga et al., 2008). Besides that, several works (Fischer & Kleinschmidt, 2018, Henke et al., 2019) have reported on galactooligosaccharides (GOS) production by Kl-β-gal. GOS are a group of substances containing two to eight monosaccharide units, one of them is a terminal glucose and the other is galactose (González-Catańo et al., 2017).

Galacto-Oligosaccharides (GOS)

Galacto-oligosaccharides (GOS) are considered prebiotics that promote the proliferation of beneficial intestinal bacteria, improving the microbiota, so they have several beneficial health effects, such as immunity increasing, reducing cholesterol levels, and preventing colon cancer. They belong to the group of short-chain carbohydrates, not being absorbed in the small intestine and still being able to improve the absorption of minerals, as there is osmotic transfer of water in the large intestine (Carneiro et al., 2018; Julio-Gonzales et al., 2020; Sijbers et al. 2020). According to Gibson and Roberfroid (1995), prebiotics are non-digestible food ingredients that promote an increase in the number of beneficial bacteria in the gastrointestinal (GI) tract, increasing the health of the same and potentially improving the host's health.

There has been a growing interest in studies in this area. Prebiotics can be obtained in two ways: through chemical or enzymatic synthesis. Studies suggest that synthesis by enzymes is more beneficial in relation to chemistry as it allows the control of regioselectivity and stereochemistry of the reaction final products (Duarte et al., 2017). They also allow the control of GOS that improve the growth of bifidobacteria and lactobacilli in the large intestine and are included as food ingredients to reduce the growth of pathogenic microorganisms. They also aid in the absorption of minerals due to the osmotic transfer of water to the large intestine (Carneiro et al., 2018).

GOS is divided into α-galacto-oligosaccharides (α-GOS) and β-galactooligosaccharides (β-GOS) due to the different galactosidic bonds attached (Tian et al., 2019; Mistry et al., 2020). To date, the most used for research and commercial production is β-GOS, which is synthesized by β-galactosidase through the lactose transgalactosylation reaction. However, there are currently few studies on α-GOS. α-GOS is an oligosaccharide widely distributed in various plants (Dai et al., 2018). GOS production methods mainly include the following (Weijers et al., 2008; Torres et al., 2010): (1) Natural extraction, GOS is mainly produced by the hydrolysis of lactose or whey. GOS has few sources in nature, mainly extracted from the seeds of leguminous plants, and does not contain charges, so it is difficult to separate and extract GOS from natural components; (2) Hydrolyzed polysaccharides, polysaccharide hydrolysis has the disadvantages of low conversion rate and complex components, and it is not easy to obtain GOS products with high content; (3) Chemical synthesis, the required reagents are toxic and easy to remain, and the synthesis process may cause pollution, so it is not suitable for food grades production of GOS; (4) Enzymatic synthesis, GOS is mainly synthesized through the transglucosidase of β-galactosidase. The enzymatic reaction has the advantages of safety and high efficiency, and does not cause environmental pollution, and is currently the main method for the GOS production (Mei et al., 2022).

GOS are galactose-containing oligosaccharides whose chemical structures vary by chain length, branching, and glycosyl linkages. Depending on the source of β-galactosidase, using lactose as a single substrate, the galactose released during the enzymatic hydrolysis of lactose can be transferred to another lactose molecule through b(1®6), b(1®4) or b(1®3) glycosidic linkage to the galactose moiety (Varney et al., 2017; Lu et al., 2020). Among the biological activity GOS enhances health-related short-chain fatty acids (SCFAs) production, growth and differentiation of colonic epithelial cells, energy transduction in colonocytes, lipid and carbohydrate metabolism, reduces the number of potentially pathogenic bacteria, and promotes intestinal normal function (Farthing, 2004). Research shows (He et al., 2021a) that the addition of GOS to a high-fat diet can effectively stimulate the growth of Bifidobacteria. Application of galacto-oligosaccharides supplementation in infant formula achieves the same gut flora as breastfeeding. GOS occurs naturally in breast milk and is an ideal ingredient in infant formula. Supplementation of low levels of GOS to infant formula improves stool frequency, reduces stool pH, and stimulates intestinal Bifidobacteria and Lactobacilli, as in breastfed infants (Ben et al., 2008). One study found that feeding infant formula with a small amount of GOS (0.24 g/dL) for 3 months stimulated the growth of Bifidobacteria and Lactobacilli but did not harm the growth of *E. coli*.

GOS is recognized as a natural ingredient with prebiotic properties, and numerous studies reveal its health-promoting and food-functional properties, making it a potential choice as a nutritional supplement or food ingredient. The popularity of prebiotics in the functional food market is rapidly rising. More recently, most prebiotic ingredients have been used in breakfast bread, baked goods, and snacks. The rise in obesity and other metabolic-related health problems is rapidly encouraging consumers to focus on low-carb, well-balanced healthy diets, which in turn is driving demand for prebiotics. GOS and other prebiotics have been recommended for various human foods such as baked goods, sweeteners, and yogurt (Mei et al., 2022).

Whey Protein

Whey proteins (WP) and their derivative components play an imperative and revolutionary role in the pharmaceuticals, healthcare products, and formulations of numerous functional foods. In addition, WPs exhibit numerous therapeutic applications in the prevention of various diseases (Brandelli et al., 2015, Deeth & Bansal, 2018). Cheese whey can be used through membrane filtration, which allows the recovery of nutritionally important proteins (Escobar, Souza & Lehn, 2014). Through this process, whey permeate is obtained, rich in lactose and mineral salts (Baldasso, Barros & Tesaro, 2011), which can be used as a culture medium for yeasts that have the ability to produce the β-galactosidase enzyme. Whey protein concentrate is also obtained as a product from the cheese whey filtration retentate. The protein portion contains approximately 50% β-lactoglobulin (BLG), 25% α-lactalbumin (ALA) and 25% other proteins (Fitzsimons et al., 2006), such as bovine serum albumin (BSA), immunoglobulins (Ig's) and glyco-macropeptides (GMP), varying in size, molecular weight and function (Kinsella & Whitehead, 1989; Aimutis, 2004), being 20% of the total milk constituent proteins.

Studies have shown that whey proteins have some advantages over milk caseins. One of the explanations for such advantages may be related to the structure of casein itself, since, unlike whey proteins, milk caseins undergo conformational changes, in which they are denatured in the presence of stomach acids and end up losing part of their function. To restore its physiological function, the peptides have to be absorbed by the intestine in their active form (Renhe, 2008). Thus, whey proteins have gastro-resistance activity in the stomach environment (Gebara et al., 2013).

The demand for cheese whey has increased to obtain its proteins, β-lactoglobulin and α-lactalbumin due to its high functional and nutritional values. Cheese whey proteins also have several applications in the pharmaceutical industry (Yadav et al., 2015). β–lactoglobulin has a medium molecular weight (18.4 to 36.8 kDa), which gives it resistance to the action of acids and proteolytic enzymes present in the stomach and is therefore absorbed in the small intestine. α–lactoalbumin with a molecular weight of 14.2 kDa has antimicrobial activity against pathogenic bacteria (Sgarbieri, 2005; Haraguchi et al., 2006; Krissansen, 2007; Hernández et al., 2008; Santos et al., 2011).

Whey proteins have emerged from dairy side-streams as widely used and high-value food product ingredients, allowing a targeted alteration of the viscosity, solubility, foaming, emulsification, and water absorption properties of food matrices (Farrokhi et al., 2019; Inthavong et al., 2019). Edible films and coatings have additionally expanded the appeal of integrating whey proteins into formulations (Ramos et al., 2012; Çakmak et al., 2020). Furthermore, whey proteins are enzymatically hydrolyzed, liberating bioactive peptides, and thereby triggering beneficial physiological effects related to the immune system, cardiovascular system, nervous system, and gastrointestinal system (Madureira et al., 2010; Dullius et al., 2018).

The uses of enzymes in industrial processes in the food area have a high cost. In order to reduce this value and enable the use of enzyme technology in different industrial segments, studies seek to develop and improve methods of enzymatic immobilization (Bon, 2008). Immobilization can be done using different supports, organic and inorganic, and by different methodologies such as microencapsulation, physical adsorption, covalent bonding, among others. Alginate is a naturally occurring anionic polymer isolated from brown algae that has the ability to form strong hydrogels. Protein-loaded alginate microcapsules can be manufactured by extruding a sodium alginate solution containing the bioactive protein into a divalent cross-linking solution, such as a calcium chloride solution. Among the microencapsulation processes is the production of alginate microcapsules, by ionic gelation, and alginate is considered practically non-toxic. In the food industry, the main application of this technology involves the protection of biologically active compounds or cells from deleterious biological and physiological environments for efficient delivery to destination sites. (Doherty et al., 2011; Paques et al., 2014).

Whey protein is an ideal wall material to encapsulate and deliver nutraceuticals due to its advantages of high nutritional value, availability, excellent techno-functional properties, and resistance to pepsin digestion (Singh et al., 2014). Many nutraceuticals have been encapsulated into whey protein-based nanoparticles to develop functional food formulations, such as astaxanthin (Shen et al., 2018), epigallocatechin-3-gallate (Shpigelman et al., 2012), curcumin (Mohammadian et al., 2020), and quercetin (Simões et al., 2020). The demand for health-promoting dairy products, functional foods, and nutraceuticals has escalated over the past several decades. Besides either fortified or enriched foods that incorporate novel bioactive components, functional foods exhibit therapeutic potentials beyond basic nutritional needs, whether from plant and animal sources (Klopčič et al., 2020). Novel biotechnology methods and sophisticated microbial manipulation techniques are key in developing high-value dairy products, infant formulas, and functional foods from milk, colostrum, and whey to meet consumer demands (Sharma, 2019).

Bioethanol

Because of the high lactose content and nutrients in the cheese whey, this by-product has been considered a potential source for biotechnological transformation into different bioproducts of industrial interest, such as bioethanol. The conversion of lactose into ethanol, through alcoholic fermentation is an alternative for the bioremediation of the polluting whey waste that would not only reduce the dairy environmental impact but might also be an alternative way in the ethanol production (Guimarães et al., 2010; Murari et al., 2019).

In the past decades, research on ethanol production has been driven by the growing demand for cleaner, more renewable energy sources (Lima et al., 2020), especially due to the drastic increase in the energy crisis and the depletion of fossil fuel reserves, such as oil and its derivatives, considered the major contributors to global warming (Oke et al., 2016; Murari et al., 2019). In addition to biofuel, cheese whey ethanol is potable, and it can be used in the food, beverage, pharmaceutical, and cosmetic industries (Guimarães et al., 2010).

There are microorganisms able to metabolize lactose as a source of carbon and energy. Lactose, the main ingredient of cheese whey, is a disaccharide consisting of galactose and glucose. Both β-galactosidase and lactose permease are indispensable for microorganisms to ferment lactose. Among the few microorganisms able to ferment lactose, only a few yeasts, such as *Kluyveromyces fragilis*, *Kluyveromyces marxianus*, and *Candida pseudotropicalis*, can ferment it directly (Kargi & Ozmıhcı, 2006; Guimarães et al., 2010). Lactic acid bacteria, such as Lactobacillus, are the representative lactose-consuming microorganisms. They are able to convert lactose to lactic acid, but not to ethanol (Liu et al., 2016). Most filamentous fungi have a low assimilative capacity for lactose (Okamoto et al., 2019).

K. marxianus has desirable attributes for industrial fermentation processes, such as thermotolerance, a high growth rate, and metabolizing capacity, and often ferments a wide variety of carbohydrates, such as pentoses, hexoses, and disaccharides (Lane & Morrissey, 2010). Studies have reported the use of *K. marxianus* with satisfactory results for the conversion of lactose into ethanol, obtaining ethanolic yield from 26 to 86% using concentrate cheese whey (Guo et al., 2010; Arora et al., 2017). Murari et al. (2019) evaluated the bioethanol production from two coproducts generated by the dairy industry (cheese whey and whey permeate) obtained by a fermentation process making use of the yeast *K. marxianus*. Both were found to be suitable and promising to convert lactose to ethanol under anaerobic conditions, having achieved a theoretical ethanolic yield of 76% in cheese whey.

A group of research from Japan isolated some basidiomycete fungi capable of producing ethanol from sugars, lignocellulosic biomass, and kitchen wastes (Okamoto et al., 2011; Okamoto et al., 2014; Okamoto et al., 2017). In 2019, the same research group examined the applicability of the brown rot fungus Neolentinus lepideus for ethanol production from cheese whey and reported the ability of this fungi to efficiently ferment whey and milk (Okamoto et al., 2019).

Saccharomyces cerevisiae is usually the first choice for industrial processes involving alcoholic fermentation, due to its good fermentative capacity, its capacity to grow rapidly under anaerobic conditions, extensive industrial and scientific knowledge accumulated that makes it one of the best studied organisms; however, wild *S. cerevisiae* strains are unable to metabolize lactose because it lacks the genes required for its transport (lactose permease) and hydrolysis (beta-galactosidase) (Guimarães et al., 2010; Álvarez-Cao et al., 2020). Therefore, the first applications of *S. cerevisiae* in cheese whey fermentation require the pre-hydrolysis of lactose using commercial lactose-degrading enzymes, however this step adds more costs for the process. To overcome this problem, several studies were developed with genetic engineering *S. cerevisiae* strains to modify this yeast for lactose consumption, like the introduction of heterologous lactose genes (Borgne, 2012; Álvarez-Cao et al., 2020).

Some of the recombinant strains are effective in converting lactose to ethanol, completely metabolizing the lactose available and reaching ethanol titres that can be economically competitive (Farkas et al., 2019; Zou et al., 2021). Therefore, *S. cerevisiae* strains are considered the best choice for ethanol production. Converting cheese whey into bioethanol is an advantageous option to increase the added value of agro-industrial processes (Murari et al., 2019).

Bioplastic

Plastic waste accumulation is another environmental concern worldwide since 80% of the total waste is accumulated in the natural environment (Geyer et al., 2017). These wastes cause harmful effects to the environment, since they are fossil fuel-based products and non-biodegradable (Schmaltz et al., 2020). Hence, alternatives such as eco-friendly plastic alternatives and renewable-based materials have been studied in order to reduce the need for conventional plastics (Tsang et al. 2019).

One of the researched areas of bioplastics are the ones produced from proteins, including dairy by-products such as whey protein (Chalermthai et al., 2019). One type of biodegradable bioplastic is called Polyhydroxyalkanoates (PHAs). PHAs are natural and intracellular polyesters produced by bacteria as a way to store energy when cells encounter nutrient limitations. This material offers an approach to carbon neutrality and sustainable industry (El-malek et al., 2020). These organisms use several substrates for PHA production and cheese whey is considered a nutritive medium, however the isolation of microorganisms that can produce PHAs effectively when growing on whey is still challenging (Tsang et al. 2019).

Ahn et al. (2001) studied the use of cheese whey as carbon source to recombinant *Escherichia coli*, since the wild type is not capable of direct metabolization of cheese whey due to the lack of β-galactosidase activity. *E.coli* containing *Alcaligenes latus* PHA synthesizing genes were able to produce a good concentration and productivity of polyhydroxybutyrate (PHB), confirming cheese whey as an adequate substrate for PHA production. *Haloferax mediterranei* has also been employed to convert cheese whey to PHA (Koller, 2015; Pais et al., 2016), suggesting cheese whey as a plausible carbon source for a different PHA accumulating microorganisms. Colombo et al. (2019) developed a two-step method for the generation of PHA and biohydrogen from the mixed microorganism culture from dairy-waste streams coming from cheese whey deproteinization. Israni et al. (2020) used crude whey as feedstock in the generation of bioplastic using *Bacillus megaterium* Ti3 species. Acid treated whey hydrolysate was also used as a medium for the growth of *Halomonas halophila* accumulating PHA after 48 h of cultivation (Kucera et al., 2018). Furthermore, some less known PHA-producers genera such as Paenibacillus and Lysinibacillus were detected in the long-term operation of cheese whey feeding (Khatami et al., 2021).

Recent studies evaluated the life cycle assessment (LCA) of bioplastic production from cheese whey protein to assess the environmental impacts of such plastics, compared to conventional plastics. Chalermthai et al. (2021) studied the production of the cheese whey plastic using free-radical copolymerization method with copolymer called polyethylene glycol methyl ether methacrylate (PEGMA). The work showed that this process had much lower environmental impact compared to the biopolymers like polylactic acid (PLA) and PHB. In addition, the energy consumption used in the production was much lower than any of the other polymers found in literature. The environmental impact score of the cheese whey plastic showed the lowest Global Warming Potential (GWP), which indicates the environmental benefits of its production compared to conventional plastics.

RECOMMENDATION AND SOLUTION

Sustainable Uses of Cheese Whey into Circular Economy Concept

The circular economy becomes increasingly fundamental to the great challenge of increasing food production allied to sustainable production (Mak et al., 2019; European Commission, 2015). Reusing, repairing, renovating, and recycling materials and products, considering waste into a resource are essential concepts of the circular economy. Thus, by-products generated by food systems that can be processed are considered the most relevant resources as they can be converted into new food ingredients and foods, thus meeting the food needs of a growing world population (Lavelli, 2021).

Cheese whey is an excellent medium for the growth of microorganisms and, therefore, there are difficulties in relation to the volumes and high costs of refrigerated transport and heat treatments that guarantee greater stability of the product (Panesar et al., 2007). The processing of cheese whey is an opportunity to avoid its inappropriate disposal in the environment and add value to this product. In addition, whey processing meets the objectives already stipulated at the last meeting of the United Nations Summit, whose theme was Transforming the World, and with goals to be achieved by 2030, focusing on sustainable production and the use of waste with greater participation of recycling and reuse (UM General Assembly, 2015). Several works were described in this chapter on the valorization of cheese whey through its various functional and technological applications and describing its beneficial effects on human health.

The great challenge is to make these strategies possible, especially among small and medium-sized dairy industries, since it is very expensive to process this by-product. The integration of biorefineries to these small and medium-sized companies can contribute (Zandona et al., 2021) with the flow of raw materials according to the concept of circular economy and zero waste generation, covering all pillars of sustainability, that includes the environment, society and economy (Romani et al., 2018). The joint search for government actions that facilitate the flow of production, the adoption of programs to improve quality in the industries, the implementation of strategically located units for pre-concentration and forwarding to a processing unit (biorefineries), can represent a solution viable and ecologically healthy solution to the problem of improper disposal of cheese whey (Lavelli & Becalli, 2022).

CONCLUSION

Recent data indicate that the dairy industry produces a great amount of cheese whey annually and of this total, 50% of it is discarded and are not re-used. This chapter described several ways of using cheese whey as a raw material for the manufacture of several products with high added value, however the implementation of these processes is still restricted. The biggest challenge in the present cheese whey processing is a deep change in the current management and processing systems, especially in redirecting on-farm production of cheese whey directed towards the industry, which will use it as a raw material. It would be necessary to implement a system in which the major actors involved in the dairy chain such as milk farms, cheese factories and rennet suppliers could send this waste directly to the food and pharmaceutical industries, in addition to ethanol refineries and bioplastic factories. Only a joint action can make cheese whey waste a product with added value, zero waste generation and, consequently, attending the environmental legislation and current public policies.

References

Addai, F. P., Lin, F., Wang, T., Kosiba, A. A., Sheng, P., Yu, F., & Shi, H. (2020). Technical integrative approaches to cheese whey valorization towards a sustainable environment. *Food & Function*, *11*(10), 8407–8423. doi:10.1039/D0FO01484B PMID:32955061

Ahmad, T., Aadil, R. M., Ahmed, H., Rahman, U., Soares, B. C., Souza, S. L., & Cruz, A. G. (2019). Treatment and utilization of dairy industrial waste: A review. *Trends in Food Science & Technology*, *88*, 361–372. doi:10.1016/j.tifs.2019.04.003

Ahmed, S., Kattimani, L., Divatar, M., Gajare, S., Shivalee, A., Farheen, A., & Irfana, M. (2016). Optimization of lactase production under submerged fermentation by Lactobacillus sp.KLSA 22. *International Journal of Pure and Applied Bioscience*, *4*(4), 212–220. doi:10.18782/2320-7051.2321

Ahn, W. S., Park, S. J., & Lee, S. Y. (2001). Production of poly (3-hydroxybutyrate) from whey by cell recycle fed-batch culture of recombinant Escherichia coli. *Biotechnology Letters*, *23*(3), 235–240. doi:10.1023/A:1005633418161

Aimutis, W. L. (2004). Bioactive properties of milk proteins with particular focus on anticariogenesis. *The Journal of Nutrition*, *134*(4), 989–995. doi:10.1093/jn/134.4.989S PMID:15051859

Almeida, K. E., Bonassi, I. A., & Roca, R. O. (2001). Physical and chemical characteristics of fermented dairy beverages using minas cheese whey. *Food Science and Technology (Campinas)*, *21*(2), 187–192. doi:10.1590/S0101-20612001000200012

Álvarez-Cao, M. E., Becerra, M., & González-Siso, M. I. (2020). Biovalorization of cheese whey and molasses wastes to galactosidases by recombinant yeasts. In N. K. Rathinam & R. Sani (Eds.), *Biovalorisation of Wastes to Renewable Chemicals and Biofuels* (pp. 149–161). Elsevier. doi:10.1016/B978-0-12-817951-2.00008-0

Anisha, G. S. (2017). β-galactosidases. In Pandey, A., Negi, S., & Soccol, C. R. (Eds.). Current developments in biotechnology and bioengineering: production, isolation and purification of industrial products (395-421). Elsevier.

Argenta, A. B., Nogueira, A., & Scheer, A. P. (2021). Hydrolysis of whey lactose: Kluyveromyces lactisβ-galactosidase immobilisation and integrated process hydrolysis-ultrafiltration. *International Dairy Journal, 117*, 105007. doi:10.1016/j.idairyj.2021.105007

Arora, R., Behera, S., Sharma, N. K., & Kumar, S. (2017). Augmentation of ethanol production through statistically designed growth and fermentation medium using novel thermotolerant yeast isolates. *Renewable Energy, 109*, 406–421. doi:10.1016/j.renene.2017.03.059

Assembly, U. G. (2015). Transforming our world: the 2030 Agenda for Sustainable Development. A/Res/70/1, *Resolution adopted by the General Assembly on 25 September 2015*.GA.

Baldasso, C., Barros, T. C., & Tessaro, I. C. (2011). Concentration and purification of whey proteins by ultrafiltration. *Desalination, 278*(1-3), 381–386. doi:10.1016/j.desal.2011.05.055

Barba, F. J. (2021). An integrated approach for the valorization of cheese whey. *Foods, 10*(3), 564. doi:10.3390/foods10030564 PMID:33803106

Becerra, M., Cerdan, E., & Gonzalez Siso, M. I. (1998). Dealing with different methods for Kluyveromyces lactisβ-galactosidase purification. *Biological Procedures Online, 1*(1), 48–58. doi:10.1251/bpo4 PMID:12734592

Ben, X. M., Li, J., Feng, Z. T., Shi, S. Y., Lu, Y. D., Chen, R., & Zhou, X.-Y. (2008). Low level of galacto-oligosaccharide in infant formula stimulates growth of intestinal Bifidobacteria and Lactobacilli. *World Journal of Gastroenterology, 14*(42), 6564–6568. doi:10.3748/wjg.14.6564 PMID:19030213

Bentahar, J., Doyen, A., Beaulieu, L., & Deschênes, J. S. (2019). Investigation of β-galactosidase production by microalga Tetradesmus obliquus in determined growth conditions. *Journal of Applied Phycology, 31*(1), 301–308. doi:10.100710811-018-1550-y

Bolognesi, L. S., Gabardo, S., Dall Cortivo, P. R., & Ayub, M. A. Z. (2021). Biotechnological production of galactooligosaccharides (GOS) using porungo cheese whey. *Food Science and Technology (Campinas), 42*, e64520. doi:10.1590/fst.64520

Bon, E. P. S., Ferrara, M. A., & Corvo, M. L. (2008). *Enzimas em biotecnologia: produção, aplicações e mercado*. Editora Interciência.

Borgne, S. L. (2012). Genetic engineering of industrial strains of Saccharomyces cerevisiae. In A. Lorence (Ed.), *Recombinant gene expression* (pp. 451–465). Humana Press. doi:10.1007/978-1-61779-433-9_24

Brandelli, A., Daroit, D. J., & Corrêa, A. P. F. (2015). Whey as a source of peptides with remarkable biological activities. *Food Research International, 73*, 149–161. doi:10.1016/j.foodres.2015.01.016

Çakmak, H., Özselek, Y., Turan, O. Y., Fıratlıgil, E., & Karbancioğlu-Güler, F. (2020). Whey protein isolate edible films incorporated with essential oils: Antimicrobial activity and barrier properties. *Polymer Degradation & Stability, 179*, 109285. doi:10.1016/j.polymdegradstab.2020.109285

Carneiro, L. A. B. C., Yu, L., Dupree, P., & Ward, R. J. (2018). Characterization of a β-galactosidase from Bacillus subtilis with transgalactosylation activity. *International Journal of Biological Macromolecules, 120*, 279–287. doi:10.1016/j.ijbiomac.2018.07.116 PMID:30036621

Carrillo-Lopez, L. M., Garcia-Galicia, I. A., Tirado-Gallegos, J. M., Sanchez-Vega, R., Huerta-Jimenez, M., Ashokkumar, M., & Alarcon-Rojo, A. D. (2021). Recent advances in the application of ultrasound in dairy products: Effect on functional, physical, chemical, microbiological and sensory properties. *Ultrasonics Sonochemistry, 73*, 105467. doi:10.1016/j.ultsonch.2021.105467 PMID:33508590

Carvalho, F., Prazeres, A. R., & Rivas, J. (2013). Cheese whey wastewater: Characterization and treatment. *The Science of the Total Environment, 445*, 385–396. doi:10.1016/j.scitotenv.2012.12.038 PMID:23376111

Çelik, K., Önür, Z. Y., Baytekin, H., & Coşkun, B. (2020). Whey processing: Utlization and Major Products. In K. Çelik (Ed.), *Whey every aspect* (pp. 33–60). Tudás Alapítvány.

Chalermthai, B., Giwa, A., Schmidt, J. E., & Taher, H. (2021). Life cycle assessment of bioplastic production from whey protein obtained from dairy residues. *Bioresource Technology Reports, 15*, 100695. doi:10.1016/j.biteb.2021.100695

Choonia, H. S., & Lele, S. S. (2013). Kinetic modeling and implementation of superior process strategies for β-galactosidase production during submerged fermentation in a stirred tank bioreactor. *Biochemical Engineering Journal, 77*, 49–57. doi:10.1016/j.bej.2013.04.021

Colombo, B., Calvo, M. V., Sciarria, T. P., Scaglia, B., Kizito, S. S., D'Imporzano, G., & Adani, F. (2019). Biohydrogen and polyhydroxyalkanoates (PHA) as products of a two-steps bioprocess from deproteinized dairy wastes. *Waste Management (New York, N.Y.)*, *95*, 22–31. doi:10.1016/j.wasman.2019.05.052 PMID:31351607

Dai, Z., Lyu, W., Xiang, X., Tang, Y., Hu, B., Ou, S., & Zeng, X. (2018). Immunomodulatory effects of enzymatic-synthesized α-galactooligosaccharides and evaluation of the structure-activity relationship. *Journal of Agricultural and Food Chemistry*, *66*(34), 9070–9079. doi:10.1021/acs.jafc.8b01939 PMID:30086236

De Almeida, M. P. G., Mockaitis, G., & Weissbrodt, D. G. (2021). *Got Whey? The significance of cheese whey at the confluence of dairying, environmental impacts, energy and resource biorecovery*. The Evolving Scholar.

Deeth, H. C., & Bansal, N. (2018). *Whey proteins: From milk to medicine*. Academic Press.

Doherty, S. B., Gee, V. L., Rossa, R. P., Stanton, C., Fitzgerald, G. F., & Brodkorb, A. (2011). Development and characterisation of whey protein micro-beads as potential matrices for probiotic protection. *Food Hydrocolloids*, *25*(6), 1604–1617. doi:10.1016/j.foodhyd.2010.12.012

Duarte, L. S., Schöffer, J. da N., Lorenzoni, A. S. G., Rodrigues, R. C., Rodrigues, E., & Hertz, P. F. (2017). A new biopocess for the production of prebiotc lactosusrose by immobilized β-galactosidase. *Process Biochemistry*, *55*, 96–103. doi:10.1016/j.procbio.2017.01.015

Dullius, A., Goettert, M. I., & Souza, C. F. V. (2018). Whey protein hydrolysates as a source of bioactive peptides for functional foods—Biotechnological facilitation of industrial scale-up. *Journal of Functional Foods*, *42*, 58–74. doi:10.1016/j.jff.2017.12.063

El-malek, F., Farag, A., Omar, S., & Khairy, H. (2020). Polyhydroxyalkanoates (PHA) from Halomonas pacifica ASL10 and Halomonas salifodiane ASL11 isolated from Mariout salt lakes. *International Journal of Biological Macromolecules*, *161*, 1318–1328. doi:10.1016/j.ijbiomac.2020.07.258 PMID:32755698

Erman, M., Demir, S., Ocak, E., Tüfenkçi, Ş., Oğuz, F., & Akköprü, A. (2011). Effects of Rhizobium, arbuscular mycorrhiza and whey applications on some properties in chickpea (Cicer arietinum L.) under irrigated and rainfed conditions. *Field Crops Research*, *122*(1), 14–24. doi:10.1016/j.fcr.2011.02.002

Escobar, G. P., Souza, C. F. V., & Lehn, D. N. (2014). Avaliação de beta-galactosidase livre e imobilizada na hidrólise da lactose do permeado de soro de queijo. *Caderno Pedagógico, 11*(1), 117–129.

Eurostat (2020). Milk and milk product statistics. https://ec.europa.eu/eurostat/statistics-explained/index.php?title=Milk_and_milk_product_statistics#:~:text=In%202021%2C%202.9%20million%20tonnes,about%2023%20%25)%20in%202021.

FAO. (2015). *Milk facts*. FAO. https://www.fao.org/dairy-production-products/production/en/

Farkas, C., Rezessy-Szabó, J. M., Gupta, V. K., Bujna, E., Pham, T. M., Pásztor-Huszár, K., & Nguyen, Q. D. (2019). Batch and fed-batch ethanol fermentation of cheese-whey powder with mixed cultures of different yeasts. *Energies, 12*(23), 4495. doi:10.3390/en12234495

Farrokhi, F., Badii, F., Ehsani, M. R., & Hashemi, M. (2019). Functional and thermal properties of nanofibrillated whey protein isolate as functions of denaturation temperature and solution pH. *Colloids and Surfaces. A, Physicochemical and Engineering Aspects, 583*, 124002. doi:10.1016/j.colsurfa.2019.124002

Farthing, M. J. (2004). Bugs and the gut: An unstable marriage. *Best Practice & Research. Clinical Gastroenterology, 18*(2), 233–239. doi:10.1016/j.bpg.2003.11.001 PMID:15123066

Fischer, C., & Kleinschmidt, T. (2018). Synthesis of galactooligosaccharides in milk and whey: A review. *Comprehensive Reviews in Food Science and Food Safety, 17*(3), 678–697. doi:10.1111/1541-4337.12344 PMID:33350129

Fitzsimons, S. M., Mulvihill, D. M., & Moris, E. R. (2006). Denaturation and aggregation processes in thermal gelation of whey proteins resolved by differential scanning calorimetry. *Food Hydrocolloids, 11*, 62–69.

Fox, P. F., Guinee, T. P., Cogan, T. M., & McSweeney, P. L. H. (2017). Microbiology of Cheese Ripening. In P. F. Fox, T. P. Guinee, T. M. Cogan, & P. L. McSweeney (Eds.), *Fundamentals of cheese science* (pp. 333–390). Springer. doi:10.1007/978-1-4899-7681-9_11

Gebara, C., Chaves, K. S., Ribeiro, M. C. E., Souza, F. N., Grosso, C. R. F., & Gigante, M. L. (2013). Viability of Lactobacillus acidophilus La5 in pectin–whey protein microparticles during exposure to simulated gastrointestinal conditions. *Food Research International, 51*(2), 872–878. doi:10.1016/j.foodres.2013.02.008

Geyer, R., Jambeck, J. R., & Law, K. L. (2017). Production, use, and fate of all plastics ever made. *Science Advances, 3*(7), e1700782. doi:10.1126ciadv.1700782 PMID:28776036

Gibson, G. R., & Roberfroid, M. B. (1995). Dietary modulation of the human colonic microbiota: Introducing the concept of prebiotics. *The Journal of Nutrition, 125*(6), 1401–1412. doi:10.1093/jn/125.6.1401 PMID:7782892

González-Cataño, F., Tovar-Castro, L., Castaño-Tostado, E., Regalado-Gonzalez, C., García-Almendarez, B., Cardador-Martínez, A., & Amaya-Llano, S. (2017). Improvement of covalent immobilization procedure of β-galactosidase from Kluyveromyces lactis for galactooligosaccharides production: Modeling and kinetic study. *Biotechnology Progress, 33*(6), 1568–1578. doi:10.1002/btpr.2509 PMID:28571115

Guerrero, C., Vera, C., Conejeros, R., & Illanes, A. (2015). Transgalactosylation and hydrolytic activities of commercial preparations of β-galactosidase for the synthesis of prebiotic carbohydrates. *Enzyme and Microbial Technology, 70*, 9–17. doi:10.1016/j.enzmictec.2014.12.006 PMID:25659627

Guimarães, P. M., Teixeira, J. A., & Domingues, L. (2010). Fermentation of lactose to bio-ethanol by yeasts as part of integrated solutions for the valorisation of cheese whey. *Biotechnology Advances, 28*(3), 375–384. doi:10.1016/j.biotechadv.2010.02.002 PMID:20153415

Guo, X., Zhou, J., & Xiao, D. (2010). Improved ethanol production by mixed immobilized cells of Kluyveromyces marxianus and Saccharomyces cerevisiae from cheese whey powder solution fermentation. *Applied Biochemistry and Biotechnology, 160*(2), 532–538. doi:10.100712010-008-8412-z PMID:19002658

Haraguchi, F. K., Abreu, W. C., & Paula, H. (2006). Proteínas do soro do leite: Composição, propriedades nutricionais, aplicações no esporte e benefícios para a saúde humana. *Revista Brasileira de Nutrição, 19*(4), 479–488. doi:10.1590/S1415-52732006000400007

He, N., Chen, H., Zhou, Z., Zhao, W., Wang, S., Lv, Z., Liu, N., Wang, H., Yang, Z., & Li, S. (2021). Enzymatically synthesized a-galactooligosaccharides attenuate metabolic syndrome in high-fat diet induced mice in association with the modulation of gut microbiota. *Food & Function, 12*(11), 4960–4971. doi:10.1039/D0FO03113E PMID:34100482

Henke, S., Shakhno, N., Botvynko, A., & Bedn, A. (2019). Production of galactooligosaccharides using various combinations of the comercial β-galactosidases. *Biochemical and Biophysical Research Communications*, *517*(4), 762–766. doi:10.1016/j.bbrc.2019.08.001 PMID:31395344

Hernández, L. B., Recio, I., & Amigo, L. (2008). β-lactoglobulin as source of bioactive peptides. *Journal of Amino Acids*, *35*(2), 257–265. doi:10.100700726-007-0585-1 PMID:17726638

IBGE. (2021). *Produção de Leite*. IBGE. https://www.ibge.gov.br/explica/producao-agropecuaria/leite/br

Inthavong, W., Chassenieux, C., & Nicolai, T. (2019). Viscosity of mixtures of protein aggregates with different sizes and morphologies. *Soft Matter*, *15*(23), 4682–4688. doi:10.1039/C9SM00298G PMID:31114829

Israni, N., Venkatachalam, P., Gajaraj, B., Varalakshmi, K. N., & Shivakumar, S. (2020). Whey valorization for sustainable polyhydroxyalkanoate production by Bacillus megaterium: Production, characterization and in vitro biocompatibility evaluation. *Journal of Environmental Management*, *255*, 109884. doi:10.1016/j.jenvman.2019.109884 PMID:32063322

Jelen, P. (2003). Whey processing. In H. Roginski, J. W. Fuquay, & P. F. Fox (Eds.), *Encylcopedia of dairy sciences*. Academic Press.

Jin, Y., Parashar, A., Mason, B., & Bressler, D. C. (2016). Simultaneous hydrolysis and cofermentation of whey lactose with wheat for ethanol production. *Bioresource Technology*, *221*, 616–624. doi:10.1016/j.biortech.2016.09.063 PMID:27693727

Julio-Gonzalez, L. C., Hernandez-Hernandez, O., Moreno, F. J., Jimeno, M. L., Doyaguez, E. G., Olano, A., & Corzo, N. (2020). Hydrolysis and transgalactosylation catalysed by β-galactosidase from brush border membrane vesicles isolated from pig small intestine: A study using lactulose and its mixtures with lactose or galactose as substrates. *Food Research International*, *129*, 108811. doi:10.1016/j.foodres.2019.108811 PMID:32036892

Kargi, F., & Ozmıhcı, S. (2006). Utilization of cheese whey powder (CWP) for ethanol fermentations: Effects of operating parameters. *Enzyme and Microbial Technology*, *38*(5), 711–718. doi:10.1016/j.enzmictec.2005.11.006

Khatami, K., Perez-Zabaleta, M., Owusu-Agyeman, I., & Cetecioglu, Z. (2021). Waste to bioplastics: How close are we to sustainable polyhydroxyalkanoates production? *Waste Management (New York, N.Y.)*, *119*, 374–388. doi:10.1016/j.wasman.2020.10.008 PMID:33139190

Kindstedt, P. S. (2014). The basic of cheesemaking. In C. W. Donnelly (Ed.), *Cheese and Microbes* (pp. 17–38). ASM Press.

Kinsella, J. E., & Whitehead, D. M. (1989). Proteins in whey: Chemical, physical and functional properties. *Advances in Food and Nutrition Research*, *33*, 343–438. doi:10.1016/S1043-4526(08)60130-8 PMID:2697234

Klopčič, M., Slokan, P., & Erjavec, K. (2020). Consumer preference for nutrition and health claims: A multi-methodological approach. *Food Quality and Preference*, *82*, 103863. doi:10.1016/j.foodqual.2019.103863

Koller, M. (2015). Recycling of waste streams of the biotechnological poly (hydroxyalkanoate) production by Haloferax mediterranei on whey. *International Journal of Polymer Science*, 2015.

Krissansen, G. W. (2007). Emerging health properties of whey proteins and their clinical implications. *Journal of the American College of Nutrition*, *26*(6), 713–723. doi:10.1080/07315724.2007.10719652 PMID:18187438

Kucera, D., Pernicová, I., Kovalcik, A., Koller, M., Mullerova, L., Sedlacek, P., & Obruca, S. (2018). Characterization of the promising poly (3-hydroxybutyrate) producing halophilic bacterium Halomonas halophila. *Bioresource Technology*, *256*, 552–556. doi:10.1016/j.biortech.2018.02.062 PMID:29478784

Lane, M. M., & Morrissey, J. P. (2010). Kluyveromyces marxianus: A yeast emerging from its sister's shadow. *Fungal Biology Reviews*, *24*(1-2), 17–26. doi:10.1016/j.fbr.2010.01.001

Lavelli, V. (2021). Circular food supply chains–Impact on value addition and safety. *Trends in Food Science & Technology*, *114*, 323–332. doi:10.1016/j.tifs.2021.06.008

Lavelli, V., & Beccalli, M. P. (2022). Cheese whey recycling in the perspective of the circular economy: Modeling processes and the supply chain to design the involvement of the small and medium enterprises. *Trends in Food Science & Technology*, *126*, 86–98. doi:10.1016/j.tifs.2022.06.013

Lima, M. A., Mendes, L. F. R., Mothé, G. A., Linhares, F. G., de Castro, M. P. P., Da Silva, M. G., & Sthel, M. S. (2020). Renewable energy in reducing greenhouse gas emissions: Reaching the goals of the Paris agreement in Brazil. *Environmental Development*, *33*, 100504. doi:10.1016/j.envdev.2020.100504

Liu, J., Dantoft, S. H., Würtz, A., Jensen, P. R., & Solem, C. (2016). A novel cell factory for efficient production of ethanol from dairy waste. *Biotechnology for Biofuels*, *9*(1), 1–11. doi:10.118613068-016-0448-7 PMID:26925162

Lu, L., Guo, L., Wang, K., Liu, Y., & Xiao, M. (2020). β-Galactosidases: A great tool for synthesizing galactose-containing carbohydrates. *Biotechnology Advances*, *39*, 107465. doi:10.1016/j.biotechadv.2019.107465 PMID:31689470

Madureira, A. R., Tavares, T., Gomes, A. M. P., Pintado, M. E., & Malcata, F. X. (2010). Invited review: Physiological properties of bioactive peptides obtained from whey proteins. *Journal of Dairy Science*, *93*(2), 437–455. doi:10.3168/jds.2009-2566 PMID:20105516

Mak, T. M., Xiong, X., Tsang, D. C., Iris, K. M., & Poon, C. S. (2020). Sustainable food waste management towards circular bioeconomy: Policy review, limitations and opportunities. *Bioresource Technology*, *297*, 122497. doi:10.1016/j.biortech.2019.122497 PMID:31818718

Martínez-Villaluenga, C., Cardelle-Cobas, A., Corzo, N., Olano, A., & Villamiel, M. (2008). Optimization of conditions for galactooligosaccharide synthesis during lactose hydrolysis by β-galactosidase from Kluyveromyces lactis (Lactozym 3000 L HP G). *Food Chemistry*, *107*(1), 258–264. doi:10.1016/j.foodchem.2007.08.011

Martins, G. N., Ureta, M. M., Tymczyszyn, E. E., Castilho, P. C., & Gomez-Zavaglia, A. (2019). Technological aspects of the production of fructo and enzymatic synthesis and hydrolysis. *Frontiers in Nutrition*, *6*, 78. doi:10.3389/fnut.2019.00078 PMID:31214595

Mei, Z., Yuan, J., & Li, D. (2022). Biological activity of galacto-oligosaccharides: A review. *Frontiers in Microbiology*, *13*, 993052. doi:10.3389/fmicb.2022.993052 PMID:36147858

Mistry, R. H., Liu, F., Borewicz, K., Lohuis, M. A. M., Smidt, H., Verkade, H. J., & Tietge, U. J. F. (2020). Long-term β-galacto-oligosaccharides supplementation decreases the development of obesity and insulin resistance in mice fed a western-type diet. *Molecular Nutrition & Food Research*, *64*(12), 1900922. doi:10.1002/mnfr.201900922 PMID:32380577

Mohammadian, M., Moghadam, M., Salami, M., Emam Djomeh, Z., Alavi, F., Momen, S., & Moosavi-Movahedi, A. A. (2020). Whey protein aggregates formed by non-toxic chemical cross-linking as novel carriers for curcumin delivery: Fabrication and characterization. *Journal of Drug Delivery Science and Technology*, *56*, 101531. doi:10.1016/j.jddst.2020.101531

Murari, C. S., da Silva, D. C. M. N., Schuina, G. L., Mosinahti, E. F., & Del Bianchi, V. L. (2019). Bioethanol production from dairy industrial coproducts. *BioEnergy Research*, *12*(1), 112–122. doi:10.100712155-018-9949-5

Murari, C. S., Machado, W. R. C., Schuina, G. L., & Del Bianchi, V. L. (2019). Optimization of bioethanol production from cheese whey using Kluyveromyces marxianus URM 7404. *Biocatalysis and Agricultural Biotechnology*, 20, 101182. https://www.oecd-ilibrary.org/sites/aa3fa6a0-en/index.html?itemId=/content/component/aa3fa6a0-en#:~:text=This%20chapter%20describes%20the%20market,whole%20milk%20powder%20are%20discussed

Okamoto, K., Nakagawa, S., Kanawaku, R., & Kawamura, S. (2019). Ethanol production from cheese whey and expired milk by the brown rot fungus Neolentinus lepideus. *Fermentation (Basel, Switzerland)*, *5*(2), 49. doi:10.3390/fermentation5020049

Okamoto, K., Nitta, Y., Maekawa, N., & Yanase, H. (2011). Direct ethanol production from starch, wheat bran and rice straw by the white rot fungus Trametes hirsuta. *Enzyme and Microbial Technology*, *48*(3), 273–277. doi:10.1016/j.enzmictec.2010.12.001 PMID:22112911

Okamoto, K., Uchii, A., Kanawaku, R., & Yanase, H. (2014). Bioconversion of xylose, hexoses and biomass to ethanol by a new isolate of the white rot basidiomycete Trametes versicolor. *SpringerPlus*, *3*(1), 1–9. doi:10.1186/2193-1801-3-121 PMID:24624317

Okamoto, K., Yoneda, M., & Fumioka, T. (2017). Isolation of a Peniophora strain capable of producing ethanol from starch and kitchen waste. *Fermentation Technology*, *6*(1), 1000140.

Oke, M. A., Annuar, M. S. M., & Simarani, K. (2016). Mixed feedstock approach to lignocellulosic ethanol production—Prospects and limitations. *BioEnergy Research*, *9*(4), 1189–1203. doi:10.100712155-016-9765-8

Oliveira, C., Guimarães, P. M. R., & Domingues, L. (2011). Recombinant microbial systems for improved β-galactosidase production and biotechnological applications. *Biotechnology Advances*, *29*(6), 600–6009. doi:10.1016/j.biotechadv.2011.03.008 PMID:21514372

Otieno, D. O. (2010). Synthesis of β-galactooligosaccharides from lactose using microbial β-galactosidases. *Comprehensive Reviews in Food Science and Food Safety*, *9*(5), 471–482. doi:10.1111/j.1541-4337.2010.00121.x PMID:33467831

Pais, J., Serafim, L. S., Freitas, F., & Reis, M. A. (2016). Conversion of cheese whey into poly (3-hydroxybutyrate-co-3-hydroxyvalerate) by Haloferax mediterranei. *New Biotechnology*, *33*(1), 224–230. doi:10.1016/j.nbt.2015.06.001 PMID:26134839

Panesar, P. S., Kennedy, J. F., Gandhi, D. N., & Bunko, K. (2007). Bioutilisation of whey for lactic acid production. *Food Chemistry*, *105*(1), 1–14. doi:10.1016/j.foodchem.2007.03.035

Panghal, A., Kumar, V., Dhull, S. B., Gat, Y., & Chhikara, N. (2017). Utilization of dairy industry waste-whey in formulation of papaya RTS beverage. *Current Research in Nutrition and Food Science Journal*, *5*(2), 168–174. doi:10.12944/CRNFSJ.5.2.14

Paques, J. P., Van der Linden, E., Van Rijn, C. J. M., & Sagis, L. M. C. (2014). Preparation methods of alginate nanoparticles. *Advances in Colloid and Interface Science*, *209*, 163–171. doi:10.1016/j.cis.2014.03.009 PMID:24745976

Pereira-Rodríguez, A., Fernandez-Leiro, R., Gonzalez Siso, M. I., Cerdan, M. E., Becerra, M., & Sanz-Aparicio, J. (2010). Crystallization and preliminary X-ray crystallographic analysis of β-galactosidase from Kluyveromyces lactis. *Acta Crystallographica*, *66*, 297–300. PMID:20208165

Pires, A. F., Marnotes, N. G., Rubio, O. D., Garcia, A. C., & Pereira, C. D. (2021). Dairy by-products: A review on the valorization of whey and second cheese whey. *Foods*, *10*(5), 1067. doi:10.3390/foods10051067 PMID:34066033

Rama, G. R., Kuhn, D., Beux, S., Maciel, M. J., & de Souza, C. F. V. (2019). Potential applications of dairy whey for the production of lactic acid bacteria cultures. *International Dairy Journal*, *98*, 25–37. doi:10.1016/j.idairyj.2019.06.012

Ramos, O. L., Fernandes, J. C., Silva, S. I., Pintado, M. E., & Malcata, F. X. (2012). Edible films and coatings from whey proteins: A review on formulation, and on mechanical and bioactive properties. *Critical Reviews in Food Science and Nutrition*, *52*(6), 533–552. doi:10.1080/10408398.2010.500528 PMID:22452733

Renhe, I. R. T. (2008). O papel do leite na nutrição. *Revista do Instituto de Latícinios Cândido Tostes*, *363*(63), 36–43.

Romani, A., Michelin, M., Domingues, L., & Teixeira, J. A. (2018). Valorization of wastes from agrofood and pulp and paper industries within the biorefinery concept: southwestern Europe scenario. In T. Bhaskar, A. Pandey, S. V. Mohan, & D. Lee (Eds.), *J, Khanal, S.K. Waste biorefinery* (pp. 487–504). Elsevier. doi:10.1016/B978-0-444-63992-9.00016-1

Santos, F. F., Queiroz, R. C. S., & Almeida Neto, J. A. (2017). Avaliação da aplicação das técnicas da Produção Mais Limpa em um laticínio no Sul da Bahia. *Gestão & Produção*, *25*(1), 117–131. doi:10.1590/0104-530x2234-16

Santos, H. C. M. Jr, Maranduba, H. L., de Almeida Neto, J. A., & Rodrigues, L. B. (2017). Life cycle assessment of cheese production process in a small-sized dairy industry in Brazil. *Environmental Science and Pollution Research International, 24*(4), 3470–3482. doi:10.100711356-016-8084-0 PMID:27873115

Santos, M. J., Teixeira, J. A., & Rodrigues, L. R. (2011). Fractionation and recovery of whey proteins by hydrophobic interaction chromatography. *Journal of Chromatography. B, Analytical Technologies in the Biomedical and Life Sciences, 879*(7-8), 475–479. doi:10.1016/j.jchromb.2011.01.003 PMID:21292570

Saquib, S. A., Akram, A. S. A., Halim, S. A., & Tassaduq, R. (2017). Sources of β-galactosidase and its applications in food industry. *Biotech, 7*(1), 79. PMID:28500401

Schmaltz, E., Melvin, E. C., Diana, Z., Gunady, E. F., Rittschof, D., Somarelli, J. A., & Dunphy-Daly, M. M. (2020). Plastic pollution solutions: emerging technologies to prevent and collect marine plastic pollution. *Environment international,* 144, 106067. Sgarbieri, V.C., Revisão: Propriedades estruturais e físico-químicas das proteínas do leite. *Brazilian Journal of Food Technology, 8*(1), 43–56.

Sharma, R. (2019). Whey proteins in functional foods. In H. Deeth & N. Bansal (Eds.), *Whey proteins* (pp. 637–663). Elsevier. doi:10.1016/B978-0-12-812124-5.00018-7

Shen, X., Zhao, C., Lu, J., & Guo, M. (2018). Physicochemical properties of whey-protein-stabilized astaxanthin nanodispersion and its transport via a Caco-2 monolayer. *Journal of Agricultural and Food Chemistry, 66*(6), 1472–1478. doi:10.1021/acs.jafc.7b05284 PMID:29345928

Shpigelman, A., Cohen, Y., & Livney, Y. (2012). Thermally-induced β-lactoglobulin-EGCG nanovehicles: Loading, stability, sensory and digestive-release study. *Food Hydrocolloids, 29*(1), 57–67. doi:10.1016/j.foodhyd.2012.01.016

Sienkiewicz, T., & Riedel, C. L. (1990). *Whey and Whey Utilization*. Verlag Th. Mann.

Sijbers, A. M., Schoemaker, R. J. W., Nauta, A., & Alkema, W. (2020). Revealing new leads for the impact of galacto-oligosaccharides on gut commensals and gut health benefts through text mining. *Beneficial Microbes, 11*(3), 283–302. doi:10.3920/BM2019.0105 PMID:32299221

Simões, L., Martins, J., Pinheiro, A., Vicente, A., & Ramos, O. (2020). β-lactoglobulin micro- and nanostructures as bioactive compounds vehicle: In vitro studies. *Food Research International, 131*, 108979. doi:10.1016/j.foodres.2020.108979 PMID:32247463

Singh, T., Øiseth, S., Lundin, L., & Day, L. (2014). Influence of heat and shear induced protein aggregation on the in vitro digestion rate of whey proteins. *Food & Function*, *5*(11), 2686–2698. doi:10.1039/C4FO00454J PMID:25205335

Tian, S., Wang, J., Yu, H., Wang, J., & Zhu, W. (2019). Changes in ileal microbial composition and microbial metabolism by an early-life galacto-oligosaccharides intervention in a neonatal porcine model. *Nutrients*, *11*(8), 1753. doi:10.3390/nu11081753 PMID:31366090

Torres, D. P. M., Gonçalves, M., Teixeira, J. A., & Rodrigues, L. R. (2010). Galacto-oligosaccharides: Production, properties, applications, and significance as prebiotics. *Comprehensive Reviews in Food Science and Food Safety*, *9*(5), 438–454. doi:10.1111/j.1541-4337.2010.00119.x PMID:33467830

Tsang, Y. F., Kumar, V., Samadar, P., Yang, Y., Lee, J., Ok, Y. S., & Jeon, Y. J. (2019). Production of bioplastic through food waste valorization. *Environment International*, *127*, 625–644. doi:10.1016/j.envint.2019.03.076 PMID:30991219

United States Department of Agriculture (USDA). (2022). *Dairy: World Markets and Trade*. USDA. https://apps.fas.usda.gov/psdonline/circulars/dairy.pdf

Varney, J., Barrett, J., Scarlata, K., Catsos, P., Gibson, P. R., & Muir, J. G. (2017). FODMAPs: Food composition, defining cutoff values and international application. *Journal of Gastroenterology and Hepatology*, *32*, 53–61. doi:10.1111/jgh.13698 PMID:28244665

Vera, C., Guerrero, C., Aburto, C., Cordova, A., & Illanes, A. (2020). Conventional and nonconventional applications of β-galactosidases. *Biochimica et Biophysica Acta. Proteins and Proteomics*, *1868*(1), 140271. doi:10.1016/j.bbapap.2019.140271 PMID:31494342

Weijers, C. A., Franssen, M. C., & Visser, G. M. (2008). Glycosyltransferasecatalyzed synthesis of bioactive oligosaccharides. *Biotechnology Advances*, *26*(5), 436–456. doi:10.1016/j.biotechadv.2008.05.001 PMID:18565714

Xavier, J. R., Ramana, K. V., & Sharma, R. K. (2018). β-galactosidase: Biotechnological applications in food processing. *Journal of Food Biochemistry*, *42*(5), 12564. doi:10.1111/jfbc.12564

Ximenes, L. F. (2021). *Lácteos*. https://www.bnb.gov.br/s482-dspace/bitstream/123456789/909/1/2021_CDS_181.pdf

Yadav, J. S. S., Bezawada, J., Ajila, C. M., Yan, S., Tyagi, R. D., & Surampalli, R. Y. (2014). Mixed culture of Kluyveromyces marxianus and Candida krusei for single-cell protein production and organic load removal from whey. *Bioresource Technology*, *164*, 119–127. doi:10.1016/j.biortech.2014.04.069 PMID:24844166

Yadav, J. S. S., Yan, S., Pilli, S., Kumar, L., Tyagi, R. D., & Surampalli, R. Y. (2015). Cheese whey: A potential resource to transform into bioprotein, functional/nutritional proteins and bioactive peptides. *Biotechnology Advances*, *33*(6), 756–774. doi:10.1016/j.biotechadv.2015.07.002 PMID:26165970

Zandona, E., Blažić, M., & Režek Jambrak, A. (2021). Whey utilization: Sustainable uses and environmental Approach. *Food Technology and Biotechnology*, *59*(2), 147–161. doi:10.17113/ftb.59.02.21.6968 PMID:34316276

Zou, J., Chen, X., Hu, Y., Xiao, D., Guo, X., Chang, X., & Zhou, L. (2021). Uncoupling glucose sensing from GAL metabolism for heterologous lactose fermentation in Saccharomyces cerevisiae. *Biotechnology Letters*, *43*(8), 1607–1616. doi:10.100710529-021-03136-8 PMID:33937967

Chapter 11
The Importance of Sustainability Indicators in Dairy Farming:
The Perceptions of Students of Agricultural Science Courses in Brazil

Mirian Fabiana da Silva
State University of Maringá, Brazil

Ferenc I. Bánkuti
State University of Maringá, Brazil

EXECUTIVE SUMMARY

The objective of this research was to assess the perception of agricultural science undergraduate students from Brazilian institutions on dairy farm sustainability indicators. The total number of survey respondents was 351. Undergraduate students from public institutions rated technical and environmental indicators as the most important. Students who had had previous contact with rural production gave lower scores (lower importance) to environmental indicators and higher scores (higher importance) to economic and technical indicators. Agricultural engineering students attributed lower importance (scores) to environmental indicators than agronomic engineering and veterinary medicine students. On the basis of these findings, the authors conclude that the plurality of students' perceptions on dairy farm sustainability indicators is mainly related to concrete on-farm experiences, differences in syllabus between agricultural science courses, and the public or private nature of the educational system.

DOI: 10.4018/978-1-6684-5472-5.ch011

INTRODUCTION

Brazilian dairy farming, as well as other production sectors, has faced several challenges and criticisms in relation to technical, economic, environmental, and social aspects. Specifically, concerns have arisen about the need to preserve natural resources, minimize environmental impacts, optimize production, increase efficiency and economic viability, improve quality of life and working conditions, meet human needs, and supply food to the population without compromising future generations (Oudshoorn et al., 2012; WCED, 1987). For this, the importance of promoting production in a sustainable manner is increasingly recognized (Herrero et al., 2015; Olde et al., 2016).

Efficient development of livestock production necessitates a systemic and multidisciplinary approach. A systemic view allows to understand the interrelation between elements and factors involved in dairy production. From this perspective, the dairy farm is analyzed as a system composed of a set of interrelated activities and operations for the production of goods (milk, animals, and others) from specific resources (feed, animals, and others), which are influenced by internal and external factors (Batalha & Silva, 2010).

Dairy production can be analyzed by using groups of indicators. Indicators are of utmost importance for systemic analysis of the current situation of production systems and for the development of actions aimed at production sustainability (Gaudino et al., 2014; Lebacq et al., 2013; Olde et al., 2017; Bánkuti et al., 2020).

The development and use of indicators should take into account the dynamic and social context of production systems, with participation of the community associated with the activity. By examining the community's perspective of indicators, it is possible to determine whether these variables hold meaning and utility for those involved, thereby providing valuable insight for sustainability assessment (King et al., 2000; Marandure et al., 2017; Munyaneza et al., 2019; Okumah et al., 2018). Understanding the perception of professionals on indicators is essential for the adoption of such indicators in the field. Participation of the actors involved in dairy farming enables efficient planning, increases the probability of successful implementation of actions and strategies, and disseminates knowledge on the activity (Almeida et al., 2017).

Few studies have been conducted to investigate which indicators professionals and in-training individuals linked to dairy farming (i.e., researchers, professors, students, technicians, farmers, and stakeholders) deem important for sustainability assessment (Bélanger et al., 2012, 2015; Meul et al., 2008; Munyaneza et al., 2019; Van Calker et al., 2005). Silva and Gameiro (2022) analyzed the perceptions of Brazilian professionals, such as researchers, professors, advisers, farmers, and others, about the importance of using indicators. The purpose of their research was to examine the views of individuals with experience in dairy farming. However, there is a need for studies assessing the perception of future professionals, that is, students in the field of Agricultural Sciences (Agronomic Engineering, Agricultural Engineering, Veterinary Medicine, Animal Sciences, and the like).

The level of knowledge and perception of individuals on indicators can influence the adoption and diffusion of assessment methods. Understanding students' views on dairy farm sustainability assessment may guide the implementation of lectures, subjects, and even courses to fill potential knowledge gaps. Thus, educational institutions and organizations will be able to introduce sustainable practices into their routines and create professionals who are equipped with the knowledge to produce milk in a more sustainable manner.

According to several authors, there is a growing need for universities to prepare undergraduate students to work toward sustainable development (Aleixo et al., 2021; Hasan et al., 2018; Rampasso et al. 2019; Rieckmann, 2012; Salovaara et al., 2021; Sidiropoulos, 2018). Adoption of a broader, transdisciplinary vision of sustainability may serve as a useful starting point for researchers, teachers, and higher education coordinators to evaluate training and prioritize the necessary improvements.

In view of the above, this study aimed to examine the perception of Agricultural Science undergraduate students from Brazilian institutions on dairy farm sustainability indicators based on their life experiences and formation.

METHODS

This descriptive study used a questionnaire-based survey method. The target audience comprised students of Agricultural Science courses attending Brazilian higher education institutions. The research was approved by the Human Research Ethics Standing Committee of the local university (COPEP protocol no. 4.660.307)

The questionnaire was created using Google Forms. This platform was chosen because it can be easily accessed by participants, as it is available online. A link to the questionnaire was sent to course coordinators and secretariats by e-mail. To obtain the e-mail addresses of coordinators and secretariats, we used a list of Brazilian higher education institutions provided by the Ministry of Education (Cadastro Nacional de Cursos e Instituições de Educação Superior Cadastro e-MEC, https://emec.mec.gov.br/). With the list of institutions in hand, we searched at university and college webpages for the contacts of course coordinators and/or secretariats. For some institutions, it was not possible to find contact details.

Coordinators and professors were contacted. Those who agreed to participate in the survey provided the link to the questionnaire to their students. For one institution, it was possible to send the link directly to students, as their e-mail addresses were made available. Participants were also invited to share the link with other students.

The questionnaire used was that developed by Silva (2020), with some modifications in the sections regarding participants' characteristics and list of indicators. The questionnaire is divided into two parts. The first part comprises questions about participants' characteristics, and the second part contains questions about the importance of indicators in the assessment of dairy farm sustainability. The social variables used to characterize the students are presented in Table 1. These variables were analyzed by descriptive statistics.

In the second part of the questionnaire, responses were rated on a five-point Likert scale (5 = very important, 4 = important, 3 = desirable, 2 = non-priority, 1 = dispensable) plus an option that indicated that the participant does not have knowledge on that item (0 = I do not know or I have no knowledge) (Likert, 1932; Sullivan & Artino JR, 2013). Several authors used the Likert scale to assess dairy production sustainability indicators (Bélanger et al., 2015; Gazola et al., 2018; Munyaneza et al., 2019; Van Calker et al., 2005). Other authors used the Likert scale to assess students' perceptions of sustainable development (Hasan et al., 2018; Saqib et al., 2020; Wang et al., 2020).

The questionnaire was prepared based on a literature review. Scientific articles published between 1999 and 2019 and using indicators to assess dairy farm sustainability were searched in the ScienceDirect, Web of Science, Scopus, and Google Scholar databases. Searches were also performed by reading the reference list of selected articles (Silva, 2020; Silva & Gameiro, 2021). The indicators evaluated in the study are presented in Table 2.

Indicator scores were analyzed by descriptive statistics and multivariate analysis. Multivariate analysis allows identifying variables that are correlated and reducing the dataset by the principal component method, achieving minimal information loss and maximum proportion of variance with a minimum number of components. In the current study, factor analysis was used to define sustainability indicators. It is an interdependence technique used to synthesize information of several original variables into a smaller set of intrinsic factors or indicators (Hair et al., 2014).

Table 1. Social variables used to characterize the sample of undergraduate students

Variable	Type
Public or private institution	Binary
Undergraduate course	Nominal
Previous contact with rural production	Binary
Lived in rural areas	Binary
Internship or work experience with dairy cattle	Binary

Table 2. Technical, economic, social, and environmental indicators evaluated in the study

Technical indicators	Social indicators
TI_1 - Annual milk production (kg milk/year)	SI_1 - Milk quality
TI_2 - Stocking rate (livestock units/ha)	SI_2 - Penalty for not meeting milk composition requirements (%)
TI_3 - Lactating cows per area (cows/ha)	SI_3 - Price bonus (%)
TI_4 - Milk production per area (kg milk/ha/year)	SI_4 - Animal welfare index
TI_5 - Milk production per permanent worker (kg milk/ worker)	SI_5 - Employment (employee/day)
TI_6 - Lactating cows per employee (cows/employee)	SI_6 - Employee turnover
TI_7 - Milk production per lactating cow (kg milk/day)	SI_7 - Accidents at work (accidents/employee/year)
TI_8 - Total milk production (kg milk/day)	SI_8 - Lost time (%)
TI_9 - Ratio of lactating cows to total cows (%)	SI_9 - Working time (h/employee/month)
TI_10 - Ratio of lactating cows to herd size (%)	SI_10 - Rest days (rest days/employee/month)
TI_11 - Reproductive index	SI_11 - Employee salary (R$/employee/month)
TI_12 - Discard rate (%)	SI_12 - Remuneration of family labor (R$/year)
TI_13 - Morbidity rate (%)	SI_13 - Quality of life
TI_14 - Mortality rate (%)	SI_14 - Education
TI_15 - Feed self-sufficiency (%)	SI_15 - Entrepreneurship
TI_16 - Concentrate consumption per lactating cow (kg dry matter/lactating cow/year)	SI_16 - Social involvement
TI_17 - Dairy production efficiency (kg milk/kg dry matter)	SI_17 - Succession
TI_18 - Milk production to concentrate feed ratio (kg milk/ kg dry matter)	SI_18 - Training and professional development (h/employee/ year)
TI_19 - Body condition score of cows per lactation phase (scale ranging from 1 to 5)	SI_19 - Diversification of farm activities (activities/farm)
	SI_20 - Payment for ecosystem services (R$/ha/year)
Economic indicators	**Environmental indicators**
ECI_1 - Average price of milk (R$/kg)	ENI_1 - Energy use per kg of milk (MJ/kg milk) and production of milk by fossil energy source
ECI_2 - Gross farm income (R$/year)	ENI_2 - Use of renewable energy (%)
ECI_3 - Gross income from milk in relation to gross farm income (%)	ENI_3 - Ammonia emission (kg ammonia/ha)
ECI_4 - Expenditure on hired labor per gross income from milk (%)	ENI_4 - Global warming potential (kg CO_2 equivalents/kg milk)
ECI_5 - Expenditure on concentrate per gross income from milk (%)	ENI_5 - Acidification potential (g sulfur dioxide equivalents/ kg milk)
ECI_6 - Total cost of dairy farm (R$/year)	ENI_6 - Eutrophication potential (g phosphate or nitrate equivalents/kg milk)
ECI_7 - Total unit cost of milk (R$/kg)	ENI_7 - Terrestrial and aquatic ecotoxicity (g 1,4-dichlorobenzene equivalents/kg milk)
ECI_8 - Total cost of milk per milk price (%)	ENI_8 - Nutrient balance (kg nitrogen, phosphorus, and potassium/year)

continues on following page

Table 2. Continued

Technical indicators	Social indicators
ECI_9 - Annual gross margin (R$/year)	ENI_9 - Nutrient use efficiency (%)
ECI_10 - Unit gross margin (R$/kg)	ENI_10 - Land occupation (m^2/kg milk)
ECI_11 - Gross margin per area (R$/ha)	ENI_11 - Air quality
ECI_12 - Annual net margin (R$/year)	ENI_12 - Water consumption (m^3 water/kg milk)
ECI_13 - Unit net margin (R$/kg)	ENI_13 - Consumption of surface water, groundwater, and rain (%)
ECI_14 - Farm profit (R$/year)	ENI_14 - Quality of water used for human and animal consumption and milking
ECI_15 - Unit profit (R$/kg)	ENI_15 - Wastewater reuse (%)
ECI_16 - Profit per area (R$/ha)	ENI_16 - Effluent production per milked cow (kg/milked cow)
ECI_17 - Capital stock per area (R$/ha)	ENI_17 - Manure treated and reused (%)
ECI_18 - Capital stock per kg of milk (R$/kg milk/day)	ENI_18 - Management of animal and human waste and manure
ECI_19 - Expenditure on feed (R$/year)	ENI_19 - Use of chemical fertilizer (kg NPK/ha)
ECI_20 - Rate of return on capital invested in land (%/year)	ENI_20 - Soil fertility
ECI_21 - Cost/benefit ratio	ENI_21 - Organic carbon in pasture soil (t/ha/year)
ECI_22 - Solvency	ENI_22 - Soil loss (t/ha/year)
ECI_23 - Degree of indebtedness	ENI_23 - Soil erosion (% of farm area)
ECI_24 - Profitability (%)	ENI_24 - Degraded areas (% of farm area)
ECI_25 - Farm leveling point (kg/day)	ENI_25 - Soil management (Yes or no)
ECI_26 - Cash flow	ENI_26 - Management of animal health, agricultural waste, and carcass disposal
	ENI_27 - Use of antibiotics (dose/animal/year)
	ENI_28 - Disposal of milk from medicated animals
	ENI_29 - Pesticide use (kg pesticides/ha)
	ENI_30 - Degree of biodiversity
	ENI_31 - Preservation area or legal reserve (% of farm area)
	ENI_32 - Protection of watercourse or permanent preservation area (% of farm area)
	ENI_33 - Grant and license
	ENI_34 - Emergy synthesis

Source: adapted from Silva and Gameiro (2022).

Orthogonal varimax rotation was performed to obtain the lowest number of factors explaining the maximum possible variance. The criteria for retention of extracted variables and factors were factor loadings equal to or greater than 0.5, eigenvalues equal to or greater than 1.0, and cumulative variance of at least 60%. Factors were labeled according to the variables (indicators) that defined them. Kaiser–Meyer–Olkin (KMO) and Bartlett's sphericity tests were performed to assess the adequacy of data for factor analysis (Hair et al., 2014). Factor scores obtained by factor analysis were saved as regression measures. With this procedure, factor scores can be used with metric variables for various statistical analyses, for example, test of means (Field, 2009).

Students were classified according to the following items: nature of higher education institution (private or public), previous contact with rural production (yes or no), experience with dairy cattle (no experience or internship/job), having lived in rural areas (yes or no), and undergraduate course (Veterinary Medicine, Animal Sciences, Agricultural Sciences, Agribusiness, Agronomic Engineering, Agricultural Engineering, or Biosystems Engineering). The mean of factor scores generated for sustainability indicators was compared between groups, and differences were assessed by analysis of variance (ANOVA) at $p < 0.05$. Differences between undergraduate courses were assessed using Tukey's test ($p < 0.05$). Data were processed and analyzed statistically using Jamovi software version 1.8.

RESULTS

A total of 351 students from 85 Brazilian higher education institutions answered the questionnaire. Of these, 70.09% were from public institutions and 29.91% from private ones,75.50% had had previous contact with rural production before entering higher education, 49.57% had lived in rural areas, and 46.72% had interned or worked with dairy cattle. Figure 1 shows the distribution frequency of undergraduate courses. Veterinary Medicine had the highest number of students.

Of the students who answered the questionnaire, 24.79% were in the fifth year of the undergraduate course, as shown in Figure 2.

Figure 3 shows the distribution of the 351 participants according to regions of the country. The majority of study participants were from the southern region (36.47%), and Paraná was the state with the highest number of participants (n = 80).

The KMO value was 0.960, indicating that the sample size and variables were adequate for factor analysis. The result of the KMO test ranges from 0 to 1: 0 indicates inadequacy of the data set, values of 0.5 and greater are acceptable, and values of 0.6 and above are recommended (Hair et al., 2014). Bartlett's test was significant ($\chi2 = 30488$, df = 3570, $p < 0.001$), indicating the presence of significant relationships between variables. Four factors were retained after applying the predetermined criteria (Hair et al., 2014). Together, the four factors explained 61.10% of the cumulative variance of the data set (Table 3).

Figure 1. Distribution of students (n = 351) according to undergraduate course

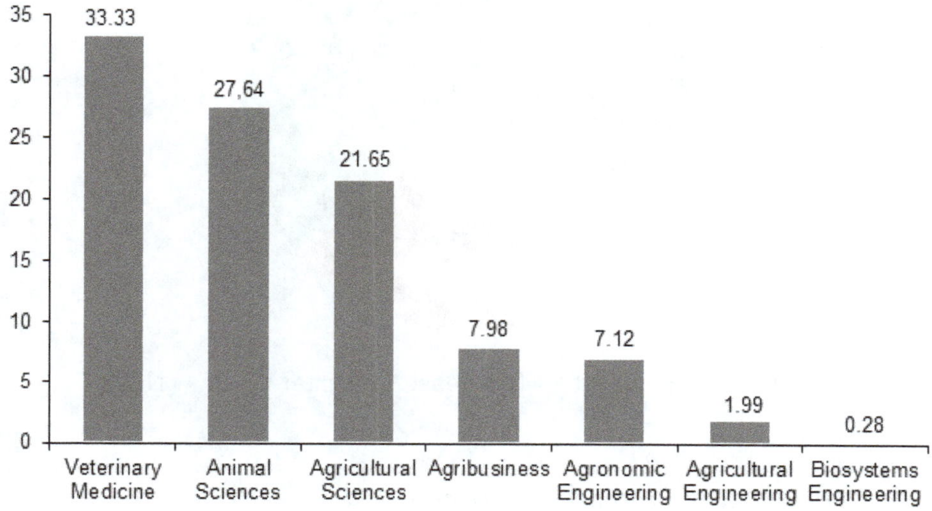

Figure 2. Distribution of students according to year of undergraduate education

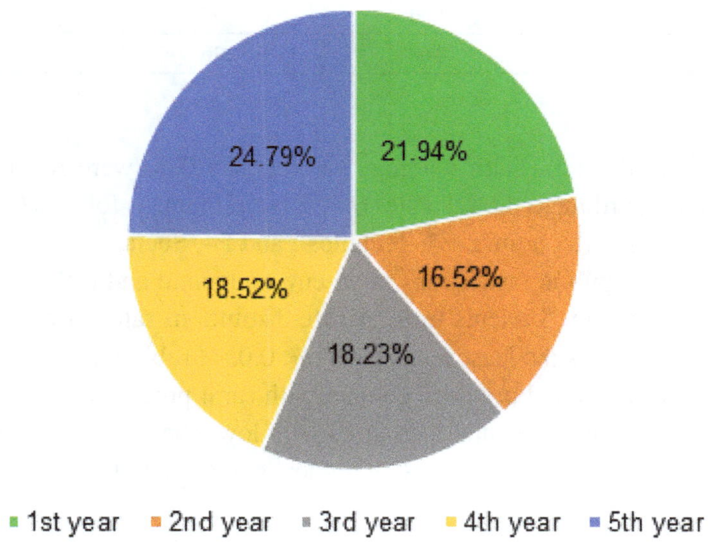

Figure 3. Frequency distribution of undergraduate students (n = 351) according to region where the institution is located

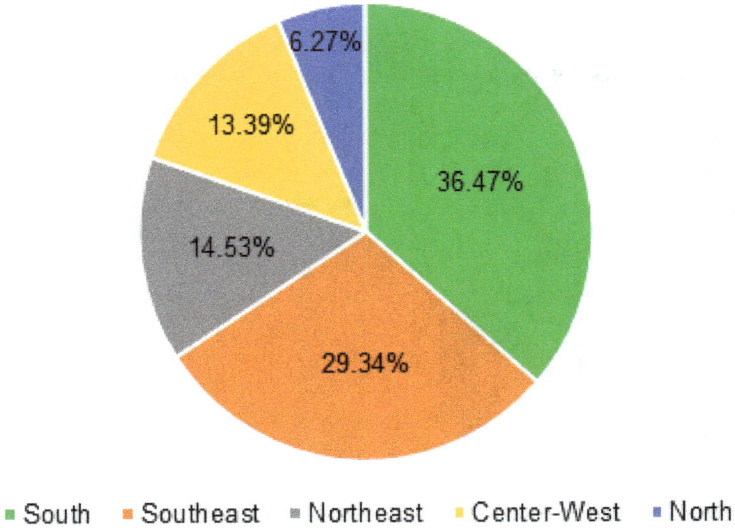

Table 3. Factor eigenvalues and variance explained

Factor	Eigenvalue	% of Variance	% Cumulative
1	16.60	19.60	19.60
2	12.80	15.00	34.60
3	11.80	13.90	48.40
4	10.80	12.70	61.10

Variables (indicators) with loadings greater than 0.50 were retained. Factors were labeled according to the variables that defined them, as follows (Table 4): F1, Environmental; F2, Economic; F3, Technical; and F4, Social.

Students from public institutions rated environmental and technical indicators as the most important. Students from private institutions rated environmental and technical indicators as the least important ($p < 0.05$) (Table 5).

Students who had had previous contact with rural production (before entering the course) rated environmental indicators with lower scores (less important) and economic and technical indicators with higher scores (more important) ($p < 0.05$) (Table 6).

Table 4. Rotated factor matrix

F1_Environmental		F2_Economic	
ENI_31 - Preservation area or legal reserve	0.776	ECI_11 - Gross margin per area	0.771
ENI_32 - Protection of watercourse or permanent preservation area	0.740	ECI_13 - Unit net margin	0.763
ENI_30 - Degree of biodiversity	0.739	ECI_10 - Unit gross margin	0.750
ENI_16 - Effluent production per milked cow	0.726	ECI_12 - Net margin	0.746
ENI_13 - Consumption of surface water, groundwater, and rain	0.717	ECI_18 - Capital stock per kg of milk	0.713
ENI_8 - Nutrient balance	0.715	ECI_25 - Farm leveling point	0.713
ENI_17 - Manure treated and reused	0.699	ECI_17 - Capital stock per area	0.693
ENI_18 - Animal and human waste and manure management	0.697	ECI_9 - Gross margin	0.683
ENI_9 - Nutrient use efficiency	0.686	ECI_19 - Expenditure on feed	0.676
ENI_24 - Degraded areas	0.683	ECI_20 - Rate of return on capital invested in land	0.666
ENI_15 - Wastewater reuse	0.682	ECI_22 - Solvency	0.655
ENI_5 - Acidification potential	0.680	ECI_23 - Degree of indebtedness	0.642
ENI_12 - Use of water per kg of milk	0.677	ECI_3 - Gross income from milk in relation to gross farm income	0.632
ENI_23 - Soil erosion	0.674	ECI_5 - Expenditure on concentrate per gross income from milk	0.623
ENI_25 - Soil management	0.672	ECI_16 - Profit per area	0.608
ENI_10 - Land occupation	0.667	ECI_26 - Cash flow	0.602
ENI_3 - Ammonia emission	0.665	ECI_4 - Expenditure on hired labor per gross income from milk	0.588
ENI_14 - Quality of water used for human and animal consumption and milking	0.646	ECI_2 - Gross income	0.570
ENI_26 - Management of animal health, agricultural waste, and carcass disposal	0.644	ECI_24 - Profitability	0.568
ENI_2 - Use of renewable energy	0.630	ECI_21 – Cost/benefit ratio	0.509
ENI_11 - Air quality	0.628		
ENI_29 - Pesticide use	0.616		
ENI_22 - Soil loss	0.609		
ENI_34 - Emergy synthesis	0.605		
ENI_6 - Eutrophication potential	0.601		
ENI_33 - Grant and license	0.600		
ENI_19 - Use of chemical fertilizer	0.586		
ENI_20 - Soil fertility	0.577		
ENI_28 - Disposal of milk from medicated animals	0.566		
ENI_27 - Use of antibiotics	0.557		

continues on following page

Table 4v. Continued

F1_Environmental		F2_Economic	
F3_Technical		**F4_ Social**	
TI_7 - Milk production per lactating cow	0.764	SI_10 - Rest days	0.734
TI_9 - Ratio of lactating cows per total cows	0.714	SI_11 - Employee salary	0.730
TI_17 - Dairy efficiency	0.689	SI_9 - Working time	0.721
TI_11 - Reproductive index	0.688	SI_18 - Training and professional development	0.706
TI_8 - Milk production per total cows	0.671	SI_14 - Education	0.674
TI_16 - Concentrate consumption per lactating cow	0.645	SI_5 - Employment	0.674
TI_12 - Discard rate	0.633	SI_15 - Entrepreneurship	0.665
TI_13 - Morbidity rate	0.626	SI_16 - Social involvement	0.658
TI_18 - Milk production to concentrate feed ratio	0.625	SI_13 - Quality of life	0.654
TI_14 - Mortality rate	0.624	SI_17 - Succession	0.651
TI_1 - Annual milk production	0.622	SI_4 - Animal welfare index	0.635
TI_10 - Ratio of lactating cows to herd size	0.606	SI_12 - Remuneration of family labor	0.616
TI_3 - Lactating cows per area	0.600	SI_7 - Accidents at work	0.612
TI_15 - Feed self-sufficiency	0.585	SI_2 - Penalty for not meeting milk composition requirements	0.605
ECI_14 - Farm profit	0.580	SI_6 - Employee turnover	0.568
TI_4 - Milk production per area	0.551	SI_8 - Lost time	0.549
TI_2 - Stocking rate	0.549	SI_19 - Diversification of farm activities	0.540
		SI_1 - Milk quality	0.534

Table 5. Environmental, economic, technical, and social factor scores of participants according to type of undergraduate institution

Factor	Type of institution	*n*	Mean	Standard deviation	*p*-value
F1_Environmental	Private	105	−0.168	0.991	0.039
	Public	246	0.072	0.997	
F2_Economic	Private	105	0.106	0.895	0.194
	Public	246	−0.045	1.040	
F3_Technical	Private	105	−0.198	1.220	0.015
	Public	246	0.085	0.879	
F4_Social	Private	105	0.021	−0.009	0.793
	Public	246	−0.009	1.270	

Table 6. Environmental, economic, technical, and social factor scores of participants according to previous contact with rural production

Factor	Previous contact with rural production	*n*	Mean	Standard deviation	*p*-value
F1_ Environmental	Yes	265	−0.079	0.975	0.009
	No	86	0.244	1.040	
F2_Economic	Yes	265	0.112	0.892	< 0.001
	No	86	−0.344	1.220	
F3_Technical	Yes	265	0.119	0.878	< 0.001
	No	86	−0.366	1.240	
F4_Social	Yes	265	0.005	0.972	0.882
	No	86	−0.014	1.090	

Students who had lived in rural areas rated technical, economic, social, and environmental indicators similarly to students who had not lived in rural areas (Table 7).

Table 7. Environmental, economic, technical, and social factor scores of participants according to previous experience living in rural areas

Factor	Lived in rural areas	*n*	Mean	Standard deviation	*p*-value
F1_Environmental	Yes	174	−0.017	0.853	0.758
	No	177	0.016	1.130	
F2_Economic	Yes	174	0.100	0.902	0.064
	No	177	−0.098	1.080	
F3_Technical	Yes	174	0.014	1.000	0.803
	No	177	−0.013	0.999	
F4_Social	Yes	174	0.041	0.819	0.453
	No	177	−0.040	1.150	

Students who had interned or worked with dairy cattle considered economic and technical indicators to be more important compared with students who did not have experience with dairy cattle ($p < 0.05$) (Table 8).

Table 8. Environmental, economic, technical, and social factor scores of participants according to internship or work experience with dairy cattle

Factor	Internship or work experience with dairy cattle	*n*	Mean	Standard deviation	*p*-value
F1_Environmental	No	187	0.045	1.030	0.370
	Yes	164	−0.051	0.959	
F2_Economic	No	187	−0.111	1.070	0.027
	Yes	164	0.126	0.902	
F3_Technical	No	187	−0.168	1.120	<0.001
	Yes	164	0.191	0.810	
F4_Social	No	187	−0.070	1.110	0.160
	Yes	164	0.080	0.860	

Differences in factor scores between courses were only observed for environmental indicators ($p < 0.05$). Only one participant was enrolled in Biosystems Engineering; thus, the course was excluded from the analysis. Students of Agricultural Engineering attributed lower scores (lower importance) to environmental indicators than Agronomic Engineering and Veterinary Medicine students ($p < 0.05$) (Table 9).

DISCUSSION

Analysis of students' perceptions of indicators should be seen as an important strategy for the adoption and implementation of indicators on dairy farms and may subsidize changes to the syllabus of courses associated with animal production or other educational activities, such as short courses and lectures. Today's students will be the main influencers and decision makers of the future (Aleixo et al., 2021).

Environmental, social, economic, and technical sustainability has been increasingly demanded by control and regulatory bodies, as well as by various markets, particularly the foreign market. Therefore, formation and training of future agricultural professionals are crucial for market performance.

Several factors may influence students' perceptions of sustainable development. Perception of indicators might be related to knowledge about the subject, beliefs, values, experiences, and interactions with society and the environment (Naderi et al., 2022). Such factors may influence decisions about how an activity should be evaluated and developed. Individuals perceive and interpret a certain subject and its problems differently (Almeida et al., 2017). Aleixo et al. (2021) and García-González et al. (2020) argued that the life experiences and formation of undergraduate students influence their knowledge and actions toward sustainable development.

Table 9. Environmental, economic, technical, and social factor scores of participants according to undergraduate course

Factor	Undergraduate course	*n*	Mean	Standard deviation	*p*-value
F1_Environmental	Agribusiness	28	−0.018ab*	0.964	0.014
	Agricultural Sciences	76	−0.064ab	0.924	
	Agronomic Sciences	25	0.100a	0.691	
	Agricultural Sciences	7	−1.154bc	1.752	
	Veterinary Medicine	117	0.170a	1.005	
	Animal Sciences	97	−0.092ab	1.013	
F2_Economic	Agribusiness	28	0.133	1.012	0.248
	Agricultural Sciences	76	0.128	0.962	
	Agronomic Sciences	25	0.255	0.633	
	Agricultural Sciences	7	0.110	0.849	
	Veterinary Medicine	117	−0.165	1.118	
	Animal Sciences	97	−0.013	0.948	
F3_Technical	Agribusiness	28	−0.250	1.027	0.082
	Agricultural Sciences	76	−0.045	0.973	
	Agronomic Sciences	25	−0.164	1.050	
	Agricultural Sciences	7	0.516	0.635	
	Veterinary Medicine	117	−0.085	1.109	
	Animal Sciences	97	0.214	0.844	
F4_Social	Agribusiness	28	−0.260	1.602	0.33
	Agricultural Sciences	76	−0.139	1.042	
	Agronomic Sciences	25	−0.003	0.674	
	Agricultural Sciences	7	0.062	1.263	
	Veterinary Medicine	117	0.143	0.987	
	Animal Sciences	97	0.008	0.785	

*Means within a column followed by different lowercase letters are significantly different by Tukey's test ($p < 0.05$).

One way to improve the perception and knowledge of students is through educational actions and practical experience of problems associated with the activity. According to Rampasso et al. (2019), institutions must insert sustainable practices into the syllabus to better prepare students. It is important to offer opportunities that develop students' capacity to think, elaborate judgments, and articulate information, expanding their critical vision and the power of reflection during academic formation (Bergmann et al., 2020).

According to several authors, higher education institutions play an important role in the formation of future professionals by generating new knowledge, contributing to skill development, increasing awareness about sustainability, and promoting problem-solving, factors that contribute to a more holistic, integrated vision of activities (Aleixo et al., 2021; Hasan et al., 2018; Rampasso et al., 2019; Rieckmann, 2012; Salovaara et al., 2021; Sidiropoulos, 2018). Incorporating the concept of sustainability in education may help to construct a sustainable society (Yadav & Prakash, 2022). Higher education institutions are responsible for preparing students, changing their attitudes, and encouraging them to think about future generations and manage social, cultural, economic, and environmental challenges (Sidiropoulos, 2018).

Students must understand and balance technical, economic, environmental, and social dimensions and their interrelationships (Watson et al., 2013). This improvement must go beyond technical training, encompassing systemic/holistic thinking, reflective and critical thinking, intellectually understanding and solving of interrelated/interconnected problems through application of multidisciplinarity, interdisciplinarity, transdisciplinarity, and skills to deal with complex challenges, organize and interpret data, and formulate questions (Acevedo-Osorio et al., 2020; Corrêa et al., 2020; Gil-Doménech et al., 2021; Holdsworth & Sandri, 2021; Karatzoglou, 2013; Lambrechts et al., 2018; Olalla & Merino, 2019; Rampasso et al., 2018; Rieckmann, 2012; Salovaara et al., 2021; Watson et al., 2013). An enriched understanding can drive students to recognize and explore new perspectives to generate innovative solutions for sustainability (Salovaara et al., 2021).

Dialog between farmers, experts, teachers, students, and educational programs can influence the perception of students and their behaviors in the search for more sustainable production techniques (Cortner et al., 2019; Souza & Araujo, 2015). It is critical that students have transparent access to information and knowledge on how to assess sustainability, so that they can analyze and apply research results.

Students from public and private higher education institutions had different views about environmental and technical indicators. According to Servant-Miklos and Noordzij (2021), such differences can be explained by differences in curricular orientation and pedagogical approaches. Sidiropoulos (2018) argued that different educational and cultural traditions can expose students to different conceptions of sustainable development. Students from private institutions considered environmental indicators to have lower importance. These results disagree with those of Wang et al. (2020), who compared public and private university students in China and found that private university students agreed more often on the importance of sustainable development and had a higher level of perception about commitment, knowledge, attitudes, and practices toward sustainability.

Students who had had previous contact with rural production (before entering the course) attributed lower scores to environmental indicators. Furthermore, this group of students and those with experience in dairy cattle activities considered economic and technical indicators to be the most important. In other words, individuals with daily experience in the activity focused on technical/economic aspects, productivity, and profit. The problems and challenges faced by dairy production are complex and interconnected in an uncertain and rapidly changing environment. These transformations require a shift in the mindset of individuals to think beyond productivity and profit.

Souza and Araujo (2015) observed that environmental education is presented in an uncritical and technical way, without political, historical, or social contextualization; that is, technical solutions are prioritized without criticism for the resolution of environmental issues. Given this, it is necessary to promote a systemic view capable of underscoring the interrelationships between social, economic, and environmental factors and the agricultural sector, thereby mitigating environmental damage, and stimulating a change in attitude toward environmental and social problems (Santos et al., 2017). Furthermore, actions that contribute to a more holistic academic formation and contemplate the specificities of rural environments, production systems, rural farmers, and their knowledge and market issues are needed. Such actions may form professionals with the ability to assess and critically interpret rural situations and develop intervention proposals that contribute to sustainable development (Marques et al., 2015).

CONCLUSION

There is a plurality of students' perceptions about sustainability indicators for assessment of dairy farming in Brazil. Such a diversity of views is mainly related to personal experience in milk production, the educational system (whether public or private), and differences in the syllabus of Agricultural Science courses.

The results highlight the importance of developing education based on systemic thinking, focused on including technical, social, economic, and environmental interrelationships in students' academic formation. These actions may contribute to the improvement of sustainable dairy production.

Incorporation of a comprehensive vision of the dairy activity demands interdisciplinarity, multidisciplinarity, and transdisciplinarity. This is a challenge for the academic system, which, in addition to educating students, should lead them to reflect on their own actions in a global perspective.

The results of this study can be used by researchers, teachers, and coordinators to guide and promote the inclusion of technical, social, economic, and environmental aspects and their interrelationships in the syllabus of Agricultural Sciences programs. The findings may also contribute to systemic thinking and empower students to act in complex situations. Both theoretical knowledge bases and practical activities are progressive and necessary for the continuous formation of students and professionals.

The results can be used to rethink the syllabus of disciplines, create new disciplines, develop actions such as programs, projects, courses and workshops, events, and service provision, in addition to working on interdisciplinary and interprofessional relationships. For the sustainable development of cattle production, it is necessary to work on the teaching, research, and extension triad, with dialogical interaction with the community, with the curricular units and with other professionals, with the development of research and teaching, generating transformations in the training of students and the community.

Nevertheless, we highlight that the results should be considered with caution before generalization. Evaluation of a larger sample of students from different cultural contexts may lead to different results.

REFERENCES

Acevedo-Osorio, Á., Hofmann-Souki, S., & Cruz Morales, J. (2020). Holistic competence orientation in sustainability-related study programmes lessons from implementing transdisciplinary studen. *Sustainability Science*, *15*(1), 233–246. doi:10.100711625-019-00687-8

Aleixo, A. M., Leal, S., & Azeiteiro, U. M. (2021). Higher education students' perceptions of sustainable development in Portugal. *Journal of Cleaner Production*, *327*, 129429. doi:10.1016/j.jclepro.2021.129429

Almeida, R., Scatena, L. M., & Luz, M. S. (2017). Percepção ambiental e políticas públicas - dicotomia e desafios no desenvolvimento da cultura de sustentabilidade. *Ambiente & Sociedade*, *XX*, 43–64. doi:10.1590/1809-4422asoc20150004r1v2012017

Bánkuti, F. I., Prizon, R. C., Damasceno, J. C., De Brito, M. M., Pozza, M. S. S., & Lima, P. G. L. (2020). Farmers' actions toward sustainability: A typology of dairy farms according to sustainability indicators. *Animal*, *14*, s417–s423. doi:10.1017/S1751731120000750 PMID:32290889

Batalha, M. O., & Silva, A. L. (2010). Gerenciamento de sistemas agroindustriais: definições, especificidades e correntes metodológicas. In *M. O. Batalha (Org.), Gestão agroindustrial: Grupo de estudos e pesquisas agroindustriais* (3rd ed., pp. 1–62). Atlas.

Bélanger, V., Vanasse, A., Parent, D., Allard, G., & Pellerin, D. (2012). Development of agri-environmental indicators to assess dairy farm sustainability in Quebec, Eastern Canada. *Ecological Indicators*, *23*, 421–430. doi:10.1016/j.ecolind.2012.04.027

Bélanger, V., Vanasse, A., Parent, D., Allard, G., & Pellerin, D. (2015). DELTA: An integrated indicator-based self-assessment tool for the evaluation of dairy farms sustainability in Quebec, Canada. *Agroecology and Sustainable Food Systems*, *39*(9), 1022–1046. doi:10.1080/21683565.2015.1069775

Bergmann, E., Lima, T. B. C., Pedroso, D. S., Debortoli, E. C., & Aguiar, V. S. L. (2020). Investigando as Análises na Ciência Animal: Uma ação extracurricular na formação de Zootecnistas. *Brazilian Journal of Development*, *6*(8), 54619–54630. doi:10.34117/bjdv6n8-031

Corrêa, M., Lima, B. V. M., Martins, V. W. B., Rampasso, I. S., Anholon, R., Quelhas, O. L. G., & Leal Filho, W. (2020). An analysis of the insertion of sustainability elements in undergraduate design courses offered by Brazilian higher education institutions: An exploratory study. *Journal of Cleaner Production*, *272*, 122733. doi:10.1016/j.jclepro.2020.122733

Cortner, O., Garrett, R. D., Valentim, J. F., Ferreira, J., Niles, M. T., Reis, J., & Gil, J. (2019). Perceptions of integrated crop-livestock systems for sustainable intensification in the Brazilian Amazon. *Land Use Policy*, *82*, 841–853. doi:10.1016/j.landusepol.2019.01.006

Field, A. (2009). *Descobrindo a estatística usando o SPSS* (2nd ed.). Artmed.

García-González, E., Jiménez-Fontana, R., & Azcárate Goded, P. (2020). Approaches to teaching and learning for sustainability: Characterizing students' perceptions. *Journal of Cleaner Production*, *274*, 122928. doi:10.1016/j.jclepro.2020.122928

Gaudino, S., Goia, I., Grignani, C., Monaco, S., & Sacco, D. (2014). Assessing agro-environmental performance of dairy farms in northwest Italy based on aggregated results from indicators. *Journal of Environmental Management*, *140*, 120–134. doi:10.1016/j.jenvman.2014.03.010 PMID:24747935

Gazola, M. G., Bánkuti, F. I., Brito, M. M., Prizon, R. C., Kuwahara, K. C., Pozza, M. S. S., & Damasceno, J. C. (2018). Development and application of a sustainability assessment model for dairy production systems. *Semina. Ciências Agrárias*, *39*, 2685–2702. doi:10.5433/1679-0359.2018v39n6p2685

Gil-Doménech, D., Magomedova, N., Sánchez-Alcázar, E. J., & Lafuente-Lechuga, M. (2021). Integrating Sustainability in the Business Administration and Management Curriculum: A Sustainability Competencies Map. *Sustainability*, *13*(16), 9458. doi:10.3390u13169458

Hair, J. F., Black, W. C., Babin, B. J., & Anderson, R. E. (2014). *Multivariate Data Analysis* (7th ed.). Pearson Education.

Hasan, S. S., Haque, M. E., Suchi, I. Z., & Hossain, M. A. (2018). Assessment of Diploma Agricultural Students' Attitude towards Educational Sustainability: A Study of Selected Agricultural Training Institutes of Bangladesh. *Journal of Education. Society and Behavioural Science*, *25*(2), 1–12. doi:10.9734/JESBS/2018/41170

Herrero, M., Wirsenius, S., Henderson, B., Rigolot, C., Thornton, P., Havlík, P., Boer, I., & Gerber, P. (2015). Livestock and the Environment: What Have We Learned in the Past Decade? *Annual Review of Environment and Resources*, *40*(1), 177–202. doi:10.1146/annurev-environ-031113-093503

Holdsworth, S., & Sandri, O. (2021). Investigating undergraduate student learning experiences using the good practice learning and teaching for sustainability education (GPLTSE) framework. *Journal of Cleaner Production*, *311*, 127532. doi:10.1016/j.jclepro.2021.127532

Karatzoglou, B. (2013). An in-depth literature review of the evolving roles and contributions of universities to Education for Sustainable Development. *Journal of Cleaner Production*, *49*, 44–53. doi:10.1016/j.jclepro.2012.07.043

King, C., Gunton, J., Freebairn, D., Coutts, J., & Webb, I. (2000). The Sustainability Indicator Industry: Where to from here? A focus group study to explore the potential of farmer participation in the development of indicators. *Australian Journal of Experimental Agriculture*, *40*(4), 631–642. doi:10.1071/EA99148

Lambrechts, W., Ghijsen, P. W. T., Jacques, A., Walravens, H., Liedekerke, L. V., & Petegem, P. V. (2018). Sustainability segmentation of business students: Toward selfregulated development of critical and interpretational competences in a post-truth era. *Journal of Cleaner Production*, *202*, 561–570. doi:10.1016/j.jclepro.2018.07.303

Lebacq, T., Baret, P. V., & Stilmant, D. (2013). Sustainability indicators for livestock farming. A review. *Agronomy for Sustainable Development*, *33*(2), 311–327. doi:10.100713593-012-0121-x

Likert, R. (1932). A technique for the measurement of attitudes. *Archives de Psychologie*, *22*, 90–95.

Marandure, T., Mapiye, C., Makombe, G., & Dzama, K. (2017). Indicator-based sustainability assessment of the smallholder beef cattle production system in South Africa. *Agroecology and Sustainable Food Systems*, *41*(1), 3–29. doi:10.1080/21683565.2016.1231152

Marques, L. G., Neumann, P. S., & Zarnott, A. V. (2015). Trabalho de campo na formação holística e crítica de estudantes de Agronomia: A experiência da disciplina de Desenvolvimento Rural – UFSM. *Cadernos de Agroecologia*, *10*, 1–5.

Meul, M., Van Passel, S., Nevens, F., Dessein, J., Rogge, E., Mulier, A., & Van Hauwermeiren, A. (2008). MOTIFS: A monitoring tool for integrated farm sustainability. *Agronomy for Sustainable Development*, *28*(2), 321–332. doi:10.1051/agro:2008001

Munyaneza, C., Kurwijila, L. R., Mdoe, N. S. Y., Baltenweck, I., & Twine, E. E. (2019). Identification of appropriate indicators for assessing sustainability of small-holder milk production systems in Tanzania. *Sustainable Production and Consumption*, *19*, 141–160. doi:10.1016/j.spc.2019.03.009

Naderi, N., Monavvarifard, F., & Salehi, L. (2022). Fostering sustainability-oriented knowledge-sharing in academic environment: A key strategic process to achieving SDGs through development of students' sustainable entrepreneurship competences. *International Journal of Management Education*, *20*(1), 100603. doi:10.1016/j.ijme.2022.100603

Okumah, M., Martin-Ortega, J., & Novo, P. (2018). Effects of awareness on farmers' compliance with diffuse pollution mitigation measures: A conditional process modelling. *Land Use Policy*, *76*, 36–45. doi:10.1016/j.landusepol.2018.04.051

Olalla, C. B., & Merino, A. (2019). Competences for sustainability in undergraduate business studies: A content analysis of value-based course syllabi in Spanish universities. *International Journal of Management Education*, *17*(2), 239–253. doi:10.1016/j.ijme.2019.02.006

Olde, E. M., Moller, H., Marchand, F., Mcdowell, R. W., Macleod, C. J., Sautier, M., Halloy, S., Barber, A., Benge, J., Bockstaller, C., Bokkers, E. A. M., Boer, I. J. M., Legun, K. A., Le Quellec, I., Merfield, C., Oudshoorn, F. W., Reid, J., Schader, C., Szymanski, E., & Manhire, J. (2017). When experts disagree: The need to rethink indicator selection for assessing sustainability of agriculture. *Environment, Development and Sustainability*, *19*(4), 1327–1342. doi:10.100710668-016-9803-x

Olde, E. M., Oudshoorn, F. W., Sørensen, C. A. G., Bokkers, E. A. M., & Boer, I. J. M. (2016). Assessing sustainability at farm-level: Lessons learned from a comparison of tools in practice. *Ecological Indicators*, *66*, 391–404. doi:10.1016/j.ecolind.2016.01.047

Oudshoorn, F. W., Kristensen, T., Van Der Zijpp, A. J., & Boer, I. J. M. (2012). Sustainability evaluation of automatic and conventional milking systems on organic dairy farms in Denmark. *NJAS Wageningen Journal of Life Sciences*, *59*(1-2), 25–33. doi:10.1016/j.njas.2011.05.003

Rampasso, I. S., Anholon, R., Silva, D., Cooper Ordonez, R. E., Quelhas, O. L. G., Leal Filho, W., & Santa-Eulália, L. A. (2018). An analysis of the difficulties associated to sustainability insertion in engineering education: Examples from HEIs in Brazil. *Journal of Cleaner Production*, *193*, 363–371. doi:10.1016/j.jclepro.2018.05.079

Rampasso, I. S., Anholon, R., Silva, D., Ordoñez, R. E. C., Santa-Eulalia, L. A., Quelhas, O. L. G., Leal Filho, W., & Aguirre, L. F. G. (2019). Analysis of the perception of engineering students regarding sustainability. *Journal of Cleaner Production*, *233*, 461–467. doi:10.1016/j.jclepro.2019.06.105

Rieckmann, M. (2012). Future-oriented higher education: Which key competencies should be fostered through university teaching and learning? *Futures*, *44*(2), 127–135. doi:10.1016/j.futures.2011.09.005

Salovaara, J. J., Pietikainen, J., & Cantell, H. (2021). Perceptions of interconnected sustainability: Students' narratives bridging transition and education. *Journal of Cleaner Production*, *281*, 125336. doi:10.1016/j.jclepro.2020.125336

Santos, A. M., Lopes, E. R. N., & Silva, M. F. Jr. (2017). Percepção ambiental de estudantes do ensino técnico federal em agropecuária e a contribuição da educação ambiental na formação profissional. *Revista Brasileira de Educação Ambiental*, *12*(2), 136–155. doi:10.34024/revbea.2017.v12.2315

Saqib, Z. A., Zhang, Q., Ou, J., Saqib, K. A., Majeed, S., & Amar Razzaq, A. (2020). Education for sustainable development in Pakistani higher education institutions: An exploratory study of students' and teachers' perceptions. *International Journal of Sustainability in Higher Education*, *21*(6), 1249–1267. doi:10.1108/IJSHE-01-2020-0036

Servant-Miklos, V. F. C., & Noordzij, G. (2021). Investigating the impact of problem-oriented sustainability education on students' identity: A comparative study of planning and liberal arts students. *Journal of Cleaner Production*, *280*, 124846. doi:10.1016/j.jclepro.2020.124846

Sidiropoulos, E. (2018). The personal context of student learning for sustainability: Results of a multi-university research study. *Journal of Cleaner Production, 181*, 537–554. doi:10.1016/j.jclepro.2018.01.083

Silva, M. F. (2020). *Indicadores de sustentabilidade para a pecuária leiteira.* [Tese Doutorado em Ciências, Universidade de São Paulo].

Silva, M. F., & Gameiro, A. H. (2021). Indicadores de sustentabilidade para a produção de leite: Uma revisão de literatura. *Revista Livre de Sustentabilidade e Empreendedorismo*, *6*, 208–237.

Silva, M. F., & Gameiro, A. H. (2022). Sustainability indicators for Brazilian dairy livestock: The perception of professionals in the sector. *Revista Brasileira de Zootecnia*, *51*, e20210049. doi:10.37496/rbz5120210049

Souza, V. M., & Araujo, J. (2015). O currículo verde: Uma discussão sobre a inserção do meio ambiente nas grades curriculares dos cursos de graduação das instituições públicas de ensino superior do estado do Rio de Janeiro. *Revista Brasileira de Educação Ambiental*, *10*(3), 147–163. doi:10.34024/revbea.2015.v10.1909

Sullivan, G. M., & Artino, A. R. Jr. (2013). Analyzing and Interpreting Data From Likert-Type Scales. *Journal of Graduate Medical Education*, *5*(4), 541–542. doi:10.4300/JGME-5-4-18 PMID:24454995

Van Calker, K. J., Berentsen, P. B. M., Giesen, G. W. J., & Huirne, R. B. M. (2005). Identifying and ranking attributes that determine sustainability in Dutch dairy farming. *Agriculture and Human Values*, *22*(1), 53–63. doi:10.100710460-004-7230-3

Wang, J., Yang, M., & Maresova, P. (2020). Sustainable Development at Higher Education in China: A comparative study of students' perception in public and private universities. *Sustainability*, *12*(6), 2158. doi:10.3390u12062158

Watson, M. K., Lozano, R., Noyes, C., & Rodgers, M. (2013). Assessing curricula contribution to sustainability more holistically: Experiences from the integration of curricula assessment and students' perceptions at the Georgia Institute of Technology. *Journal of Cleaner Production*, *61*, 106–116. doi:10.1016/j.jclepro.2013.09.010

World Commission on Environment and Development - WCED. (1987). Report of the World Commission on Environment and Development: Our Common Future. Geneva.

Yadav, A., & Prakash, A. (2022). Factors influencing sustainable development integration in management education: An Empirical Assessment of management education institutions in India. *International Journal of Management Education*, *20*(1), 100604. doi:10.1016/j.ijme.2022.100604

Compilation of References

Abdalhamed, A. M., Zeedan, G. S. G., & Zeina, H. A. A. A. (2018). Isolation and identification of bacteria causing mastitis in small ruminants and their susceptibility to antibiotics, honey, essential oils, and plant extracts. *Veterinary World, 11*(3), 355–362. doi:10.14202/vetworld.2018.355-362 PMID:29657429

Abreu, P. S. (2010). *Desenvolvimento da Voz do Cliente no Desdobramento da Função Qualidade (QFD): Aplicação em Projetos de Cadeira de Rodas [Development of the Voice of the Customer in Quality Function Deployment (QFD): Application in Wheelchair Projects].* [Doctoral dissertation. UNICAMP, Brazil].

Abreu, F. D. S. (1997). QFD-desdobramento da função qualidade-estruturando a satisfação do cliente [QFD-deployment of the quality function-structuring customer satisfaction]. *Revista de Administração de Empresas, 37*(2), 47–55. doi:10.1590/S0034-75901997000200005

Abril, A. G., Villa, T. G., Barros-Velázquez, J., Cañas, B., Sánchez-Pérez, A., Calo-Mata, P., & Carrera, M. (2020). Staphylococcus aureus Exotoxins and Their Detection in the Dairy Industry and Mastitis. *Toxins, 12*(9), 1–18. doi:10.3390/toxins12090537 PMID:32825515

Acevedo-Osorio, Á., Hofmann-Souki, S., & Cruz Morales, J. (2020). Holistic competence orientation in sustainability-related study programmes lessons from implementing transdisciplinary studen. *Sustainability Science, 15*(1), 233–246. doi:10.100711625-019-00687-8

Addai, F. P., Lin, F., Wang, T., Kosiba, A. A., Sheng, P., Yu, F., & Shi, H. (2020). Technical integrative approaches to cheese whey valorization towards a sustainable environment. *Food & Function, 11*(10), 8407–8423. doi:10.1039/D0FO01484B PMID:32955061

Addis, M. F., Maffioli, E. M., Ceciliani, F., Tedeschi, G., Zamarian, V., Tangorra, F., Albertini, M., Piccinini, R., & Bronzo, V. (2020). Influence of subclinical mastitis and intramammary infection by coagulase-negative staphylococci on the cow milk peptidome. *Journal of Proteomics, 226*(June), 103885. doi:10.1016/j.jprot.2020.103885 PMID:32645476

Ahmad, T., Aadil, R. M., Ahmed, H., Rahman, U., Soares, B. C., Souza, S. L., & Cruz, A. G. (2019). Treatment and utilization of dairy industrial waste: A review. *Trends in Food Science & Technology, 88*, 361–372. doi:10.1016/j.tifs.2019.04.003

Ahmed, H., Tamminen, L. M., & Emanuelson, U. (2022). Temperature, productivity, and heat tolerance: Evidence from Swedish dairy production. *Climatic Change*, *175*(1-2), 10. doi:10.100710584-022-03461-5

Ahmed, I., Kumar, S., & Aggarwal, D. (2020, March). Assessment of Knowledge and Practices of Hygienic Milk Production among Dairy Farmworkers, Southwest Delhi. *Indian Journal of Community Medicine*, *45*(5, Suppl 1), S26–S30. doi:10.4103/ijcm.IJCM_366_19 PMID:32476737

Ahmed, S., Kattimani, L., Divatar, M., Gajare, S., Shivalee, A., Farheen, A., & Irfana, M. (2016). Optimization of lactase production under submerged fermentation by Lactobacillus sp.KLSA 22. *International Journal of Pure and Applied Bioscience*, *4*(4), 212–220. doi:10.18782/2320-7051.2321

Ahn, W. S., Park, S. J., & Lee, S. Y. (2001). Production of poly (3-hydroxybutyrate) from whey by cell recycle fed-batch culture of recombinant Escherichia coli. *Biotechnology Letters*, *23*(3), 235–240. doi:10.1023/A:1005633418161

Aimutis, W. L. (2004). Bioactive properties of milk proteins with particular focus on anticariogenesis. *The Journal of Nutrition*, *134*(4), 989–995. doi:10.1093/jn/134.4.989S PMID:15051859

Akao, Y., & Mazur, G. H. (2003). The leading edge in QFD: Past, present and future. *International Journal of Quality & Reliability Management*, *20*(1), 20–35. doi:10.1108/02656710310453791

Akkermans, J. P., & Vecht, U. (1994). Streptococcal infections as cause of death in pigs brought in for necropsy. *Tijdschrift voor Diergeneeskunde*, *119*, 123–128. PMID:8134911

Alatas, M. S., Citil, O. B., Kahraman, O., & Ozbilgin, A. (2015). Causes of milk fat depression in dairy cows. Scientific Papers. *Series D. Animal Science*, 80-85. http://animalsciencejournal.usamv.ro/pdf/2015/Art11.pdf

Alba, D. F., da Rosa, G., Hanauer, D., Saldanha, T. F., Souza, C. F., Baldissera, M. D., da Silva dos Santos, D., Piovezan, A. P., Girardini, L. K., & Schafer Da Silva, A. (2019). Subclinical mastitis in Lacaune sheep: Causative agents, impacts on milk production, milk quality, oxidative profiles, and treatment efficacy of ceftiofur. *Microbial Pathogenesis*, *137*(September), 103732. doi:10.1016/j.micpath.2019.103732 PMID:31505264

Albright, J. L., Tuckey, S. L., & Woods, G. T. (1961). Antibióticos no Leite—Uma Revisão. *Journal of Dairy Science*, *44*(5), 779–807. doi:10.3168/jds.S0022-0302(61)89819-6

Aleixo, A. M., Leal, S., & Azeiteiro, U. M. (2021). Higher education students' perceptions of sustainable development in Portugal. *Journal of Cleaner Production*, *327*, 129429. doi:10.1016/j.jclepro.2021.129429

Alekish, M. O., Ismail, Z. B., Awawdeh, M. S., & Shatnawi, S. (2017). Effects of intramammary infusion of sage (Salvia officinalis) essential oil on milk somatic cell count, milk composition parameters and selected hematology and serum biochemical parameters in Awassi sheep with subclinical mastitis. *Veterinary World*, *10*(8), 895–900. doi:10.14202/vetworld.2017.895-900 PMID:28919679

Algharib, S. A., Dawood, A., & Xie, S. (2020). Nanoparticles for treatment of bovine Staphylococcus aureus mastitis. *Drug Delivery*, *27*(1), 292–308. doi:10.1080/10717544.2020.1724209 PMID:32036717

Ali, Z., Dimri, U., & Jhambh, R. (2015). Prevalence and antibiogram of bacterial pathogens from subclinical mastitis in buffaloes. *Buffalo Bulletin*, *34*(1), 41–44.

Allen, J. C. (1990). Milk Synthesis and Secretion Rates in Cows with Milk Composition Changed by Oxytocin 1, 2. *Journal of Dairy Science*, *73*(4), 975–984. doi:10.3168/jds.S0022-0302(90)78755-3 PMID:2345208

Allen, J. D., Hall, L. W., Collier, R. J., & Smith, J. F. (2015). Effect of core body temperature, time of day, and climate conditions on behavioral patterns of lactating dairy cows experiencing mild to moderate heat stress. *Journal of Dairy Science*, *98*(1), 1–10. doi:10.3168/jds.2013-7704 PMID:25468707

Almeida, T. V. Parâmetros de qualidade do leite cru bovino: contagem bacteriana total e contagem de células somáticas. *Dissertação* (Mestrado em Ciência Animal) – Universidade Federal de Goiás, Goiânia, 2013.

Almeida, K. E., Bonassi, I. A., & Roca, R. O. (2001). Physical and chemical characteristics of fermented dairy beverages using minas cheese whey. *Food Science and Technology (Campinas)*, *21*(2), 187–192. doi:10.1590/S0101-20612001000200012

Almeida, R., Scatena, L. M., & Luz, M. S. (2017). Percepção ambiental e políticas públicas - dicotomia e desafios no desenvolvimento da cultura de sustentabilidade. *Ambiente & Sociedade*, *XX*, 43–64. doi:10.1590/1809-4422asoc20150004r1v2012017

Álvarez-Cao, M. E., Becerra, M., & González-Siso, M. I. (2020). Biovalorization of cheese whey and molasses wastes to galactosidases by recombinant yeasts. In N. K. Rathinam & R. Sani (Eds.), *Biovalorisation of Wastes to Renewable Chemicals and Biofuels* (pp. 149–161). Elsevier. doi:10.1016/B978-0-12-817951-2.00008-0

Alves, E. C., Dahmer, A. M., & Borges, A. F. (2014). Total bacterial count and somatic cell count in refrigerated raw milk stored in communal tanks. *Brazilian Journal of Food Technology*, *17*(3), 221–225. doi:10.1590/1981-6723.1114

Anderson, E. S., & Datta, N. (1965). Resistance To Penicillins And Its Transfer In Enterobacteriaceae. *Lancet*, *285*(7382), 407–409. doi:10.1016/S0140-6736(65)90004-8 PMID:14238093

Andrietta, J. M., & Miguel, P. A. C. (2007). Aplicação do programa Seis Sigma no Brasil: Resultados de um levantamento tipo survey exploratório-descritivo e perspectivas para pesquisas futuras [A study on the application of the Six Sigma programme in Brazil: results of an exploratory-descriptive survey and perspectives for future research]. *Gestão & Produção*, *14*(2), 203–219. doi:10.1590/S0104-530X2007000200002

Anisha, G. S. (2017). β-galactosidases. In Pandey, A., Negi, S., & Soccol, C. R. (Eds.). Current developments in biotechnology and bioengineering: production, isolation and purification of industrial products (395-421). Elsevier.

Antanaitis, R., Juozaitienė, V., Jonike, V., Baumgartner, W., & Paulauskas, A. (2021). Milk lactose as a biomarker of subclinical mastitis in dairy cows. *Animals (Basel)*, *11*(6), 1–11. doi:10.3390/ani11061736 PMID:34200862

Antanaitis, R., Juozaitienė, V., Malašauskienė, D., Televičius, M., Urbutis, M., Rutkaukas, A., Šertvytytė, G., & Baumgartner, W. (2022). Identification of Changes in Rumination Behavior Registered with an Online Sensor System in Cows with Subclinical Mastitis. *Veterinary Sciences*, *9*(9), 454. doi:10.3390/vetsci9090454 PMID:36136670

Argenta, A. B., Nogueira, A., & Scheer, A. P. (2021). Hydrolysis of whey lactose: Kluyveromyces lactisβ-galactosidase immobilisation and integrated process hydrolysis-ultrafiltration. *International Dairy Journal*, *117*, 105007. doi:10.1016/j.idairyj.2021.105007

Arnott, G., Ferris, C. P., & O'Connell, N. E. (2017). Review: Welfare of dairy cows in continuously housed and pasture-based production systems. *Animal*, *11*(2), 261–273. doi:10.1017/S1751731116001336 PMID:27364762

Arora, R., Behera, S., Sharma, N. K., & Kumar, S. (2017). Augmentation of ethanol production through statistically designed growth and fermentation medium using novel thermotolerant yeast isolates. *Renewable Energy*, *109*, 406–421. doi:10.1016/j.renene.2017.03.059

Assembly, U. G. (2015). Transforming our world: the 2030 Agenda for Sustainable Development. A/Res/70/1, *Resolution adopted by the General Assembly on 25 September 2015.*GA.

Asseng, S., Spänkuch, D., Hernandez-Ochoa, I. M., & Laporta, J. (2021, June). The upper temperature thresholds of life. [Erratum in: Lancet Planet Health. 2021 Sep;5] [9] [: e578. PMID: 34119012. Doi: 10.1016/S2542-5196] [21]. *The Lancet. Planetary Health*, *5*(6), e378–e385. doi:10.1016/S2542-5196(21)00079-6 PMID:34119012

Assis, J., Ferreira, J. D., Martins, H. H., & Schneider, M. B. (2016). Cadeia produtiva do leite no Brasil no contexto do comércio internacional. *Revista de Ciências Empresariais*, *17*(1), 63–93.

Baldasso, C., Barros, T. C., & Tessaro, I. C. (2011). Concentration and purification of whey proteins by ultrafiltration. *Desalination*, *278*(1-3), 381–386. doi:10.1016/j.desal.2011.05.055

Balemi, A., Gumi, B., Amenu, K., Girma, S., Gebru, M., Tekle, M., Rius, A. A., D'souza, D. H., Agga, G. E., & Dego, O. K. (2021). Prevalence of mastitis and antibiotic resistance of bacterial isolates from cmt positive milk samples obtained from dairy cows, camels, and goats in two pastoral districts in southern ethiopia. *Animals (Basel)*, *11*(6), 1530. doi:10.3390/ani11061530 PMID:34073967

Bánkuti, F. I., Prizon, R. C., Damasceno, J. C., De Brito, M. M., Pozza, M. S. S., & Lima, P. G. L. (2020). Farmers' actions toward sustainability: A typology of dairy farms according to sustainability indicators. *Animal*, *14*(2), 417–423. doi:10.1017/S1751731120000750 PMID:32290889

Barba, F. J. (2021). An integrated approach for the valorization of cheese whey. *Foods*, *10*(3), 564. doi:10.3390/foods10030564 PMID:33803106

Barros, L. (2001). Transtornos metabólicos que afetam a qualidade do leite. In F. H. D. González, J. W. Dürr, & R. S. Fontaneli (Eds.), *Uso do leite para monitorar a nutrição e metabolismo de vacas leiteiras* (pp. 46–60). Ed. UFRGS.

Batalha, M. O., & Silva, A. L. (2010). Gerenciamento de sistemas agroindustriais: definições, especificidades e correntes metodológicas. In *M. O. Batalha (Org.), Gestão agroindustrial: Grupo de estudos e pesquisas agroindustriais* (3rd ed., pp. 1–62). Atlas.

Bauman, D. E., McGuire, M. A., & Harvatine, K. J. (2022). Mammary Gland, Milk Biosynthesis and Secretion: Milk Fat. In P.L.H. McSweeney (Ed.), J. P. McNamara (Ed.), Encyclopedia of Dairy Sciences (3rd Edition), (pp. 190-197). Academic Press. doi:10.1016/B978-0-12-818766-1.00049-0

Beauchemin, K. A. (2018). Invited review: Current perspectives on eating and rumination activity in dairy cows. *Journal of Dairy Science*, *101*(6), 4762–4784. doi:10.3168/jds.2017-13706 PMID:29627250

Becerra, M., Cerdan, E., & Gonzalez Siso, M. I. (1998). Dealing with different methods for Kluyveromyces lactisβ-galactosidase purification. *Biological Procedures Online*, *1*(1), 48–58. doi:10.1251/bpo4 PMID:12734592

Beede, D. K., & Collier, R. J. (1986). Potential nutritional strategies for intensively managed cattle during thermal stress. *Journal of Animal Science*, *62*(2), 543–554. doi:10.2527/jas1986.622543x

Bélanger, V., Vanasse, A., Parent, D., Allard, G., & Pellerin, D. (2012). Development of agri-environmental indicators to assess dairy farm sustainability in Quebec, Eastern Canada. *Ecological Indicators*, *23*, 421–430. doi:10.1016/j.ecolind.2012.04.027

Bélanger, V., Vanasse, A., Parent, D., Allard, G., & Pellerin, D. (2015). DELTA: An integrated indicator-based self-assessment tool for the evaluation of dairy farms sustainability in Quebec, Canada. *Agroecology and Sustainable Food Systems*, *39*(9), 1022–1046. doi:10.1080/21683565.2015.1069775

Belhadj Slimen, I., Najar, T., Ghram, A., & Abdrrabba, M. (2016). Heat stress effects on livestock: Molecular, cellular and metabolic aspects, a review. *Journal of Animal Physiology and Animal Nutrition*, *100*(3), 401412. doi:10.1111/jpn.12379 PMID:26250521

Beloti, V., Dias, J. A., & Oliveira, A. M. Ordenha e boas práticas de produção. In: Pecuária Leiteira na Amazônia, Embrapa. Brasília, DF: Embrapa, p. 105-130, 2020.

Beloti, V. (2015). *Leite: Obtenção, Inspeção e Qualidade*. Editora Planta.

Beltrán, M. C., Althaus, R. L., Molina, A., Berruga, M. I., & Molina, M. P. (2015). Analytical strategy for the detection of antibiotic residues in sheep and goat's milk. *Spanish Journal of Agricultural Research*, *13*(1), e0501. https://revistas.inia.es/index.php/sjar/article/ view/6522. doi:10.5424jar/2015131-6522

Bentahar, J., Doyen, A., Beaulieu, L., & Deschênes, J. S. (2019). Investigation of β-galactosidase production by microalga Tetradesmus obliquus in determined growth conditions. *Journal of Applied Phycology, 31*(1), 301–308. doi:10.100710811-018-1550-y

Ben, X. M., Li, J., Feng, Z. T., Shi, S. Y., Lu, Y. D., Chen, R., & Zhou, X.-Y. (2008). Low level of galacto-oligosaccharide in infant formula stimulates growth of intestinal Bifidobacteria and Lactobacilli. *World Journal of Gastroenterology, 14*(42), 6564–6568. doi:10.3748/wjg.14.6564 PMID:19030213

Bergmann, E., Lima, T. B. C., Pedroso, D. S., Debortoli, E. C., & Aguiar, V. S. L. (2020). Investigando as Análises na Ciência Animal: Uma ação extracurricular na formação de Zootecnistas. *Brazilian Journal of Development, 6*(8), 54619–54630. doi:10.34117/bjdv6n8-031

Berman, A. (2011). Invited review: Are adaptations present to support dairy cattle productivity in warm climates? *Journal of Dairy Science, 94*(5), 2147–2158. doi:10.3168/jds.2010-3962 PMID:21524505

Bernabucci, U., N. Lacetera, L.H. Baumgard, P.P. Rhoads, B. Ronchi, and A. Nardone, (2010). Metabolic and hormonal acclimation to heat stress in domesticated ruminants. *Animal: an international journal of animal bioscience, 4*, 1167. doi:10.1017/S175173111000090X

Bernabucci, U., Basiricò, L., Morera, P., Dipasquale, D., Vitali, A., Cappelli, F. P., & Calamari, L. (2015). Effect of summer season on milk protein fractions in Holstein cows. *Journal of Dairy Science, 98*(3), 1–13. doi:10.3168/jds.2014-8788 PMID:25547301

Berry, D. P., Friggens, N. C., Lucy, M., & Roche, J. R. (2016). Milk production and fertility in cattle. *Annual Review of Animal Biosciences, 4*(1), 269290. doi:10.1146/annurev-animal-021815-111406 PMID:26526546

Bertipaglia, E. C. A., Silva, R. G., & Maia, A. S. C. (2018). Fertility and hair coat characteristics of Holstein cows in a tropical environment. *Animal Reproduction (AR), 2*, 187194.

Beyene, T. (2015). Veterinary Drug Residues in Food-animal Products: Its Risk Factors and Potential Effects on Public Health. *Journal of Veterinary Science & Technology, 7*(01), 1–7. doi:10.4172/2157-7579.1000285

Bisutti, V., Vanzin, A., Toscano, A., Pegolo, S., Giannuzzi, D., Tagliapietra, F., Schiavon, S., Gallo, L., Trevisi, E., Negrini, R., & Cecchinato, A. (2022). Impact of somatic cell count combined with differential somatic cell count on milk protein fractions in Holstein cattle. *Journal of Dairy Science, 105*(8), 6447–6459. doi:10.3168/jds.2022-22071 PMID:35840397

Blowey, R. W., & Edmondson, P. (2010). *Mastitis Control in Dairy Herds* (2nd ed.). CAB International. doi:10.1079/9781845935504.0000

Bohmanova, J., Misztal, I., & Cole, J. B. (2007). Temperaturehumidity indices as indicators of milk production losses due to heat stress. *Journal of Dairy Science, 90*(4), 19471956. doi:10.3168/jds.2006-513 PMID:17369235

Bolognesi, L. S., Gabardo, S., Dall Cortivo, P. R., & Ayub, M. A. Z. (2021). Biotechnological production of galactooligosaccharides (GOS) using porungo cheese whey. *Food Science and Technology (Campinas)*, *42*, e64520. doi:10.1590/fst.64520

Bon, E. P. S., Ferrara, M. A., & Corvo, M. L. (2008). *Enzimas em biotecnologia: produção, aplicações e mercado*. Editora Interciência.

Boora, A., Yadav, S., Devi, P., Singh, K. P., Cheema, P. S., Muley, V., Sehrawat, V., Dhamanaskar, K., & Singh, I. (2021). an Open Label Study To Assess the Efficacy of Ceftiofur in Treatment of Clinical Mastitis in Buffaloes. *Buffalo Bulletin*, *40*(4), 545–555.

Borgne, S. L. (2012). Genetic engineering of industrial strains of Saccharomyces cerevisiae. In A. Lorence (Ed.), *Recombinant gene expression* (pp. 451–465). Humana Press. doi:10.1007/978-1-61779-433-9_24

Bradley, A. J. (2002). Bovine mastitis: An evolving disease. *Veterinary Journal (London, England)*, *164*(2), 116–128. doi:10.1053/tvjl.2002.0724 PMID:12359466

Brandelli, A., Daroit, D. J., & Corrêa, A. P. F. (2015). Whey as a source of peptides with remarkable biological activities. *Food Research International*, *73*, 149–161. doi:10.1016/j.foodres.2015.01.016

Brasil. (2002). Ministério da Agricultura, Pecuária e Abastecimento. Instrução Normativa n°51, de 19 de setembro de 2002 [Normative Instruction No. 51, of September 19, 2002]. Brasília: Mapa, Brazil.

Brasil. (2011). Ministério da Agricultura, Pecuária e Abastecimento. Instrução normativa n° 62, de 26 de dezembro de 2011 [Normative Instruction No. 62, of December 26, 2011]. Brasília, Mapa, Brazil Brasil. (2016). Ministério da Agricultura, Pecuária e Abastecimento. Instrução Normativa n°07, de 03 de maio de 2016 [Normative Instruction No. 07, of May 03, 2016]. Brasília: Mapa, Brazil.

Brasil. (2018a). Ministério da Agricultura, Pecuária e Abastecimento. Instrução Normativa N° 76, de 26 de Novembro de 2018. [Normative Instruction No. 76, of November 26, 2018]. Brasília: Mapa, Brazil.

Brasil. (2018b). Ministério da Agricultura, Pecuária e Abastecimento. Instrução Normativa N° 77, de 26 de Novembro de 2018. [Normative Instruction No. 77, of November 26, 2018]. Brasília: Mapa, Brazil.

Brasil. (2018c). Ministério da Agricultura, Pecuária e Abastecimento. Instrução Normativa n°31, de 29 de junho de 2018 [Normative Instruction No. 29, of June 29, 2018]. Brasília: Mapa, Brazil.

Brasil, R. B., Silva, M. A. P., Carvalho, T. S., Cabral, J. F., Nicolau, E. S., & Neves, R. B. S. (2012, December). *Revista do Instituto de Laticínios Cândido Tostes*. Avaliação da qualidade do leite cru em função do tipo de ordenha e das condições de transporte e armazenamento. *Santa Terezinha*, *67*(389), 34–42.

Brito, L. G., da Silva Netto, F. G., Salman, A. K. D., & da Silva, W. C. (2005). Cartilha para o produtor de leite de Rondônia. Vol. I: Instalações, sanidade animal e a Instrução Normativa 51 [Booklet for the Rondônia milk producer. vol. I: Facilities, animal health and Normative Instruction 51]. Embrapa Rondônia-Documentos (INFOTECA-E).

Britt, J. H., Cushman, R. A., Dechow, C. D., Dobson, H., Humblot, P., Hutjens, M. F., Jones, G. A., Mitloehner, F. M., Ruegg, P. L., Sheldon, I. M., & Stevenson, J. S. (2021). Review: Perspective on high-performing dairy cows and herds. *Animal*, *xxxx*, 100298. doi:10.1016/j.animal.2021.100298 PMID:34266782

Brogden, K. A. (2005). Antimicrobial peptides: Pore formers or metabolic inhibitors in bacteria? *National Review*, *3*(3), 238–250. doi:10.1038/nrmicro1098 PMID:15703760

Brown-Brandl, T. M. (2018). Understanding heat stress in beef cattle. *Revista Brasileira de Zootecnia*, *47*(0), e20160414. doi:10.1590/rbz4720160414

Brown-Brandl, T. M., Chitko-McKown, C. G., Eigenberg, R. A., Mayer, J. J., Welsh, T. H. Jr, Davis, J. D., & Purswell, J. L. (2017). Physiological responses of feedlot heifers provided access to different levels of shade. *Animal*, *11*(8), 1344–1353. doi:10.1017/S1751731116002664 PMID:28007043

Brown, K., Uwiera, R. R. E., Kalmokoff, M. L., Brooks, S. P. J., & Inglis, G. D. (2016). Antimicrobial growth promoter use in livestock: A requirement to understand their modes of action to develop effective alternatives. *International Journal of Antimicrobial Agents*, *49*(1), 12–24. doi:10.1016/j.ijantimicag.2016.08.006 PMID:27717740

Çakmak, H., Özselek, Y., Turan, O. Y., Fıratlıgil, E., & Karbancioğlu-Güler, F. (2020). Whey protein isolate edible films incorporated with essential oils: Antimicrobial activity and barrier properties. *Polymer Degradation & Stability*, *179*, 109285. doi:10.1016/j.polymdegradstab.2020.109285

Calegari, F., Calamari, L., & Frazzi, E. (2014). Fan cooling of the resting area in a free stalls dairy barn. *International Journal of Biometeorology*, *58*(6), 1225–1236. doi:10.100700484-013-0716-1 PMID:24122338

Callaway, T. R., & .... (2003). "Probiotics, Prebiotics and Competitive Exclusion for Prophylaxis Against Bacterial Disease," Animal Health Research Reviews 9, no. 02 (2008): 217-25; Roberto M. La Ragione and Martin J. Woodward, "Competitive Exclusion by Bacillus subtilis Spores of Salmonella enterica Serotype Enteritidis and Clostridium perfringens in Young Chickens. *Veterinary Microbiology*, *94*(3), 245–256. PMID:12814892

Campra, N. A., Reinoso, E. B., Montironi, I. D., Moliva, M. V., Raviolo, J., Ruiz Moreno, F., Marin, C., Camacho, N. M., Paredes, A. J., Morán, M. C., Estein, S. M., Maletto, B. A., Palma, S. D., & Cariddi, L. N. (2022). Spray-drying-microencapsulated *Minthostachys verticillata* essential oil and limonene as innovative adjuvant strategy to bovine mastitis vaccines. *Research in Veterinary Science*, *149*, 136–150. doi:10.1016/j.rvsc.2022.04.014 PMID:35792419

Carneiro, L. A. B. C., Yu, L., Dupree, P., & Ward, R. J. (2018). Characterization of a β-galactosidase from Bacillus subtilis with transgalactosylation activity. *International Journal of Biological Macromolecules*, *120*, 279–287. doi:10.1016/j.ijbiomac.2018.07.116 PMID:30036621

Carrillo-Lopez, L. M., Garcia-Galicia, I. A., Tirado-Gallegos, J. M., Sanchez-Vega, R., Huerta-Jimenez, M., Ashokkumar, M., & Alarcon-Rojo, A. D. (2021). Recent advances in the application of ultrasound in dairy products: Effect on functional, physical, chemical, microbiological and sensory properties. *Ultrasonics Sonochemistry*, *73*, 105467. doi:10.1016/j.ultsonch.2021.105467 PMID:33508590

Carvalho, G. F. (2002). Milk yield, somatic cell count and physicochemical characteristics of raw milk collected from dairy cows in Minas Gerais state. In: Congresso Panamericano de Qualidade do Leite e Controle da Mastite. Ribeirão Preto.

Carvalho, F., Prazeres, A. R., & Rivas, J. (2013). Cheese whey wastewater: Characterization and treatment. *The Science of the Total Environment*, *445*, 385–396. doi:10.1016/j.scitotenv.2012.12.038 PMID:23376111

Castillejos, L., Calsamiglia, S., & Ferret, A. (2006). Effect of Essential Oil Active Compounds on Rumen Microbial Fermentation and Nutrient Flow in In Vitro Systems. *Journal of Dairy Science*, *89*(7), 2649–2658. doi:10.3168/jds.S0022-0302(06)72341-4 PMID:16772584

Castro, I., Alba, C., Aparicio, M., Arroyo, R., Jiménez, L., Fernández, L., Arias, R., & Rodríguez, J. M. (2019). Metataxonomic and immunological analysis of milk from ewes with or without a history of mastitis. *Journal of Dairy Science*, *102*(10), 9298–9311. doi:10.3168/jds.2019-16403 PMID:31421883

Catozzi, C., Sanchez Bonastre, A., Francino, O., Lecchi, C., De Carlo, E., Vecchio, D., Martucciello, A., Fraulo, P., Bronzo, V., Cuscó, A., D'Andreano, S., & Ceciliani, F. (2017). The microbiota of water buffalo milk during mastitis. *PLoS One*, *12*(9), 1–20. doi:10.1371/journal.pone.0184710 PMID:28926595

Ceciliani, F., Audano, M., Addis, M. F., Mitro, N., Lecchi, C., Ghaffari, M. H., Albertini, M., De Carlo, E., Vecchio, D., Di Vuolo, G., Cappelli, G., Tangorra, F., Piccinini, R., Bronzo, V., & Caruso, D. (2022). Changes in the lipidome of water buffalo milk during intramammary infection by non-aureus Staphylococci. *Scientific Reports*, *12*(1), 1–12. doi:10.103841598-022-13400-0 PMID:35690599

Çelik, K., Önür, Z. Y., Baytekin, H., & Coşkun, B. (2020). Whey processing: Utlization and Major Products. In K. Çelik (Ed.), *Whey every aspect* (pp. 33–60). Tudás Alapítvány.

CFIS - Canadian Food Inspection System. (2006). Agriculture and Agri-Food Canada. http://www.cfis.agr.ca/english/indexe.shtml.

Chaiyabutr, N., Chanpongsang, S., & Suadsong, S. (2008). Effects of evaporative cooling on the regulation of body water and milk production in crossbred Holstein cattle in a tropical environment. *International Journal of Biometeorology*, *52*(7), 575–585. doi:10.100700484-008-0151-x PMID:18427839

Chalermthai, B., Giwa, A., Schmidt, J. E., & Taher, H. (2021). Life cycle assessment of bioplastic production from whey protein obtained from dairy residues. *Bioresource Technology Reports*, *15*, 100695. doi:10.1016/j.biteb.2021.100695

Chang-Fung-Martel, J., Harrison, M. T., Brown, J. N., Rawnsley, R., Smith, A. P., & Meinke, H. (2021). Negative relationship between dry matter intake and the temperature-humidity index with increasing heat stress in cattle: A global meta-analysis. *International Journal of Biometeorology*, *65*(12), 2099–2109. doi:10.100700484-021-02167-0 PMID:34283273

Chibisa, G. E., Gorka, P., Penner, G. B., Berthiaume, R., & Mutsvangwa, T. (2015). Effects of partial replacement of dietary starch from barley or corn with lactose on ruminal function, short-chain fatty acid absorption, nitrogen utilization, and production performance of dairy cows. *Journal of Dairy Science*, *98*(4), 2627–2640. doi:10.3168/jds.2014-8827 PMID:25704977

Chiesa, L. M., DeCastelli, L., Nobile, M., Martucci, F., Mosconi, G., Fontana, M., Castrica, M., Arioli, F., & Panseri, S. (2020). Analysis of antibiotic residues in raw bovine milk and their impact toward food safety and on milk starter cultures in cheese-making process. *Lebensmittel-Wissenschaft + Technologie*, *131*, 109783. doi:10.1016/j.lwt.2020.109783

Chinnappan, R., Al Attas, S., Kaman, W. E., Bikker, F. J., & Zourob, M. (2017). Development of magnetic nanoparticle based calorimetric assay for the detection of bovine mastitis in cow milk. *Analytical Biochemistry*, *523*, 58–64. doi:10.1016/j.ab.2017.02.009 PMID:28219684

Chiquette, J. (1995). Saccharomyces cerevisiae and Aspergillus oryzae, Used Alone or In Combination, as a Feed Supplement for Beef and Dairy Cattle. *Canadian Journal of Animal Science*, *75*(3), 405–415. doi:10.4141/cjas95-060

Choonia, H. S., & Lele, S. S. (2013). Kinetic modeling and implementation of superior process strategies for β-galactosidase production during submerged fermentation in a stirred tank bioreactor. *Biochemical Engineering Journal*, *77*, 49–57. doi:10.1016/j.bej.2013.04.021

Cogan, T. M. (1972). Susceptibility of Cheese and Yogurt Starter Bacteria to Antibiotics. *Applied Microbiology*, *23*(5), 960–965. doi:10.1128/am.23.5.960-965.1972 PMID:4624215

Collier, R. J., Collier, J. L., Rhoads, R. P., & Baumgard, L. H. (2008). Invited Review: Genes involved in the bovine heat stress response. *Journal of Dairy Science*, *91*(2), 445–454. doi:10.3168/jds.2007-0540 PMID:18218730

Colombo, B., Calvo, M. V., Sciarria, T. P., Scaglia, B., Kizito, S. S., D'Imporzano, G., & Adani, F. (2019). Biohydrogen and polyhydroxyalkanoates (PHA) as products of a two-steps bioprocess from deproteinized dairy wastes. *Waste Management (New York, N.Y.)*, *95*, 22–31. doi:10.1016/j.wasman.2019.05.052 PMID:31351607

Companhia Nacional de Abastecimento – CONAB. (2018). O Brasil é responsável por cerca de 7% do leite produzido no mundo [Brazil is responsible for about 7% of the milk produced in the world]. https://www.conab.gov.br/ultimas-noticias/2634-brasil-e-responsavel-por-cerca-de-7-do-leite-produzido-no-mundo

Corrêa, M., Lima, B. V. M., Martins, V. W. B., Rampasso, I. S., Anholon, R., Quelhas, O. L. G., & Leal Filho, W. (2020). An analysis of the insertion of sustainability elements in undergraduate design courses offered by Brazilian higher education institutions: An exploratory study. *Journal of Cleaner Production*, *272*, 122733. doi:10.1016/j.jclepro.2020.122733

Cortner, O., Garrett, R. D., Valentim, J. F., Ferreira, J., Niles, M. T., Reis, J., & Gil, J. (2019). Perceptions of integrated crop-livestock systems for sustainable intensification in the Brazilian Amazon. *Land Use Policy*, *82*, 841–853. doi:10.1016/j.landusepol.2019.01.006

Costa, A. R. V. (2018). Método híbrido de desenvolvimento de produto com utilização de mock-up digital e a voz do cliente no contexto da indústria automotiva [Hybrid method of product development using digital mock-up and the voice of the customer in the context of the automotive industry]. Master's thesis, Universidade Tecnológica Federal do Paraná, Brazil.

Costa, C. C. M., de Almeida, A. L. T., Ferreira, M. A. M., & Silva, E. A. (2013). Determinantes do desenvolvimento do setor agropecuário nos municípios [Determinants of the development of the agricultural sector in the municipalities]. *Revista ADM*, 295–309.

Côté-Gravel, J., & Malouin, F. (2019). Symposium review: Features of Staphylococcus aureus mastitis pathogenesis that guide vaccine development strategies. *Journal of Dairy Science*, *102*(5), 4727–4740. doi:10.3168/jds.2018-15272 PMID:30580940

Cui, D., Wang, S., Wang, L., Wang, H., Li, X., & Liu, Y. (2015). Prophylactic strategy with herbal remedy to reduce puerperal metritis risk in dairy cows: A randomized clinical trial. *Livestock Science*, *181*, 231–235. doi:10.1016/j.livsci.2015.09.005

Cunha, A. F. (2010). Evolução anual da qualidade do leite cru de tanques individuais e comunitários do Vale do Rio Doce (MG). *Boletim do Centro de Pesquisa de Processamento de Alimentos*, [S.l.], *34*(1).

Cunha, A. Silva, L. B. G. da, Pinheiro Júnior, J. W., Silva, D. R. da, Oliveira, A. A. da F., Silva, K. P. C. da, & Mota, R. A. (2006). Perfil De Sensibilidade Antimicrobiana De Agentes Contagiosos E Ambientais Isolados De Mastite Clínica E Subclínica De Búfalas. *Arquivos do Instituto Biológico*, *73*(1), 17–21. doi:10.1590/1808-1657v73p0172006

CurbeloRodríguez, J. E., V. RodríguezCruz and A. AlmeidaMontenegro. (2016). Evaluación de la capacidad termoregulatoria en bovinos lecheros Holstein pelona puertorriqueña, Holstein normal y Jersey. *The Journal of Agriculture of the University of Puerto Rico*, *100*, 112. doi:10.1007105840172 1101

da Silva, J. G., Alves, B. H. L. S., Melo, R. P. B., Kim, P. C. P., Neto, O. L. S., Bezerra, M. J. G., Sá, S. G., & Mota, R. A. (2015). Occurrence of anti-Toxoplasma gondii antibodies and parasite DNA in raw milk of sheep and goats of local breeds reared in Northeastern Brazil. *Acta Tropica*, *142*, 145–148. doi:10.1016/j.actatropica.2014.11.011 PMID:25438258 da Silva, J. G., Camargo, A. C., de Melo, R. P. B., Aragão, B. B., de Oliveira, J. M. B., de Sena, M. J., Nero, L. A., & Mota, R. A. (2022). Meca positive staphylococcus spp. In bovine mastitis, milkers, milking environment, and the circulation of different MRSA clones at dairy cows' farms in the northeast region of Brazil. *Ciência Rural*, *52*(3), e20210008. doi:10.1590/0103-8478cr20210008

Da Silva, J. G., De Barros, M., Santos, N. D. D. L., Paiva, P. M. G., Napoleão, T. H., De Sena, M. J., Da Costa, M. M., De Oliveira, H. P., Moreira, M. A. S., & Mota, R. A. (2020). Antimicrobial activity of polypyrrole nanoparticles and aqueous extract of Moringa oleifera against Staphylococcus spp. carriers of multi-drug efflux system genes isolated from dairy farms. *The Journal of Dairy Research*, *87*(3), 309–314. doi:10.1017/S0022029920000874 PMID:32958093

Dadgostar, P. (2019). Antimicrobial Resistance: Implications and Costs. *Infection and Drug Resistance*, *2012*, 3903–3910. doi:10.2147/IDR.S234610 PMID:31908502

Dafale N.A., Srivastava, S., & Purohit, H.J. (2020). Zoonose: um link emergente para a resistência aos antibióticos sob "uma abordagem de saúde. *Indian Journal of Microbiology*, *60* (2), 139-152. . PMC 7105526. PMID 32255846 doi:10.1007/s12088-020-00860-z

Dai, Z., Lyu, W., Xiang, X., Tang, Y., Hu, B., Ou, S., & Zeng, X. (2018). Immunomodulatory effects of enzymatic-synthesized α-galactooligosaccharides and evaluation of the structure-activity relationship. *Journal of Agricultural and Food Chemistry*, *66*(34), 9070–9079. doi:10.1021/acs.jafc.8b01939 PMID:30086236

Damasceno, F. A. (2012). Compost bedded pack barns system and computational simulation of airflow through naturally ventilated reduced model.

Dänicke, S., Meyer, U., Kersten, S., & Frahm, J. (2018). Animal models to study the impact of nutrition on the immune system of the transition cow. *Research in Veterinary Science*, *116*, 15–27. doi:10.1016/j.rvsc.2018.01.023 PMID:29428254

Dash, S., Chakravarty, A. K., Singh, A., Upadhyay, A., Singh, M., & Yousuf, S. (2016). Effect of heat stress on reproductive performances of dairy cattle and buffaloes: A review. *Veterinary World*, *9*(3), 235–244. doi:10.14202/vetworld.2016.235-244 PMID:27057105

Das, R., Sailo, L., Verma, N., Bharti, P., Saikia, J., Imtiwati, & Kumar, R. (2016). Impact of heat stress on health and performance of dairy animals: A review. *Veterinary World*, *9*(3), 260–268. doi:10.14202/vetworld.2016.260-268 PMID:27057109

Dastmalchi Saei, H., & Panahi, M. (2020). Genotyping and antimicrobial resistance of Staphylococcus aureus isolates from dairy ruminants: Differences in the distribution of clonal types between cattle and small ruminants. *Archives of Microbiology*, *202*(1), 115–125. doi:10.100700203-019-01722-z PMID:31501949

De Albuquerque Fernandes, S. A., Magnavita, A. P., Ferrao, S. P., Gualberto, S. A., Faleiro, A. S., Figueiredo, A. J., & Matarazzo, S. V. (2014). Daily ingestion of tetracycline residue present in pasteurized milk: A public health problem. *Environmental Science and Pollution Research International*, *21*(5), 3427–3434. https://link.springer.com/article/10.1007/s11356-013-2286-5. doi:10.100711356-013-2286-5 PMID:24243094

De Almeida, M. P. G., Mockaitis, G., & Weissbrodt, D. G. (2021). *Got Whey? The significance of cheese whey at the confluence of dairying, environmental impacts, energy and resource biorecovery*. The Evolving Scholar.

De Rensis, F., & Scaramuzzi, R. J. (2003). Heat stress and seasonal effects on reproduction in the dairy cow—A review. *Theriogenology*, *60*(6), 11391151. doi:10.1016/S0093-691X(03)00126-2 PMID:12935853

Deeth, H. C., & Bansal, N. (2018). *Whey proteins: From milk to medicine*. Academic Press.

Defante, L.; Damasceno, J. C.; Bánkuti, F. I.; Ramos, C. E. C. D. O. (2019). Typology of dairy production systems that meet Brazilian standards for milk quality. *Revista Brasileira de Zootecnia*.

DelCurto-Wyffels, H. M., Dafoe, J. M., Parsons, C. T., Boss, D. L., DelCurto, T., Wyffels, S. A., Van Emon, M. L., & Bowman, J. G. P. (2021). The Influence of Environmental Conditions on Intake Behavior and Activity by Feedlot Steers Fed Corn or Barley-Based Diets. *Animals (Basel)*, *11*(5), 1261. doi:10.3390/ani11051261 PMID:33925628

DeVries, T. J., Deming, J. A., Rodenburg, J., Seguin, G., Leslie, K. E., & Barkema, H. W. (2011). Association of standing and lying behavior patterns and incidence of intramammary infection in dairy cows milked with an automatic milking system. *Journal of Dairy Science*, *94*(8), 3845–3855. doi:10.3168/jds.2010-4032 PMID:21787921

Diez-Gonzalez, F. (2010). Applications of Bacteriocins in Livestock. *Current Issues in Intestinal Microbiology 8*. PMID:24031528

Dikmen, S., Alava, E., Pontes, E., Fear, J. M., Dikmen, B. Y., Olson, T. A., & Hansen, P. J. (2008). Differences in thermoregulatory ability between slick-haired and wild-type lactating Holstein cows in response to acute heat stress. *Journal of Dairy Science*, *91*(9), 3395–3402. doi:10.3168/jds.2008-1072 PMID:18765598

Dikmen, S., Khan, F. A., Huson, H. J., Sonstegard, T. S., Moss, J. I., Dahl, G. E., & Hansen, P. J. (2014). The SLICK hair locus derived from Senepol cattle confers thermotolerance to intensively managed lactating Holstein cows. *Journal of Dairy Science*, *97*, 55085520. doi:10.3168/jds.20148087

Doherty, S. B., Gee, V. L., Rossa, R. P., Stanton, C., Fitzgerald, G. F., & Brodkorb, A. (2011). Development and characterisation of whey protein micro-beads as potential matrices for probiotic protection. *Food Hydrocolloids*, *25*(6), 1604–1617. doi:10.1016/j.foodhyd.2010.12.012

Dore, S., Liciardi, M., Amatiste, S., Bergagna, S., Bolzoni, G., Caligiuri, V., Cerrone, A., Farina, G., Montagna, C. O., Saletti, M. A., Scatassa, M. L., Sotgiu, G., & Cannas, E. A. (2016). Survey on small ruminant bacterial mastitis in Italy, 2013–2014. *Small Ruminant Research*, *141*, 91–93. doi:10.1016/j.smallrumres.2016.07.010

Drescher, G., Mattiello, S. P., Peixoto, R. M., Vargas, A. C., Maciel, M. N., & Costa, M. M. (2010). Caracterização Bioquímica E Perfil De Sensibilidade Aos Antimicrobianos De Agentes Bacterianos Isolados De Mastite Subclínica Ovina Na Região Oeste De Santa Catarina. *Ciência Animal Brasileira*, *11*(1), 188–193. doi:10.5216/cab.v11i1.5189

Duarte, L. S., Schöffer, J. da N., Lorenzoni, A. S. G., Rodrigues, R. C., Rodrigues, E., & Hertz, P. F. (2017). A new biopocess for the production of prebiotc lactosusrose by immobilized β-galactosidase. *Process Biochemistry*, *55*, 96–103. doi:10.1016/j.procbio.2017.01.015

Dullius, A., Goettert, M. I., & Souza, C. F. V. (2018). Whey protein hydrolysates as a source of bioactive peptides for functional foods—Biotechnological facilitation of industrial scale-up. *Journal of Functional Foods*, *42*, 58–74. doi:10.1016/j.jff.2017.12.063

Dunkelberger, E. (1995). The statutory basis for the FDA's food safety assurance programs: From GMP, to emergency permit control, to HACCP. *Food and Drug Law Journal*, *50*(3), 357–383. PMID:10343005

El-malek, F., Farag, A., Omar, S., & Khairy, H. (2020). Polyhydroxyalkanoates (PHA) from Halomonas pacifica ASL10 and Halomonas salifodiane ASL11 isolated from Mariout salt lakes. *International Journal of Biological Macromolecules*, *161*, 1318–1328. doi:10.1016/j.ijbiomac.2020.07.258 PMID:32755698

ElTarabany, M. S., & ElTarabany, A. A. (2015). Impact of maternal heat stress at insemination on the subsequent reproductive performance of Holstein, Brown Swiss, and their crosses. *Theriogenology*, *84*(9), 15231529. doi:10.1016/j.theriogenology.2015.07.040 PMID:26324113

EMA Committee for Medicinal Products for Veterinary Use (CVMP) and EFSA Panel on Biological Hazards (BIOHAZ), Murphy, D., Ricci, A., Auce, Z., Beechinor, J. G., Bergendahl, H., & Jukes, H. (2017). EMA and EFSA Joint Scientific Opinion on measures to reduce the need to use antimicrobial agents in animal husbandry in the European Union, and the resulting impacts on food safety (RONAFA). EFSA Journal, 15(1), e04666. doi:10.1111/jam.12081 PMID:23167818

EMBRAPA. Anuário Leite 2020. 1. Ed., Embrapa gado de leite, Elyon, 2020.

Erman, M., Demir, S., Ocak, E., Tüfenkçi, Ş., Oğuz, F., & Akköprü, A. (2011). Effects of Rhizobium, arbuscular mycorrhiza and whey applications on some properties in chickpea (Cicer arietinum L.) under irrigated and rainfed conditions. *Field Crops Research*, *122*(1), 14–24. doi:10.1016/j.fcr.2011.02.002

Escobar, G. P., Souza, C. F. V., & Lehn, D. N. (2014). Avaliação de beta-galactosidase livre e imobilizada na hidrólise da lactose do permeado de soro de queijo. *Caderno Pedagógico*, *11*(1), 117–129.

Eurostat (2020). Milk and milk product statistics. https://ec.europa.eu/eurostat/statistics-explained/index.php?title=Milk_and_milk_product_statistics#:~:text=In%202021%2C%202.9%20million%20tonnes,about%2023%20%25)%20in%202021.

FAO. (2015). *Milk facts*. FAO. https://www.fao.org/dairy-production-products/production/en/

FariaA. P. A.PennaC. F. A. M.PintoM. S.EndoÉ. (2020). Influência do leite com elevada contagem de células somáticas sobre características físico-químicas e processo de fermentação de iogurte. *Ciência Animal Brasileira*, v. 21. doi:10.1590/1809-6891v21e-44773

Farkas, C., Rezessy-Szabó, J. M., Gupta, V. K., Bujna, E., Pham, T. M., Pásztor-Huszár, K., & Nguyen, Q. D. (2019). Batch and fed-batch ethanol fermentation of cheese-whey powder with mixed cultures of different yeasts. *Energies*, *12*(23), 4495. doi:10.3390/en12234495

Farrokhi, F., Badii, F., Ehsani, M. R., & Hashemi, M. (2019). Functional and thermal properties of nanofibrillated whey protein isolate as functions of denaturation temperature and solution pH. *Colloids and Surfaces. A, Physicochemical and Engineering Aspects*, *583*, 124002. doi:10.1016/j.colsurfa.2019.124002

Farthing, M. J. (2004). Bugs and the gut: An unstable marriage. *Best Practice & Research. Clinical Gastroenterology*, *18*(2), 233–239. doi:10.1016/j.bpg.2003.11.001 PMID:15123066

Ferrazza, R. A., Garcia, H. D. M., Aristizábal, V. H. V., Nogueira, C. H., Veríssimo, C. J., Sartori, J. R., Sartori, R., & Ferreira, J. C. P. (2017). Thermoregulatory responses of Holstein cows exposed to experimentally induced heat stress. *Journal of Thermal Biology*, *66*, 68–80. doi:10.1016/j.jtherbio.2017.03.014 PMID:28477912

Ferreira, B. H. A., & Ribeiro, L. F. (2022). Mastites causadas por Escherichia coli, Klebsiella spp. e Streptococcus uberis relacionadas ao sistema de produção Compost Barn e o impacto na qualidade do leite. Https://Medium.Com/, 1–18. https://medium.com/@arifwicaksanaa/pengertian-use-case-a7e576e1b6bf

Field, A. (2009). *Descobrindo a estatística usando o SPSS* (2nd ed.). Artmed.

Fischer, C., & Kleinschmidt, T. (2018). Synthesis of galactooligosaccharides in milk and whey: A review. *Comprehensive Reviews in Food Science and Food Safety*, *17*(3), 678–697. doi:10.1111/1541-4337.12344 PMID:33350129

Fitzsimons, S. M., Mulvihill, D. M., & Moris, E. R. (2006). Denaturation and aggregation processes in thermal gelation of whey proteins resolved by differential scanning calorimetry. *Food Hydrocolloids*, *11*, 62–69.

Fontoura, A. B. P., Javaid, A., Sáinz de la Maza-Escolà, V., Salandy, N. S., Fubini, S. L., Grilli, E., & McFadden, J. W. (2022). Heat stress develops with increased total-tract gut permeability, and dietary organic acid and pure botanical supplementation partly restores lactation performance in Holstein dairy cows. *Journal of Dairy Science*, *105*(9), 7842–7860. doi:10.3168/jds.2022-21820 PMID:35931486

Food, U.N. and Agriculture Organization. (2012). "Probiotics in Animal Nutrition. Research in Veterinary Science, 93(1), 250-58. doi:10.1111/j.1863-2378.2007.01059.x PMID:17803515

Foroutan, A., Guo, A. C., Vazquez-Fresno, R., Lipfert, M., Zhang, L., Zheng, J., Badran, H., Budinski, Z., Mandal, R., Ametaj, B. N., & Wishart, D. S. (2019). Chemical Composition of Commercial Cow's Milk. *Journal of Agricultural and Food Chemistry*, *67*(17), 4897–4914. doi:10.1021/acs.jafc.9b00204 PMID:30994344

Forza, C. (2002). Survey research in operations management: A process-based perspective. *International Journal of Operations & Production Management*, *22*(2), 152–194. doi:10.1108/01443570210414310

Fox, P. F., Guinee, T. P., Cogan, T. M., & McSweeney, P. L. H. (2017). Microbiology of Cheese Ripening. In P. F. Fox, T. P. Guinee, T. M. Cogan, & P. L. McSweeney (Eds.), *Fundamentals of cheese science* (pp. 333–390). Springer. doi:10.1007/978-1-4899-7681-9_11

Frutos, P., Hervás, G., Giráldez, F. J., & Mantecón, A. R. (2004). Review. Tannins and ruminant nutrition. *Spanish Journal of Agricultural Research*, *2*(2), 191–202. doi:10.5424jar/2004022-73

Furlani, R. DE SOUSA, M. M., Rocha, G. N. da S. A. O., Vilar, F. C. R., Ramalho, R. C., & Peixoto, R. D. M. (2021). Antibacterial activity of essential oils against pathogens of importance in caprine and ovine mastitis. *Revista Caatinga*, *34*(3), 702–708. doi:10.1590/1983-21252021v34n322rc

Gaggìa, F., Mattarelli, P., & Biavati, B. (2010, July 31). Probiotics and prebiotics in animal feeding for safe food production. *International Journal of Food Microbiology*, *141*(Suppl 1), S15–S28. doi:10.1016/j.ijfoodmicro.2010.02.031 PMID:20382438

Gallo, R. L., Murakami, M., Ohtake, T., & Zaiou, M. (2002). Biology and clinical relevance of naturally occurring antimicrobial peptides. *The Journal of Allergy and Clinical Immunology*, *110*(6), 823–831. doi:10.1067/mai.2002.129801 PMID:12464945

Garcia, A. B., Angeli, N., Machado, L., Cardoso, F. C., & Gonzalez, F. (2015). Relationships between heat stress and metabolic and milk parameter in dairy cows in southern Brazil. *Tropical Animal Health and Production*, *47*(5), 889–894. doi:10.100711250-015-0804-9 PMID:25851928

García-Galán, A., García-Romero, E., Sánchez, A., Corrales, J. C., Contreras, A., & De la Fe, C. (2020). Isolation of Mycoplasma auris from milk of goats with clinical mastitis. *Small Ruminant Research*, *185*(March), 106089. doi:10.1016/j.smallrumres.2020.106089

García-González, E., Jiménez-Fontana, R., & Azcárate Goded, P. (2020). Approaches to teaching and learning for sustainability: Characterizing students' perceptions. *Journal of Cleaner Production*, *274*, 122928. doi:10.1016/j.jclepro.2020.122928

GarinoJunior, F., Camboim, E. K. A., Neves, P. B., Sá, A. V. V, & Almeida, A. P. (2011). Suscetibilidade a antimicrobianos e produção de betalactamase em amostras de staphylococcus isolados de mastite caprina no semiárido paraibano. *Arquivos do Instituto Biológico*, *28*(1), 103–107.

Gasques, J. G., Rezende, G. C. D., Villa Verde, C. M., Salerno, M. S., Conceição, J. C. P. R. D., & Carvalho, J. C. D. S. (2004). Desempenho e crescimento do agronegócio no Brasil [Performance and growth of agribusiness in Brazil]. Instituto de Pesquisa Econômica Aplicada - IPEA, 07–38.

Gaudino, S., Goia, I., Grignani, C., Monaco, S., & Sacco, D. (2014). Assessing agro-environmental performance of dairy farms in northwest Italy based on aggregated results from indicators. *Journal of Environmental Management*, *140*, 120–134. doi:10.1016/j.jenvman.2014.03.010 PMID:24747935

Gazola, M. G., Bánkuti, F. I., Brito, M. M., Prizon, R. C., Kuwahara, K. C., Pozza, M. S. S., & Damasceno, J. C. (2018). Development and application of a sustainability assessment model for dairy production systems. *Semina. Ciências Agrárias*, *39*, 2685–2702. doi:10.5433/1679-0359.2018v39n6p2685

GDP - Global Dairy Platform. (2016). *Global Dairy Platform Annual Review 2016*. GDP. https://www.globaldairyplatform.com/wp-content/uploads/2018/04/2016-annual-review-final.pdf

GDP - Global Dairy Platform. (2018). *Global Dairy Platform Annual Review 2018*. GDP. https://globaldairyplatform.com/wp-content/uploads/2019/02/gdp-annual-report-2018-compressed.pdf

Gebara, C., Chaves, K. S., Ribeiro, M. C. E., Souza, F. N., Grosso, C. R. F., & Gigante, M. L. (2013). Viability of Lactobacillus acidophilus La5 in pectin–whey protein microparticles during exposure to simulated gastrointestinal conditions. *Food Research International*, *51*(2), 872–878. doi:10.1016/j.foodres.2013.02.008

Gelasakis, A. I., Angelidis, A. S., Giannakou, R., Filioussis, G., Kalamaki, M. S., & Arsenos, G. (2016). Bacterial subclinical mastitis and its effect on milk yield in low-input dairy goat herds. *Journal of Dairy Science*, *99*(5), 3698–3708. doi:10.3168/jds.2015-10694 PMID:26898280

Geyer, R., Jambeck, J. R., & Law, K. L. (2017). Production, use, and fate of all plastics ever made. *Science Advances*, *3*(7), e1700782. doi:10.1126ciadv.1700782 PMID:28776036

Ghosh, S., & Mehla, R. K. (2012). Influence of dietary supplementation of prebiotics (mannanoligosaccharide) on the performance of crossbred calves. *Tropical Animal Health and Production*, *44*(3), 617–622. doi:10.100711250-011-9944-8 PMID:21805307

Gibson, G. R., & Roberfroid, M. B. (1995). Dietary modulation of the human colonic microbiota: Introducing the concept of prebiotics. *The Journal of Nutrition*, *125*(6), 1401–1412. doi:10.1093/jn/125.6.1401 PMID:7782892

Gil-Doménech, D., Magomedova, N., Sánchez-Alcázar, E. J., & Lafuente-Lechuga, M. (2021). Integrating Sustainability in the Business Administration and Management Curriculum: A Sustainability Competencies Map. *Sustainability*, *13*(16), 9458. doi:10.3390u13169458

Gomes, S. T. (1999). Diagnóstico e perspectivas da produção de leite no Brasil. In: Restrições técnicas, econômicas e institucionais ao desenvolvimento da cadeia produtiva do leite no Brasil, Brasília, p. 19-35.

Gómez-Gascón, L., Galán-Relaño, A., Cardoso-Toset, F., Barrero-Domínguez, B., Astorga, R. J., Luque, I., Tarradas, C., & Gómez-Laguna, J. (2022). Lactate dehydrogenase: Detecting high bacterial and somatic cells counts in goats from whole milk samples. *Small Ruminant Research*, 208. doi:10.1016/j.smallrumres.2022.106632

Gonçalves, J. L., Kamphuis, C., Martins, C. M. M. R., Barreiro, J. R., Tomazi, T., Gameiro, A. H., Hogeveen, H., & dos Santos, M. (2018). Bovine subclinical mastitis reduces milk yield and economic return. *Livestock Science*, *210*(January), 25–32. doi:10.1016/j.livsci.2018.01.016

González, F. H. D., Noro, G. (2011). Variações na composição do leite no subtrópico brasileiro. In: González, F.H.D.; Pinto, A.T.; Zanella, M.B.; Fischer, V.; Bondan, C. Qualidade do leite bovino: variações no trópico e no subtrópico, Passo Fundo: UPF Editora, 2, 28-53.

González-Cataño, F., Tovar-Castro, L., Castaño-Tostado, E., Regalado-Gonzalez, C., García-Almendarez, B., Cardador-Martínez, A., & Amaya-Llano, S. (2017). Improvement of covalent immobilization procedure of β-galactosidase from Kluyveromyces lactis for galactooligosaccharides production: Modeling and kinetic study. *Biotechnology Progress*, *33*(6), 1568–1578. doi:10.1002/btpr.2509 PMID:28571115

Gorniak, T., Meyer, U., Südekum, K., & Dänicke, S. (2014). Impact of mild heat stress on dry matter intake, milk yield and milk composition in mid-lactation Holstein dairy cows in a temperate climate. *Archives of Animal Nutrition*, *68*(5), 358–369. doi:10.1080/1745039X.2014.950451 PMID:25176045

Gouveia, B. G. (2013) Good Manufacturing Practices for Medicinal Products for Human Use [Dissertation, Lisbon: Lusófona University of Humanities and Technologies.]

Gouveia, B. G., Rijo, P., Gonçalo, T. S., & Reis, C. P. (2015). Good manufacturing practices for medicinal products for human use. *Journal of Pharmacy & Bioallied Sciences*, *7*(2), 87. doi:10.4103/0975-7406.154424 PMID:25883511

Govil, M., & Proth, J.-M. (2002). *Supply chain design and management: strategic and tactical perspectives.* Elsevier. https://www.sciencedirect.com/science/article/pii/B9780122941511500023

Grand, E., Respondek, F., Martineau, C., Detilleux, J., & Bertrand, G. (2013). Effects of short-chain fructooligosaccharides on growth performance of preruminant veal calves. *Journal of Dairy Science*, *96*(2), 1094–1101. doi:10.3168/jds.2011-4949 PMID:23200477

Grazziotin, R. C. B., Halfen, J., Rosa, F., Schmitt, E., Anderson, J. L., Ballard, V., & Osorio, J. S. (2020). Altered rumen fermentation patterns in lactating dairy cows supplemented with phytochemicals improve milk production and efficiency. *Journal of Dairy Science*, *103*(1), 301–312. doi:10.3168/jds.2019-16996 PMID:31733851

Guerrero, C., Vera, C., Conejeros, R., & Illanes, A. (2015). Transgalactosylation and hydrolytic activities of commercial preparations of β-galactosidase for the synthesis of prebiotic carbohydrates. *Enzyme and Microbial Technology*, *70*, 9–17. doi:10.1016/j.enzmictec.2014.12.006 PMID:25659627

Guimarães, P. M., Teixeira, J. A., & Domingues, L. (2010). Fermentation of lactose to bio-ethanol by yeasts as part of integrated solutions for the valorisation of cheese whey. *Biotechnology Advances*, *28*(3), 375–384. doi:10.1016/j.biotechadv.2010.02.002 PMID:20153415

Guo, X., Zhou, J., & Xiao, D. (2010). Improved ethanol production by mixed immobilized cells of Kluyveromyces marxianus and Saccharomyces cerevisiae from cheese whey powder solution fermentation. *Applied Biochemistry and Biotechnology*, *160*(2), 532–538. doi:10.100712010-008-8412-z PMID:19002658

Gurjar, A. A., Klaessig, S., Salmon, S. A., Schukken, R. J., & Yancey, Y. H. (2013). Evaluation of an alternative dosing regimen of a J-5 mastitis vaccine against intramammary Escherichia coli challenge in nonlactating late-gestation dairy cows. *Journal of Dairy Science*, *96*(8), 5053–5063. doi:10.3168/jds.2012-6456 PMID:23726420

Hair, J. F., Black, W. C., Babin, B. J., & Anderson, R. E. (2014). *Multivariate Data Analysis* (7th ed.). Pearson Education.

Hambali, I. U. (2018). *Development of prototype killed vaccine against Staphylococcus aureus mastitis in dairy cows.* [Doctoral thesis, Universiti Putra Malaysia].

Hambali, I. U., Bhutto, K. R., Jesse, F. F. A., Lawan, A., Odhah, M. N., Wahid, A. H., Azmi, M. L. M., Zakaria, Z., Arsalan, M., Muhammad, N. A., & Jefri, M. N. (2018). Clinical responses in cows vaccinated with a developed prototype killed Staphylococcus aureus mastitis vaccine. *Microbial Pathogenesis*, *124*, 101–105. doi:10.1016/j.micpath.2018.08.017 PMID:30114463

Hancock, R. E., & Sahl, H. G. (2006). Antimicrobial and host-defense peptides as new anti-infective therapeutic strategies. *Nature Biotechnology*, *24*(12), 1551–1557. doi:10.1038/nbt1267 PMID:17160061

Hansen, P. J. (2009). Effects of heat stress on mammalian reproduction. *Philosophical Transactions of the Royal Society of London. Series B, Biological Sciences*, *364*(1534), 33413350. doi:10.1098/rstb.2009.0131 PMID:19833646

Hansen, P. J. (2019). Reproductive physiology of the heatstressed dairy cow: Implications for fertility and assisted reproduction. *Animal Reproduction*, *16*(3), 497507. doi:10.21451/1984-3143-AR2019-0053 PMID:32435293

Haraguchi, F. K., Abreu, W. C., & Paula, H. (2006). Proteínas do soro do leite: Composição, propriedades nutricionais, aplicações no esporte e benefícios para a saúde humana. *Revista Brasileira de Nutrição*, *19*(4), 479–488. doi:10.1590/S1415-52732006000400007

Harvatine, K. J., & Allen, M. S. (2006). Effects of fatty acid supplements on milk yield and energy balance of lactating dairy cows. *Journal of Dairy Science*, *89*(3), 1081–1091. doi:10.3168/jds.S0022-0302(06)72176-2 PMID:16507705

Hasan, S. S., Haque, M. E., Suchi, I. Z., & Hossain, M. A. (2018). Assessment of Diploma Agricultural Students' Attitude towards Educational Sustainability: A Study of Selected Agricultural Training Institutes of Bangladesh. *Journal of Education. Society and Behavioural Science*, *25*(2), 1–12. doi:10.9734/JESBS/2018/41170

Hashemzadeh-Cigari, F., Khorvash, M., Ghorbani, G. R., Kadivar, M., Riasi, A., & Zebeli, Q. (2014). Effects of supplementation with a phytobiotics-rich herbal mixture on performance, udder health, and metabolic status of Holstein cows with various levels of milk somatic cell counts. *Journal of Dairy Science*, *97*(12), 7487–7497. doi:10.3168/jds.2014-7989 PMID:25306268

Hasunuma, T., Kawashima, K., Nakayama, H., Murakami, T., Kanagawa, H., Ishii, T., Akiyama, K., Yasuda, K., Terada, F., & Kushibiki, S. (2011). Effect of cellooligosaccharide or symbiotic feeding on growth performance, fecal condition, and hormone concentrations in Holstein calves. *Animal Science Journal = Nihon Chikusan Gakkaiho, 82*(4), 543–548. doi:10.1111/j.1740-0929.2010.00861.x

Heikkila, A. M., Nousiainen, J. I., & Pyorala, S. (2012). Costs of clinical mastitis with special reference to premature culling. *Journal of Dairy Science*, *95*(1), 139–150. doi:10.3168/jds.2011-4321 PMID:22192193

Heinrichs, A. J., Jones, C. M., & Heinrichs, B. S. (2003). Effects of Mannan Oligosaccharide or Antibiotics in Neonatal Diets on Health and Growth of Dairy Calves [Uyeno et al., "Effect of Probiotics/Prebiotics.]. *Journal of Dairy Science*, *86*(12), 4064–4069. doi:10.3168/jds.S0022-0302(03)74018-1 PMID:14740845

He, N., Chen, H., Zhou, Z., Zhao, W., Wang, S., Lv, Z., Liu, N., Wang, H., Yang, Z., & Li, S. (2021). Enzymatically synthesized a-galactooligosaccharides attenuate metabolic syndrome in high-fat diet induced mice in association with the modulation of gut microbiota. *Food & Function*, *12*(11), 4960–4971. doi:10.1039/D0FO03113E PMID:34100482

Henke, S., Shakhno, N., Botvynko, A., & Bedn, A. (2019). Production of galactooligosaccharides using various combinations of the comercial β-galactosidases. *Biochemical and Biophysical Research Communications*, *517*(4), 762–766. doi:10.1016/j.bbrc.2019.08.001 PMID:31395344

Hernández, L. B., Recio, I., & Amigo, L. (2008). β-lactoglobulin as source of bioactive peptides. *Journal of Amino Acids*, *35*(2), 257–265. doi:10.100700726-007-0585-1 PMID:17726638

Herrero, M., Wirsenius, S., Henderson, B., Rigolot, C., Thornton, P., Havlík, P., Boer, I., & Gerber, P. (2015). Livestock and the Environment: What Have We Learned in the Past Decade? *Annual Review of Environment and Resources*, *40*(1), 177–202. doi:10.1146/annurev-environ-031113-093503

Hill, C., Guarner, F., Reid, G., Gibson, G. R., Merenstein, D. J., Pot, B., Morelli, L., Canani, R. B., Flint, H. J., Salminen, S., Calder, P. C., & Sanders, M. E. (2014, August). Expert consensus document. The International Scientific Association for Probiotics and Prebiotics consensus statement on the scope and appropriate use of the term probiotic. *Nature Reviews. Gastroenterology & Hepatology*, *11*(8), 506–514. doi:10.1038/nrgastro.2014.66 PMID:24912386

Hillman, P.E., Lee, C.N., & Willard, S.T. (2005). Thermoregulatory responses associated with lying and standing in heat-stressed dairy cows. *American Society of agricultural and Biological Engineers*, *48*, 795-801.

Hogan, J., & Smith, K. L. (2012). Managing Environmental Mastitis. *The Veterinary Clinics of North America. Food Animal Practice*, *28*(2), 217–224. doi:10.1016/j.cvfa.2012.03.009 PMID:22664204

Holdsworth, S., & Sandri, O. (2021). Investigating undergraduate student learning experiences using the good practice learning and teaching for sustainability education (GPLTSE) framework. *Journal of Cleaner Production*, *311*, 127532. doi:10.1016/j.jclepro.2021.127532

Holko, I., Tančin, V., Vršková, M., & Tvarožková, K. (2019). Prevalence and antimicrobial susceptibility of udder pathogens isolated from dairy cows in Slovakia. *The Journal of Dairy Research*, *86*(4), 436–439. doi:10.1017/S0022029919000694 PMID:31727191

Huan Y., Kong Q., Mou H., Yi H. (2020). Antimicrobial peptides: Classification, design, application, and research progress in multiple fields. *Frontiers in Microbiology*, *11*, 2559. PMID:33178164

Hut, P. R., Scheurwater, J., Nielen, M., van den Broek, J., & Hostens, M. M. (2022). Heat stress in a temperate climate led to adapted sensor-based behavioral patterns of dairy cows. *Journal of Dairy Science*, *105*(8), 6909–6922. doi:10.3168/jds.2021-21756 PMID:35787319

IBGE. (2021). *Produção de Leite*. IBGE. https://www.ibge.gov.br/explica/producao-agropecuaria/leite/br

IEA - Instituto de Economia Agrícola. (2018). Diagnóstico da Produção e Consumo de Leite no Estado de São Paulo [Diagnosis of Milk Production and Consumption in the State of São Paulo]. *Análise e Indicadores do Agrongócio*, *13*, 1–10.

Igono, M. O., Johnson, H. D., Steevens, B. J., Krause, G. F., & Shanklin, M. D. (1987). Physiological, productive, and economic benefits of shade, spray, and fan systems versus shade for Holstein cows during summer heat. *Journal of Dairy Science*, *70*(5), 1069–1079. doi:10.3168/jds.S0022-0302(87)80113-3 PMID:3597932

Innes, G. K., Randad, P. R., Korinek, A., Davis, M. F., Price, L. B., So, A. D., & Heaney, C. D., (2020). External societal costs of antimicrobial resistance in humans attributable to antimicrobial use in livestock. *Annual review of public health*, *41*, 141-157.

Instituto Adolfo Lutz (São Paulo). (2008). *Métodos físico-químicos para análise de alimentos /coordenadores Odair Zenebon, Neus Sadocco Pascuet e Paulo Tiglea — São Paulo*. Instituto Adolfo Lutz.

Instituto Brasileiro de Geografia e Estatística - IBGE. (2019). Produção da Pecuária Municipal 2019 [Production of Municipal Livestock 2019]. https://biblioteca.ibge.gov.br/visualizacao/periodicos/84/ppm_2019_v47_br_informativo.pdf

Inthavong, W., Chassenieux, C., & Nicolai, T. (2019). Viscosity of mixtures of protein aggregates with different sizes and morphologies. *Soft Matter*, *15*(23), 4682–4688. doi:10.1039/C9SM00298G PMID:31114829

Israni, N., Venkatachalam, P., Gajaraj, B., Varalakshmi, K. N., & Shivakumar, S. (2020). Whey valorization for sustainable polyhydroxyalkanoate production by Bacillus megaterium: Production, characterization and in vitro biocompatibility evaluation. *Journal of Environmental Management*, *255*, 109884. doi:10.1016/j.jenvman.2019.109884 PMID:32063322

Izadpanah, A. E., & Gallo, R. L. (2005). Antimicrobial peptides. *Journal of the American Academy of Dermatology*, *52*, 381–390.

Jana, U. K., Suryawanshi, R. K., Prajapati, B. P., & Kango, N. (2021). Prebiotic mannooligosaccharides: Synthesis, characterization, and bioactive properties. *Food Chemistry*, *342*, 128328. doi:10.1016/j.foodchem.2020.128328 PMID:33257024

Jansen, J. B. H. P., Van Den Borne, B. H. P., Renes, R. J., Van Schaik, G., Lam, T. J. G. M., & Leeuwis, C. (2009, November). Explaining mastitis incidence in Dutch dairy farming: The influence of farmers' attitudes and behavior. *Preventive Veterinary Medicine, Amsterdam*, *92*(3), 210–223. doi:10.1016/j.prevetmed.2009.08.015 PMID:19800700

Jelen, P. (2003). Whey processing. In H. Roginski, J. W. Fuquay, & P. F. Fox (Eds.), *Encylcopedia of dairy sciences*. Academic Press.

Jin, Y., Parashar, A., Mason, B., & Bressler, D. C. (2016). Simultaneous hydrolysis and cofermentation of whey lactose with wheat for ethanol production. *Bioresource Technology*, *221*, 616–624. doi:10.1016/j.biortech.2016.09.063 PMID:27693727

Julio-Gonzalez, L. C., Hernandez-Hernandez, O., Moreno, F. J., Jimeno, M. L., Doyaguez, E. G., Olano, A., & Corzo, N. (2020). Hydrolysis and transgalactosylation catalysed by β-galactosidase from brush border membrane vesicles isolated from pig small intestine: A study using lactulose and its mixtures with lactose or galactose as substrates. *Food Research International*, *129*, 108811. doi:10.1016/j.foodres.2019.108811 PMID:32036892

Kaczorowski, Ł., Powierska-Czarny, J., Wolko, Ł., Piotrowska-Cyplik, A., Cyplik, P., & Czarny, J. (2022). The Influence of Bacteria Causing Subclinical Mastitis on the Structure of the Cow's Milk Microbiome. *Molecules (Basel, Switzerland)*, *27*(6), 1–11. doi:10.3390/molecules27061829 PMID:35335192

Kadzere, C. T., Murphy, M. R., Silanikove, N., & Maltz, E. (2002). Heat stress in lactating dairy cows: A review. *Livestock Production Science*, *77*(1), 5991. doi:10.1016/S0301-6226(01)00330-X

Kala, S. R., & Rani, N. L. (2022). Efficacy of antioxidant in the therapy of subclinical mastitis in buffaloes. *Original Article Buffalo Bulletin (January-March 2022)*, *41*(1).

Kala, S. R., Rani, N. L., Rao, V. V., & Subramanyam, K. V. (2021). Comparison of Different Diagnostic Tests for the Detection of Subclinical Mastitis in Buffaloes. *Buffalo Bulletin*, *40*(4), 653–659.

Kandeel, S. A., Morin, D. E., Calloway, C. D., & Constable, P. D. (2018). Association of California Mastitis Test Scores with Intramammary Infection Status in Lactating Dairy Cows Admitted to a Veterinary Teaching Hospital. *Journal of Veterinary Internal Medicine*, *32*(1), 497–505. doi:10.1111/jvim.14876 PMID:29222843

Karatzoglou, B. (2013). An in-depth literature review of the evolving roles and contributions of universities to Education for Sustainable Development. *Journal of Cleaner Production*, *49*, 44–53. doi:10.1016/j.jclepro.2012.07.043

Kargi, F., & Ozmıhcı, S. (2006). Utilization of cheese whey powder (CWP) for ethanol fermentations: Effects of operating parameters. *Enzyme and Microbial Technology*, *38*(5), 711–718. doi:10.1016/j.enzmictec.2005.11.006

Kartheek, M., Smith, A. A., Muthu, A. K., & Manavalan, R. (2011). Determination of adulterants in food: A Review. *Journal of Chemical and Pharmaceutical Research*, *3*(2), 629–636.

Kayitsinga, J., Schewe, R. L., Contreras, G. A., & Erskine, R. J. (2017). Antimicrobial treatment of clinical mastitis in the eastern United States: The influence of dairy farmers' mastitis management and treatment behavior and attitudes. *Journal of Dairy Science*, *100*(2), 1388–1407. doi:10.3168/jds.2016-11708 PMID:27939551

Kazancoglu, Y., Ozkan-Ozen, Y. D., & Ozbiltekin, M. (2018). Minimizing losses in milk supply chain with sustainability: An example from an emerging economy. *Resources, Conservation and Recycling*, *139*, 270–279. doi:10.1016/j.resconrec.2018.08.020

Kekana, T. W., Marume, U., Muya, M. C., & Nherera-Chokuda, F. V. (2020). Periparturient antioxidant enzymes, hematological profile and milk production of dairy cows supplemented with Moringa oleifera leaf meal. *Animal Feed Science and Technology*, *268*, 114606. doi:10.1016/j.anifeedsci.2020.114606

Khardori, N. (2022). Vaccines and vaccine resistance: Past, present, and future. *Indian Journal of Medical Microbiology*, *40*(2), 187–192. doi:10.1016/j.ijmmb.2021.12.008 PMID:34961641

Khatami, K., Perez-Zabaleta, M., Owusu-Agyeman, I., & Cetecioglu, Z. (2021). Waste to bioplastics: How close are we to sustainable polyhydroxyalkanoates production? *Waste Management (New York, N.Y.)*, *119*, 374–388. doi:10.1016/j.wasman.2020.10.008 PMID:33139190

Kilpatrick, S. (2000). Education, and training impacts on farm management practice. *Journal of Agricultural Education and Extension, Abingdon*, *7*(2), 105–116. doi:10.1080/13892240008438811

Kim, H., Min, Y., & Choi, B. (2019). Real-time temperature monitoring for the early detection of mastitis in dairy cattle: Methods and case research. *Computers and Electronics in Agriculture*, *162*(June 2018), 119–125. doi:10.1016/j.compag.2019.04.004

Kindstedt, P. S. (2014). The basic of cheesemaking. In C. W. Donnelly (Ed.), *Cheese and Microbes* (pp. 17–38). ASM Press.

King, C.C., Dschaak, C.M., Eun, J.-S., Fellner, V., Young, A.J. (2011). Quantitative analysis of microbial fermentation under normal or high ruminal temperature in continuous cultures.*The Professional Animal Scientist, 27* (4), 319-327.. doi:10.15232/S1080-7446(15)30495-2

King, C., Gunton, J., Freebairn, D., Coutts, J., & Webb, I. (2000). The Sustainability Indicator Industry: Where to from here? A focus group study to explore the potential of farmer participation in the development of indicators. *Australian Journal of Experimental Agriculture, 40*(4), 631–642. doi:10.1071/EA99148

Kinsella, J. E., & Whitehead, D. M. (1989). Proteins in whey: Chemical, physical and functional properties. *Advances in Food and Nutrition Research, 33*, 343–438. doi:10.1016/S1043-4526(08)60130-8 PMID:2697234

Klopčič, M., Slokan, P., & Erjavec, K. (2020). Consumer preference for nutrition and health claims: A multi-methodological approach. *Food Quality and Preference, 82*, 103863. doi:10.1016/j.foodqual.2019.103863

Koch L.E. and G.J. Lascano, (2018). Milk Fat Depression: Etiology, theories, and soluble carbohydrate interactions. *J. Anim. Res. Nutr 3*. doi:10.21767/2572-5459.100046

Koller, M. (2015). Recycling of waste streams of the biotechnological poly (hydroxyalkanoate) production by Haloferax mediterranei on whey. *International Journal of Polymer Science*, 2015.

Krishnamoorthy, P., Goudar, A. L., Suresh, K. P., & Roy, P. (2021). Global and countrywide prevalence of subclinical and clinical mastitis in dairy cattle and buffaloes by systematic review and meta-analysis. *Research in Veterinary Science, 136*(6450), 561–586. doi:10.1016/j.rvsc.2021.04.021 PMID:33892366

Krissansen, G. W. (2007). Emerging health properties of whey proteins and their clinical implications. *Journal of the American College of Nutrition, 26*(6), 713–723. doi:10.1080/07315724.2007.10719652 PMID:18187438

Kucera, D., Pernicová, I., Kovalcik, A., Koller, M., Mullerova, L., Sedlacek, P., & Obruca, S. (2018). Characterization of the promising poly (3-hydroxybutyrate) producing halophilic bacterium Halomonas halophila. *Bioresource Technology, 256*, 552–556. doi:10.1016/j.biortech.2018.02.062 PMID:29478784

Kuchtík, J., Šustová, K., Sýkora, V., Kalhotka, L., Pavlata, L., & Konečná, L. (2021). Changes in the somatic cells counts and total bacterial counts in raw goat milk during lactation and their relationships to selected milk traits. *Italian Journal of Animal Science, 20*(1), 911–917. doi:10.1080/1828051X.2021.1913077

Kumar, S., Mukherjee, M. M., & Varela, M. F. (2013). Modulation of Bacterial Multidrug Resistance Efflux Pumps of the Major Facilitator Superfamily. *International Journal of Bacteriology, 2013*, 1–15. doi:10.1155/2013/204141 PMID:25750934

Lacetera, N. (2019). Impact of climate change on animal health and welfare. *Animal Frontiers, 9*(1), 26–31. doi:10.1093/af/vfy030 PMID:32002236

Lambrechts, W., Ghijsen, P. W. T., Jacques, A., Walravens, H., Liedekerke, L. V., & Petegem, P. V. (2018). Sustainability segmentation of business students: Toward selfregulated development of critical and interpretational competences in a post-truth era. *Journal of Cleaner Production, 202*, 561–570. doi:10.1016/j.jclepro.2018.07.303

Lammoglia, M. A., Garcez, N., Cabrera, A., López, R. D., Carmen, I., Renteria, D., & Rojas-Ronquillo, R. (2016) Behavior affected by routine oxytocin injection in crossbred cows in the tropics. R. Bras. Zootec. V.45, n.08, 478-482. doi:10.1590/S1806-92902016000800008

Lane, M. M., & Morrissey, J. P. (2010). Kluyveromyces marxianus: A yeast emerging from its sister's shadow. *Fungal Biology Reviews, 24*(1-2), 17–26. doi:10.1016/j.fbr.2010.01.001

Lange, A., Grzenia, A., Wierzbicki, M., Strojny-Cieslak, B., Kalińska, A., Gołębiewski, M., Radzikowski, D., Sawosz, E., & Jaworski, S. (2021). Silver and copper nanoparticles inhibit biofilm formation by mastitis pathogens. *Animals (Basel), 11*(7), 1–11. doi:10.3390/ani11071884 PMID:34202806

Langoni, H., Da, D., Penachio, S., Borin, D., Felipe De Freitas Guimarães, N. I., Baldini, S., & Ii, L. (2012). 095 Somaticell ® as a screening method for somatic cell count from bovine milk. Somaticell ® as a screening method for somatic cell count from bovine milk *Avaliação do Somaticell ® como método de triagem para contagem de células somáticas do leite de bovi. 42*(6), 1095–1101.

Langoni, H. (2000). Tendências de modernização do setor lácteo: Monitoramento da qualidade do leite pela contagem de células somáticas. *Revista Educação Continuada. CMRV-SP., 3*, 57–64.

Langoni, H., Salina, A., Oliveira, G. C., Junqueira, N. B., Menozzi, B. D., & Joaquim, S. F. (2017). Considerações sobre o tratamento das mastites. *Pesquisa Veterinaria Brasileira, 37*(11), 1261–1269. doi:10.15900100-736x2017001100011

Lavelli, V. (2021). Circular food supply chains–Impact on value addition and safety. *Trends in Food Science & Technology, 114*, 323–332. doi:10.1016/j.tifs.2021.06.008

Lavelli, V., & Beccalli, M. P. (2022). Cheese whey recycling in the perspective of the circular economy: Modeling processes and the supply chain to design the involvement of the small and medium enterprises. *Trends in Food Science & Technology, 126*, 86–98. doi:10.1016/j.tifs.2022.06.013

Lebacq, T., Baret, P. V., & Stilmant, D. (2013). Sustainability indicators for livestock farming. A review. *Agronomy for Sustainable Development, 33*(2), 311–327. doi:10.100713593-012-0121-x

Lees, P., Pelligand, L., Giraud, E., & Toutain, P. (2020). A history of antimicrobial drugs in animals: Evolution and revolution. *Journal of Veterinary Pharmacology and Therapeutics*. doi:10.1111/jvp.12895 PMID:32725687

LeitnerG.LavonY.MerinU.JacobyS.BlumS. E.KrifucksO.SilanikoveN. (2019). Increasing the value of raw bulk milk quality based on mammary glands as production units vs. the udder in dairy cows with mastitis. bioRxiv. doi:10.1101/567271

Leitner, G., Lubashevsky, E., Glickman, A., Winkler, M., Saran, A., & Trainin, Z. (2003). Development of a Staphylococcus aureus vaccine against mastitis in dairy cows: I. Challenge trials. *Veterinary Immunology and Immunopathology*, *93*(1–2), 31–38. doi:10.1016/S0165-2427(03)00051-5 PMID:12753773

Leitner, G., Yadlin, N., Lubashevsy, E., Ezra, E., Glickman, A., Chaffer, M., Winkler, M., Saran, A., & Trainin, Z. (2003). Development of a Staphylococcus aureus vaccine against mastitis in dairy cows. II. Field trial. *Veterinary Immunology and Immunopathology*, *93*(3–4), 153–158. doi:10.1016/S0165-2427(03)00062-X PMID:12814700

Leonel, S. G. (2007). Um estudo do processo de planejamento tecnológico de uma empresa nascente: alinhando tecnologia, produto e mercado com foco na necessidade do cliente [A study of the technological planning process of a nascent company: aligning technology, product and market with a focus on customer needs.]. Master's thesis, Universidade Federal de Minas Gerais, Brazil.

Li, G., Chen, S., Chen, J., Peng, D., & Gu, X. (2020). Predicting rectal temperature and respiration rate responses in lactating dairy cows exposed to heat stress. *Journal of Dairy Science*, *103*(6), 5466–5484. doi:10.3168/jds.2019-16411 PMID:32278558

Likert, R. (1932). A technique for the measurement of attitudes. *Archives de Psychologie*, *22*, 90–95.

Lima, M. A., Mendes, L. F. R., Mothé, G. A., Linhares, F. G., de Castro, M. P. P., Da Silva, M. G., & Sthel, M. S. (2020). Renewable energy in reducing greenhouse gas emissions: Reaching the goals of the Paris agreement in Brazil. *Environmental Development*, *33*, 100504. doi:10.1016/j.envdev.2020.100504

Li, N., Richoux, R., Boutinaud, M., Martin, P., & Gagnaire, V. (2014). Role of somatic cells on dairy processes and products: A review. *Dairy Science & Technology*, *1*(94), 517–538. doi:10.100713594-014-0176-3 PMID:25309683

Liu, J., Dantoft, S. H., Würtz, A., Jensen, P. R., & Solem, C. (2016). A novel cell factory for efficient production of ethanol from dairy waste. *Biotechnology for Biofuels*, *9*(1), 1–11. doi:10.118613068-016-0448-7 PMID:26925162

Lock, A. L., Bauman, D. E., & Jenkins, T. C. (2008). Understanding the biology of milk fat depression: from basic concepts to practical application. *Proc. Intermountain Nutr. Conf.* (pp. 27-44). IEEE.

Lock, A. L., Overton, T. R., Harvatine, K. J., Giesy, J. G., & Bauman, D. E. (2006). Milk fat depression: impact of dietary components and their interaction during rumen fermentation. *Proc. Cornell Nutr. Conf.* (pp. 75-85). IEEE.

Lopes, M. A., & Santos, G. (2012). Comparativo econômico entre propriedades leiteiras em sistema intensivo de produção: Um estudo multicasos. *Revista Brasileira de Saúde e Produção Animal*, *13*(3), 591–603. doi:10.1590/S1519-99402012000300001

Lorenzo, S., Mira, J., Olarte, M., Guerrero, J., & Moyano, S. (2004). Análisis matricial de la voz del cliente: QFD aplicado a la gestión sanitária [Matrix analysis of the voice of the customer: QFD applied to health management]. *Gaceta Sanitaria, 18*(6), 464–471. doi:10.1016/S0213-9111(04)72034-8 PMID:15625045

Lu, L., Guo, L., Wang, K., Liu, Y., & Xiao, M. (2020). β-Galactosidases: A great tool for synthesizing galactose-containing carbohydrates. *Biotechnology Advances, 39*, 107465. doi:10.1016/j.biotechadv.2019.107465 PMID:31689470

M'Hamdi, N., Darej, C., Attia, K., Znaidi, I. L., Khattab, R., Djelailia, H., Bouraoui, R., Taboubi, R., Marzouki, L., & Ayadi, M. (2021). Modelling THI effects on milk production and lactation curve parameters of Holstein dairy cows. *Journal of Thermal Biology, 99*, 102917. doi:10.1016/j.jtherbio.2021.102917 PMID:34420599

Machado, N. A. F., Da Costa, L. B. S., Barbosa-Filho, J. A. D., De Oliveira, K. P. L., De Sampaio, L. C., Peixoto, M. S. M., & Damasceno, F. A. (2021). Using infrared thermography to detect subclinical mastitis in dairy cows in compost barn systems. *Journal of Thermal Biology, 97*(February), 102881. doi:10.1016/j.jtherbio.2021.102881 PMID:33863445

Madureira, A. R., Tavares, T., Gomes, A. M. P., Pintado, M. E., & Malcata, F. X. (2010). Invited review: Physiological properties of bioactive peptides obtained from whey proteins. *Journal of Dairy Science, 93*(2), 437–455. doi:10.3168/jds.2009-2566 PMID:20105516

Ma, F. T., Shan, Q., Jin, Y. H., Gao, D., Li, H. Y., Chang, M. N., & Sun, P. (2020). Effect of Lonicera japonica extract on lactation performance, antioxidant status, and endocrine and immune function in heat-stressed mid-lactation dairy cows. *Journal of Dairy Science, 103*(11), 10074–10082. doi:10.3168/jds.2020-18504 PMID:32896406

Magana, M., Pushpanathan, M., Santos, A. L., Leanse, L., Fernandez, M., Ioannidis, A., Giulianotti, M. A., Apidianakis, Y., Bradfute, S., Ferguson, A. L., Cherkasov, A., Seleem, M. N., Pinilla, C., de la Fuente-Nunez, C., Lazaridis, T., Dai, T., Houghten, R. A., Hancock, R. E. W., & Tegos, G. P. (2020). The value of antimicrobial peptides in the age of resistance. *The Lancet. Infectious Diseases, 20*(9), e216–e230. doi:10.1016/S1473-3099(20)30327-3 PMID:32653070

Mak, T. M., Xiong, X., Tsang, D. C., Iris, K. M., & Poon, C. S. (2020). Sustainable food waste management towards circular bioeconomy: Policy review, limitations and opportunities. *Bioresource Technology, 297*, 122497. doi:10.1016/j.biortech.2019.122497 PMID:31818718

Mantovani, H. C., Hu, H., Warobo, R. W., & Russell, J. B. (2002). Bovicin HC5, a bacteriocin from Streptococcus bovis HC5. *Microbiology, 148*(11), 3347–3352. doi:10.1099/00221287-148-11-3347 PMID:12427926

Mantovani, H. C., Kam, D. K., Ha, J. K., & Russell, J. B. (2001). The antibacterial activity and sensitivity of Streptococcus bovis strains isolated from the rumen of cattle. *FEMS Microbiology Ecology, 37*(3), 223–229. doi:10.1016/S0168-6496(01)00166-0 PMID:11311432

Marai, I. F. M., El-Darawanya, A. A., Fadielc, A., & Abdel-Hafezb, M. A. M. (2008). Reproductive performance traits as affected by heat stress and its alleviation in sheep. *Tropical and Subtropical Agroecosystems, 8*, 209–234.

Marandure, T., Mapiye, C., Makombe, G., & Dzama, K. (2017). Indicator-based sustainability assessment of the smallholder beef cattle production system in South Africa. *Agroecology and Sustainable Food Systems, 41*(1), 3–29. doi:10.1080/21683565.2016.1231152

Marion, J. C. (2007). *Contabilidade Rural: Contabilidade Agrícola, contabilidade da Pecuária, Imposto de Renda*. Pessoa Jurídica.

Marion, J. C., & Segatti, S. (2010). *Contabilidade da pecuária. 9*. Atlas.

Marques, L. G., Neumann, P. S., & Zarnott, A. V. (2015). Trabalho de campo na formação holística e crítica de estudantes de Agronomia: A experiência da disciplina de Desenvolvimento Rural – UFSM. *Cadernos de Agroecologia, 10*, 1–5.

Marth, E. H., & Ellickson, B. E. (1959). Problems Created by the Presence of Antibiotics in Milk—A Review. *Journal of Milk and Food Technology, 22*(9), 266–272. doi:10.4315/0022-2747-22.9.266

Martí-De Olives, A., Peris, C., & Molina, M. P. (2020). Effect of subclinical mastitis on the yield and cheese-making properties of ewe's milk. *Small Ruminant Research, 184*(April 2019), 106044. doi:10.1016/j.smallrumres.2019.106044

Martínez-Villaluenga, C., Cardelle-Cobas, A., Corzo, N., Olano, A., & Villamiel, M. (2008). Optimization of conditions for galactooligosaccharide synthesis during lactose hydrolysis by β-galactosidase from Kluyveromyces lactis (Lactozym 3000 L HP G). *Food Chemistry, 107*(1), 258–264. doi:10.1016/j.foodchem.2007.08.011

Martin, P., & Bateson, P. (1993). *Measuring behavior: an introductory guide*. Cambridge University Press. doi:10.1017/CBO9781139168342

Martins, G. N., Ureta, M. M., Tymczyszyn, E. E., Castilho, P. C., & Gomez-Zavaglia, A. (2019). Technological aspects of the production of fructo and enzymatic synthesis and hydrolysis. *Frontiers in Nutrition, 6*, 78. doi:10.3389/fnut.2019.00078 PMID:31214595

Marumo, J. L., Lusseau, D., Speakman, J. R., Mackie, M., & Hambly, C. (2022). Influence of environmental factors and parity on milk yield dynamics in barn-housed dairy cattle. *Journal of Dairy Science, 105*(2), 1225–1241. doi:10.3168/jds.2021-20698 PMID:34802739

Massote, V. P., Zanateli, B. M., Alvez, G. V., Gonçalves, E. S., & Guedes, E. (2019). *Diagnóstico e controle de mastite bovina: uma revisão de literatura*. Revista Agroveterinária do Sul de Minas.

Matsubara, M. T., Beloti, V., Tamanini, R., Fagnani, R., Da Silva, L. C. C., Monteiro, A. A., & Barros, M. D. A. F. (2011). Good practices in milking to reduce microbiological contamination of milk in agreste of Pernambuco. Semina. Ciencias Agrarias, Londrina.

Mazo, S. Z. (2012). Incorporação da voz do cliente nas etapas iniciais do desenvolvimento de sistemas produto-serviço (PSS) [Incorporation of the customer's voice in the initial stages of the development of product-service systems (PSS]. Master's thesis. Universidade Tecnológica Federal do Paraná, Brazil. Miguel, P.A.C., & Ho. L. L. (2010). Levantamento Tipo Survey [Survey Method].São Paulo: Elsevier Editora Ltda.

Medeiros, E. S., França, C. A., da Krewer, C. C., de Peixoto, R. M., de Souza Júnior, A. F., Cavalcante, M. B., da Costa, M. M., & Mota, R. A. (2011). Antimicrobial resistance of staphylococcus spp. isolates from cases of mastitis in buffalo in brazil. *Journal of Veterinary Diagnostic Investigation, 23*(4), 793–796. doi:10.1177/1040638711407882 PMID:21908326

Mei, Z., Yuan, J., & Li, D. (2022). Biological activity of galacto-oligosaccharides: A review. *Frontiers in Microbiology, 13*, 993052. doi:10.3389/fmicb.2022.993052 PMID:36147858

Melvin, J. M., Heuwieser, W., Virkler, P. D., Nydam, D. V., & Wieland, M. (2019). Machine milking–induced changes in teat canal dimensions as assessed by ultrasonography. *Journal of Dairy Science, 102*(3), 2657–2669. doi:10.3168/jds.2018-14968 PMID:30639026

Mercer, H. D., Baggot, J. D., & Sams, R. A. (1977). Application of pharmacokinetic methods to the drug residue profile. *Journal of Toxicology and Environmental Health, 2*(4), 787–801. doi:10.1080/15287397709529479 PMID:857041

Merz, A., Stephan, R., & Johler, S. (2016). Staphylococcus aureus isolates from goat and sheep milk seem to be closely related and differ from isolates detected from bovine milk. *Frontiers in Microbiology, 7*(MAR), 1–7. doi:10.3389/fmicb.2016.00319 PMID:27014240

Meul, M., Van Passel, S., Nevens, F., Dessein, J., Rogge, E., Mulier, A., & Van Hauwermeiren, A. (2008). MOTIFS: A monitoring tool for integrated farm sustainability. *Agronomy for Sustainable Development, 28*(2), 321–332. doi:10.1051/agro:2008001

Mimoune, N., Saidi, R., Benadjel, O., Khelef, D., & Kaidi, R. (2021). Alternative treatment of bovine mastitis. *Veterinarska Stanica, 52*(6), 639–649. doi:10.46419/vs.52.6.9

Ministério da Agricultura. Pecuária e Abastecimento – MAPA. (2020). Brasil Projeções do Agronegócio 2019/2020 a 2029/2030 [Brazil Agribusiness Projections 2019/2020 to 2029/2030]. 11° edição. Brasília: MAPA. https://www.gov.br/agricultura/pt-br/assuntos/politica-agricola/todas-publicacoes-de-politica-agricola/projecoes-do-agronegocio/projecoes-do-agronegocio_2019_20-a-2029_30.pdf

Ministro Do Estado Da Agricultura Pecuaria E Abstecimento. (2018). *Mapa 76 e 77.* Disponível em: https://saudeinspecaoanimal.com.br/in76-77.

Misra, N., Wines, T. F., Knopp, C. L., Hermann, R., Bond, L., Mitchell, B., McGuire, M. A., & Tinker, J. K. (2018). Immunogenicity of a Staphylococcus aureus-cholera toxin A2/B vaccine for bovine mastitis. *Vaccine, 36*(24), 3513–3521. doi:10.1016/j.vaccine.2018.04.067 PMID:29739718

Mistry, R. H., Liu, F., Borewicz, K., Lohuis, M. A. M., Smidt, H., Verkade, H. J., & Tietge, U. J. F. (2020). Long-term β-galacto-oligosaccharides supplementation decreases the development of obesity and insulin resistance in mice fed a western-type diet. *Molecular Nutrition & Food Research, 64*(12), 1900922. doi:10.1002/mnfr.201900922 PMID:32380577

Mohammadian, M., Moghadam, M., Salami, M., Emam Djomeh, Z., Alavi, F., Momen, S., & Moosavi-Movahedi, A. A. (2020). Whey protein aggregates formed by non-toxic chemical cross-linking as novel carriers for curcumin delivery: Fabrication and characterization. *Journal of Drug Delivery Science and Technology, 56*, 101531. doi:10.1016/j.jddst.2020.101531

Molina, C. H. A., Centenaro, G. S., & Furlan, V. J. M. (2015). Qualidade do leite cru comercializado informalmente no município de Itaqui-RS. *Vigilância Sanitária em Debate: Sociedade. Ciência & Tecnologia, 3*(4), 106–113.

Molina-Fernández, J. F. (2001). Origin and development of the Dairy Industry of Puerto Rico. [Origen y desarrollo de la industria lechera en Puerto Rico. Página 30 en La industria lechera en Puerto Rico]. NUPRESS of Miami, Inc. Miami, FL.

Molineri, A. I., Camussone, C., Zbrun, M. V., Suárez Archilla, G., Cristiani, M., Neder, V., Calvinho, L., & Signorini, M. (2021). Antimicrobial resistance of Staphylococcus aureus isolated from bovine mastitis: Systematic review and meta-analysis. *Preventive Veterinary Medicine, 188*, 105261. doi:10.1016/j.prevetmed.2021.105261 PMID:33508662

Molineri, A. I., Signorini, M. L., Cuatrín, A. L., Canavesio, V. R., Neder, V. E., Russi, N. B., & Bonazza, J. C. (2012). Calvinho, L.F. Association between milking practices and psychrotrophicbacterial counts in bulk tank milk. *Revista Argentina de Microbiologia, 44*, 187–194. PMID:23102468

Moraes, B. M. M., & Bender, R. (2017). Mercado brasileiro de lácteos: Análise do impacto de políticas de estímulo à produção [Brazilian dairy market: analysis of the impact of policies to stimulate production]. *Revista de Economia e Sociologia Rural, 55*(4), 783–800. doi:10.1590/1234-56781806-94790550410

Moroni, P., Nydam, D. V., Ospina, P. A., Scillieri-Smith, J. C., Virkler, P. D., Watters, R. D., Welcome, F. L., Zurakowski, J., Ducharme, N. G., & Yeager, A. E. (2018). 8 diseases of the Teats and Udder. In Peek, S.F., & Thomas J. Divers (Eds), Rebhun's Diseases of Dairy Cattle (389-465). Elsevier. doi:10.1016/B978-0-323-39055-2.00008-5

Morton, J. M., Tranter, W. P., Mayer, D. G., & Jonsson, N. N. (2007). Effects of environmental heat on conception rates in lactating dairy cows: Critical periods of exposure. *Journal of Dairy Science, 90*(5), 22712278. doi:10.3168/jds.2006-574 PMID:17430927

Muhammad, G., Naureen, A., Asi, M. N., Saqib, M., & Fazal-ur-Rehman. (2010). Evaluation of a 3% surf solution (surf field mastitis test) for the diagnosis of subclinical bovine and bubaline mastitis. *Tropical Animal Health and Production, 42*(3), 457–464. doi:10.100711250-009-9443-3 PMID:19731065

Mulani MS, Kamble EE, Kumkar SN, Tawre MS, Pardesi KR., (2019). Emerging Strategies to Combat ESKAPE Pathogens in the Era of Antimicrobial Resistance: A Review. *Front Microbiol.* . doi:10.3389/fmicb.2019.00539

Munyaneza, C., Kurwijila, L. R., Mdoe, N. S. Y., Baltenweck, I., & Twine, E. E. (2019). Identification of appropriate indicators for assessing sustainability of small-holder milk production systems in Tanzania. *Sustainable Production and Consumption, 19*, 141–160. doi:10.1016/j.spc.2019.03.009

Murari, C. S., Machado, W. R. C., Schuina, G. L., & Del Bianchi, V. L. (2019). Optimization of bioethanol production from cheese whey using Kluyveromyces marxianus URM 7404. *Biocatalysis and Agricultural Biotechnology*, 20, 101182. https://www.oecd-ilibrary.org/sites/aa3fa6a0-en/index.html?itemId=/content/component/aa3fa6a0-en#:~:text=This%20chapter%20describes%20the%20market,whole%20milk%20powder%20are%20discussed

Murari, C. S., da Silva, D. C. M. N., Schuina, G. L., Mosinahti, E. F., & Del Bianchi, V. L. (2019). Bioethanol production from dairy industrial coproducts. *BioEnergy Research, 12*(1), 112–122. doi:10.100712155-018-9949-5

Murphy, S. C., Martin, N. H., Barbano, D. M., & Wiedmann, M. (2016). Influence of raw milk quality on processed dairy products: How do raw milk quality test results relate to product quality and yield? *Journal of Dairy Science, 99*(12), 10128–10149. doi:10.3168/jds.2016-11172 PMID:27665134

Mylostyvyi, R. V., & Sejian, V. (2019). Welfare of dairy cattle in conditions of global climate change. *Theoretical and Applied Veterinary Medicine., 7*(1), 47–55. doi:10.32819/2019.71009

Naderi, N., Monavvarifard, F., & Salehi, L. (2022). Fostering sustainability-oriented knowledge-sharing in academic environment: A key strategic process to achieving SDGs through development of students' sustainable entrepreneurship competences. *International Journal of Management Education, 20*(1), 100603. doi:10.1016/j.ijme.2022.100603

Nardone, A., Ronchi, B., Lacetera, N., Ranieri, M. S., & Bernabucci, U. (2010). Effects of climate changes on animal production and sustainability of livestock systems. *Livestock Science, 130*(1-3), 57–69. doi:10.1016/j.livsci.2010.02.011

National Research Council. (2001). *Nutrient Requirements of Dairy Cattle. 7th rev. ed.* Natl. Acad. Sci., Washington, DC. https://www.nap.edu/catalog/9825/nutrient-requirements-of-dairy-cattle-seventh-revised-edition-2001

Nelder, J. A., & Wedderburn, R. W. M. (1972). Generalized Linear Model. *Journal of the Royal Statistical Society. Series A (General), 135*(3), 370–384. doi:10.2307/2344614

Nobrega, D. B., Naqvi, S. A., Dufour, S., Deardon, R., Kastelic, J. P., De Buck, J., & Barkema, H. W. (2020). Critically important antimicrobials are generally not needed to treat non-severe clinical mastitis in lactating dairy cows: Results from a network meta-analysis. *Journal of Dairy Science, 103*(11), 10585–10603. doi:10.3168/jds.2020-18365 PMID:32896405

Nonaka, I., Takusari, N., Tajima, K., Suzuki, T., Higuchi, K., & Kurihara, M. (2008). Effects of high environmental temperatures on physiological and nutritional status of prepubertal Holstein heifers. *Livestock Science, 113*(1), 14–23. doi:10.1016/j.livsci.2007.02.010

Nostrand, S. D., Galton, D. M., Erb, H. N., & Bauman, D. E. (1991). Effects of Daily Exogenous Oxytocin on Lactation Milk Yield and Composition. *Journal of Dairy Science, 74*(7), 2119–2127. doi:10.3168/jds.S0022-0302(91)78384-7 PMID:1894807

O'NEIL J. (2016). In A. Ro (Ed.), *Tackling drug-resistant infections globally: final report and recommendations* (Vol. 1, p. 84). Resistance.

Okamoto, K., Nakagawa, S., Kanawaku, R., & Kawamura, S. (2019). Ethanol production from cheese whey and expired milk by the brown rot fungus Neolentinus lepideus. *Fermentation (Basel, Switzerland), 5*(2), 49. doi:10.3390/fermentation5020049

Okamoto, K., Nitta, Y., Maekawa, N., & Yanase, H. (2011). Direct ethanol production from starch, wheat bran and rice straw by the white rot fungus Trametes hirsuta. *Enzyme and Microbial Technology, 48*(3), 273–277. doi:10.1016/j.enzmictec.2010.12.001 PMID:22112911

Okamoto, K., Uchii, A., Kanawaku, R., & Yanase, H. (2014). Bioconversion of xylose, hexoses and biomass to ethanol by a new isolate of the white rot basidiomycete Trametes versicolor. *SpringerPlus, 3*(1), 1–9. doi:10.1186/2193-1801-3-121 PMID:24624317

Okamoto, K., Yoneda, M., & Fumioka, T. (2017). Isolation of a Peniophora strain capable of producing ethanol from starch and kitchen waste. *Fermentation Technology, 6*(1), 1000140.

Oke, M. A., Annuar, M. S. M., & Simarani, K. (2016). Mixed feedstock approach to lignocellulosic ethanol production—Prospects and limitations. *BioEnergy Research, 9*(4), 1189–1203. doi:10.100712155-016-9765-8

Okumah, M., Martin-Ortega, J., & Novo, P. (2018). Effects of awareness on farmers' compliance with diffuse pollution mitigation measures: A conditional process modelling. *Land Use Policy, 76*, 36–45. doi:10.1016/j.landusepol.2018.04.051

Olalla, C. B., & Merino, A. (2019). Competences for sustainability in undergraduate business studies: A content analysis of value-based course syllabi in Spanish universities. *International Journal of Management Education, 17*(2), 239–253. doi:10.1016/j.ijme.2019.02.006

Olde, E. M., Moller, H., Marchand, F., Mcdowell, R. W., Macleod, C. J., Sautier, M., Halloy, S., Barber, A., Benge, J., Bockstaller, C., Bokkers, E. A. M., Boer, I. J. M., Legun, K. A., Le Quellec, I., Merfield, C., Oudshoorn, F. W., Reid, J., Schader, C., Szymanski, E., & Manhire, J. (2017). When experts disagree: The need to rethink indicator selection for assessing sustainability of agriculture. *Environment, Development and Sustainability, 19*(4), 1327–1342. doi:10.100710668-016-9803-x

Olde, E. M., Oudshoorn, F. W., Sørensen, C. A. G., Bokkers, E. A. M., & Boer, I. J. M. (2016). Assessing sustainability at farm-level: Lessons learned from a comparison of tools in practice. *Ecological Indicators, 66*, 391–404. doi:10.1016/j.ecolind.2016.01.047

Oliveira, R. P., Bezerra, B. A., de Melo, R. P. B., da Silva, D. M. S., de Carvalho, R. G., Juliano, M. A., Farias, M. P. O., de Lira, N. S. C., & Mota, R. A. (2022). Bovine mastitis in northeastern Brazil: Occurrence of emergent bacteria and their phenotypic and genotypic profile of antimicrobial resistance. *Comparative Immunology, Microbiology, and Infectious Diseases*, *85*(November 2021). doi:10.1016/j.cimid.2022.101802

Oliveira, C., Guimarães, P. M. R., & Domingues, L. (2011). Recombinant microbial systems for improved β-galactosidase production and biotechnological applications. *Biotechnology Advances*, *29*(6), 600–6009. doi:10.1016/j.biotechadv.2011.03.008 PMID:21514372

Oliveira, L. F. T., & Silva, S. P. (2012). Mudanças institucionais e produção familiar na cadeia produtiva do leite no Oeste Catarinense [Institutional changes and family production in the milk production chain in western Santa Catarina]. *Revista de Economia e Sociologia Rural*, *50*(4), 705–720. doi:10.1590/S0103-20032012000400007

Oliveira, R. P., Silva, J. G., Aragão, B. B., Carvalho, R. G., Juliano, M. A., Frazzon, J., Farias, M. P. O., & Mota, R. A. (2022). Diversity and emergence of multi-resistant Staphylococcus spp. isolated from subclinical mastitis in cows in of the state of Piauí, Brazil. *Brazilian Journal of Microbiology*, *0123456789*(4), 2215–2222. doi:10.100742770-022-00822-1 PMID:36074251

Oliveira, S. C. C., Nishi, L., Mantovani, D., Pisano Mateus, G. A., Tonial dos Santos, T. R., Takaoka Alves Baptista, A., Guttierres Gomes, R., & Bergamasco, R. (2019). EXTRATOS DE PLANTAS BRASILEIRAS NO CONTROLE DA BACTÉRIA Staphylococcus aureus CAUSADORA DA MASTITE CONTAGIOSA EM BOVINOS LEITEIROS. *Revista Tecnológica*, *27*(1), 48–58. doi:10.4025/revtecnol.v27i1.43745

Oliver, & ... . (2007). "Asas Centennial Paper: Developments and Future Outlook"; Els N.T. Meeusen et al., "Current Status of Veterinary Vaccines. *Clinical Microbiology Reviews*, *20*(3), 489–510. PMID:17630337

Olson, T. A., Lucena, C., Chase, C. C. Jr, & Hammond, A. C. (2003). Evidence of a major gene influencing hair length and heat tolerance in Bos taurus cattle. *Journal of Animal Science*, *81*(1), 8090. doi:10.2527/2003.81180x PMID:12597376

OMS. Centro de Mídia, (2017). Comunicado de Notícias. OMS publica lista de bactérias para as quais novos antibióticos são urgentemente necessários. 2017 [citado em março]. *Disponível* em: https://www.who.int/mediacentre/news/releases/2017/bacteria-antibiotics-needed/en/

Orozco, A. J., Angulo, L. M., Pérez, A. P., & Liodoro, J. H. (2012). Aspectos fisiológicos y bromatológicos de Brachiaria humidicola. *CES Medicina Veterinaria y Zootecnia*, *7*, 87–98. http://www.scielo.org.co/scielo.php?script=sci_arttext&pid=S1900-96072012000100008

Ortiz-Colón, G., Fain, S. J., Parés, I. K., Curbelo-Rodríguez, J., Jiménez-Cabán, E., Pagán-Morales, M., & Gould, W. A.OrtizColón. (2018). Assessing climate vulnerabilities and adaptive strategies for resilient beef and dairy operations in the tropics. *Climatic Change*, *146*(1-2), 4758. doi:10.100710584-017-2110-1

OrtizUriarte. (2020). B. P., N. L. RosaPadilla, R. LópezLópez, J. CurbeloRodríguez, V. M. NegrónPérez and G. OrtizColón, (2020). Comparison of milk production and calving intervals between Slick and WildType Holsteins in a tropical grazing production system. *Archivos Latinoamericanos de Producción Animal, 28*(34), 145–153. https://ojs.alpa.uy/index.php/ojs_files/article/view/2817

Otieno, D. O. (2010). Synthesis of β-galactooligosaccharides from lactose using microbial β-galactosidases. *Comprehensive Reviews in Food Science and Food Safety, 9*(5), 471–482. doi:10.1111/j.1541-4337.2010.00121.x PMID:33467831

Oudshoorn, F. W., Kristensen, T., Van Der Zijpp, A. J., & Boer, I. J. M. (2012). Sustainability evaluation of automatic and conventional milking systems on organic dairy farms in Denmark. *NJAS Wageningen Journal of Life Sciences, 59*(1-2), 25–33. doi:10.1016/j.njas.2011.05.003

Pagani Netto, C., Fontes, J. L., Pimentel, J. C. C., & Martins, S. E. (2017). *Mais Leite, Mais Renda – plano de desenvolvimento da bovinocultura leiteira paulista* [Mais Leite, Mais Renda – development plan for dairy cattle in São Paulo, Brazil]. CATI.

Pais, J., Serafim, L. S., Freitas, F., & Reis, M. A. (2016). Conversion of cheese whey into poly (3-hydroxybutyrate-co-3-hydroxyvalerate) by Haloferax mediterranei. *New Biotechnology, 33*(1), 224–230. doi:10.1016/j.nbt.2015.06.001 PMID:26134839

Pakhtunkhwa, K., Khan, I., Wazir, A. A., Khan, S., & Hassan, S. I. (2022). Prevalence of Sub-Clinical Mastitis in Buffaloes at Livestock Research and iMedPub Journals Prevalence of Sub-Clinical Mastitis in Buffaloes at Livestock Research and Development Station. *Paharpur and Surroundings of Tehsil Paharpur District D., I*(June). doi:10.5281/zenodo.6626290

Palmquist, D. L., & Jenkins, T. C. (2017). A 100-Year Review: Fat feeding of dairy cows. *Journal of Dairy Science, 100*(12), 10061–10077. doi:10.3168/jds.2017-12924 PMID:29153155

Panahi, M., & Saei, H. D. (2019). Genetic diversity and methicillin resistance of Staphylococcus aureus originating from buffaloes with mastitis in Iran. *Comparative Immunology, Microbiology, and Infectious Diseases, 62*(November 2018), 19–24. doi:10.1016/j.cimid.2018.11.014

Panesar, P. S., Kennedy, J. F., Gandhi, D. N., & Bunko, K. (2007). Bioutilisation of whey for lactic acid production. *Food Chemistry, 105*(1), 1–14. doi:10.1016/j.foodchem.2007.03.035

Panghal, A., Kumar, V., Dhull, S. B., Gat, Y., & Chhikara, N. (2017). Utilization of dairy industry waste-whey in formulation of papaya RTS beverage. *Current Research in Nutrition and Food Science Journal, 5*(2), 168–174. doi:10.12944/CRNFSJ.5.2.14

Pantoja, J. T. Olson, T. Ruiz, Á. Custodio, B. Vallejo and M. Pagán, (2005). Evaluation of factors that influence dairy productivity in dairy cows with short hair in the tropics. *Proceedings of the Annual Meeting of the Puerto Rican Society of Agricultural Sciences*. IEEE.

Paques, J. P., Van der Linden, E., Van Rijn, C. J. M., & Sagis, L. M. C. (2014). Preparation methods of alginate nanoparticles. *Advances in Colloid and Interface Science, 209*, 163–171. doi:10.1016/j.cis.2014.03.009 PMID:24745976

Patel, R., Kunjadia, P., Koringa, P., Joshi, C., & Kunjadiya, A. (2019). Microbiological profiles in clinical and subclinical cases of mastitis in milking Jafarabadi buffalo. *Research in Veterinary Science*, *125*(August 2016), 94–99. doi:10.1016/j.rvsc.2019.05.012

Peixoto, R. D. M., Aparecido, R., Mateus, M., Mastite, B., Pesquisa, B., Brasileira, V., Agrárias, C. C., & Federal, U. (2010). Mastite Em Pequenos Ruminantes: *Etiologia E Epidemiologia. 30*(9), 754–762.

Peixoto, R. de M., Silva, W. E. L., Almeida, J. R. G. S., Branco, A., & Costa, M. M. (2016). ANTIBACTERIAL POTENTIAL OF NATIVE PLANTS FROM THE CAATINGA BIOME AGAINST Staphylococcus spp. ISOLATES FROM SMALL RUMINANTS WITH MASTITIS. *Revista Caatinga*, *29*(3), 758–763. doi:10.1590/1983-21252016v29n328rc

Pengov, A., & Kirbis, A. (2009). Risks of antibiotic residues in milk following intramammary and intramuscular treatments in dairy sheep. *Analytica Chimica Acta*, *637*(1-2), 13–17. doi:10.1016/j.aca.2008.09.021 PMID:19286006

Pereira, A. M. F., Titto, E. A. L., & Almeida, J. A. A. (2019). *Adaptação dos ruminantes aos climas quentes*. Appris Editora.

Pereira-Rodríguez, A., Fernandez-Leiro, R., Gonzalez Siso, M. I., Cerdan, M. E., Becerra, M., & Sanz-Aparicio, J. (2010). Crystallization and preliminary X-ray crystallographic analysis of β-galactosidase from Kluyveromyces lactis. *Acta Crystallographica*, *66*, 297–300. PMID:20208165

Peres, J. R. (2001). O leite como ferramenta do monitoramento nutricional. In F. H. D. González, J. W. Dürr, & R. S. Fontaneli (Eds.), *Uso do leite para monitorar a nutrição e metabolismo de vacas leiteiras* (pp. 29–43). Ed. UFRGS.

Philpot, N. W., & Nickerson, S. C. (2002). *Vencendo a luta contra a mastite*. Westfalia Surge/Westfalia Landtechnik do Brasil.

Picinin, L. C. A., Bordignon-Luiz, M. T., Cerqueira, M. M. O. P., Toaldo, I. M., Souza, F. N., Leite, M. O., Fonseca, L. M., & Lana, A. M. Q. (2019). Effect of seasonal conditions and milk management practices on bulk milk quality in Minas Gerais State – Brazil. *Arquivo Brasileiro de Medicina Veterinária e Zootecnia*, *71*(4), 1355–1363. doi:10.1590/1678-4162-10063

Pieterse, R., & Todorov, S. (2017). Bacteriocins: Exploring Alternatives to Antibiotics in Mastitis Treatment. *Brazilian Journal of Microbiology*, *41*(3), 542–562.

Pieterse, R., & Todorov, S. D. (2010). Bacteriocins: exploring alternatives to antibiotics in mastitis treatment. *Brazilian Journal of Microbiology, 41*, 542-562. Plastridge, WN., (1958). Bovine Mastitis: A Review. *Journal of Dairy Science*, *41*(9), 1141–1181.

Pinheiro, F. E.; Lopes, M. A.; Mendes, J. R. R. (2015). Efeito do pagamento por qualidade do leite na rentabilidade da atividade leiteira. *Revista do Instituto de Laticínios Cândido Tostes*, [S.l.], *70*(1), 24-34.

Pires, A. F., Marnotes, N. G., Rubio, O. D., Garcia, A. C., & Pereira, C. D. (2021). Dairy by-products: A review on the valorization of whey and second cheese whey. *Foods*, *10*(5), 1067. doi:10.3390/foods10051067 PMID:34066033

Pisanu, S., Cacciotto, C., Pagnozzi, D., Uzzau, S., Pollera, C., Penati, M., Bronzo, V., & Addis, M. F. (2020). Impact of Staphylococcus aureus infection on the late lactation goat milk proteome: New perspectives for monitoring and understanding mastitis in dairy goats. *Journal of Proteomics*, *221*(January), 103763. doi:10.1016/j.jprot.2020.103763 PMID:32275959

Pol, M., & Ruegg, P. L. (2007). Treatment Practices and Quantification of Antimicrobial Drug Usage in Conventional and Organic Dairy Farms in Wisconsin. *Journal of Dairy Science*, *90*(1), 249–261. doi:10.3168/jds.S0022-0302(07)72626-7 PMID:17183093

Ponsano, E. H. G., Pinto, M. F., Grassi, T. F., Avanço, S. V., & Lima, L. K. F. (2011). Capacitação de produtores rurais para a melhoria da qualidade do leite cru produzido na região de Araçatuba-SP. *Revista Ciência em Extensão*, *7*(1), 91–101.

Portugal, J. A. B., Pires, M. F. A., & Durães, M. C. (2000). Effect of air temperature and humidity on frequency of feeding, water ingestion and rumination in Holstein cows. *Arquivo Brasileiro de Medicina Veterinária e Zootecnia*, *52*, 154–159. doi:10.1590/S0102-09352000000200013

Prado, R. M., Paulin, M. F., Prado, I. N., Santos, G. T., Benchaar, C., & Petiti, H. V. (2016). Milk yield, milk composition, and hepatic lipid metabolismo in transition dairy cows fed flaxseed or linola. DOI: . 11. ed., Sherbrook, American Dairy Science Association, 99, 8831-8846. doi:10.3168/jds.2016-11003

Prescott, JF., (2019). History of antimicrobial use in agriculture: an overview. *Antimicrobial resistance in bacteria of animal origin,* 19-27.

Prescott, J. F. (2017). History and Current Use of Antimicrobial Drugs in Veterinary Medicine. *Microbiology Spectrum*, *5*(6), 5.6.05. doi:10.1128/microbiolspec.ARBA-0002-2017 PMID:29271338

Puppel, K., Kalińska, A., Kot, M., Slósarz, J., Kunowska-Slósarz, M., Grodkowski, G., Kuczyńska, B., Solarczyk, P., Przysucha, T., & Gołębiewski, M. (2020). The effect of staphylococcus spp., streptococcus spp. and enterobacteriaceae on the development of whey protein levels and oxidative stress markers in cows with diagnosed mastitis. *Animals (Basel)*, *10*(9), 1–11. doi:10.3390/ani10091591 PMID:32906645

Quigley, J. D. III, Drewry, J. J., Murray, L. M., & Ivey, S. J. (1997). Body Weight Gain, Feed Efficiency, and Fecal Scores of Dairy Calves in Response to Galactosyl-Lactose or Antibiotics in Milk Replacers. *Journal of Dairy Science*, *80*(8), 1751–1754. doi:10.3168/jds.S0022-0302(97)76108-3 PMID:9276816

Quintanilla, P., Doménech, E., Escriche, I., Beltrán, M. C., & Molina, M. P. (2019). Food Safety Margin Assessment of Antibiotics: Pasteurized Goat's Milk and Fresh Cheese. *Journal of Food Protection*, *82*(9), 1553–1559. doi:10.4315/0362-028X.JFP-18-434 PMID:31424292

Rafiee, H., Alikhani, M., & Ghorbani, G. R. (2022). Effect of dietary protein level and corn processing on behavior activity of high producing dairy cows. *Acta Scientiarum. Animal Sciences*, *44*, e54603. doi:10.4025/actascianimsci.v44i1.54603

Rainard, P., Foucras, G., Fitzgerald, J. R., Watts, J. L., Koop, G., & Middleton, J. R. (2018). Knowledge gaps and research priorities in Staphylococcus aureus mastitis control. *Transboundary and Emerging Diseases*, *65*, 149–165. doi:10.1111/tbed.12698 PMID:28984427

Rainard, P., Gilbert, F. B., Germon, P., & Foucras, G. (2021). Invited review: A critical appraisal of mastitis vaccines for dairy cows. *Journal of Dairy Science*, *104*(10), 10427–10448. doi:10.3168/jds.2021-20434 PMID:34218921

Rall, V. L. M., Miranda, E. S., Castilho, I. G., Camargo, C. H., Langoni, H., Guimarães, F. F., Araújo Júnior, J. P., & Fernandes Júnior, A. (2014). Diversity of Staphylococcus species and prevalence of enterotoxin genes isolated from milk of healthy cows and cows with subclinical mastitis. *Journal of Dairy Science*, *97*(2), 829–837. doi:10.3168/jds.2013-7226 PMID:24359821

Rama, G. R., Kuhn, D., Beux, S., Maciel, M. J., & de Souza, C. F. V. (2019). Potential applications of dairy whey for the production of lactic acid bacteria cultures. *International Dairy Journal*, *98*, 25–37. doi:10.1016/j.idairyj.2019.06.012

Ramos, O. L., Fernandes, J. C., Silva, S. I., Pintado, M. E., & Malcata, F. X. (2012). Edible films and coatings from whey proteins: A review on formulation, and on mechanical and bioactive properties. *Critical Reviews in Food Science and Nutrition*, *52*(6), 533–552. doi:10.1080/10408398.2010.500528 PMID:22452733

Rampasso, I. S., Anholon, R., Silva, D., Cooper Ordonez, R. E., Quelhas, O. L. G., Leal Filho, W., & Santa-Eulália, L. A. (2018). An analysis of the difficulties associated to sustainability insertion in engineering education: Examples from HEIs in Brazil. *Journal of Cleaner Production*, *193*, 363–371. doi:10.1016/j.jclepro.2018.05.079

Rampasso, I. S., Anholon, R., Silva, D., Ordoñez, R. E. C., Santa-Eulalia, L. A., Quelhas, O. L. G., Leal Filho, W., & Aguirre, L. F. G. (2019). Analysis of the perception of engineering students regarding sustainability. *Journal of Cleaner Production*, *233*, 461–467. doi:10.1016/j.jclepro.2019.06.105

Rashamol, V. P., Sejian, V., Bagath, M., Krishnan, G., Archana, P. R., & Bhatta, R. (2018). Physiological adaptability of livestock to heat stress: An updated review. *Journal of Animal Behaviour and Biometeorology*, *6*(3), 62–71. doi:10.31893/2318-1265jabb.v6n3p62-71

Ravagnolo O. and I. Misztal, (2000). Genetic component of heat stress in dairy cattle, parameter estimation. *J Dairy Sci. 83*, 2126–2130. doi: . S0022-0302(00)75095-8 doi:10.3168/jds

Reis, G. L., Alves, A. A., Lana, A. M. Q., Coelho, S. G., Souza, M. R., Cerqueira, M. M. P., Penna, C. F. A. M., & Mendes, E. D. M. (2007). Procedimentos de coleta de leite cru individual e sua relação com a composição físico-química e a contagem de células somáticas. Ciência Rural, Santa Maria.

Renhe, I. R. T. (2008). O papel do leite na nutrição. *Revista do Instituto de Laticínios Cândido Tostes*, *363*(63), 36–43.

Report, P. E. W. (2017). *Alternatives to Antibiotics in Animal Agriculture*. PEW trusts. https://www.pewtrusts.org/~/media/assets/2017/07/alternatives_to_antibiotics_in_animal_agriculture.pdf. 2017.

Ribeiro, B.D., Pereira, K.S., Nascimento, R.P., Coelho, M.A.Z. (2018) Alimentos. *Microbiologia Industrial*. Elsevier, Rio de Janeiro.

Ribeiro, M. G., Costa, E. O., Leite, D. S., Langoni, H., Garino, F., Victória, C., & Listoni, F. J. P. (2006). Fatores de virulência em linhagens de Escherichia coli isoladas de mastite bovina. *Arquivo Brasileiro de Medicina Veterinária e Zootecnia*, *58*(5), 724–731. doi:10.1590/S0102-09352006000500004

Rieckmann, M. (2012). Future-oriented higher education: Which key competencies should be fostered through university teaching and learning? *Futures*, *44*(2), 127–135. doi:10.1016/j.futures.2011.09.005

RíosSolís, C. G, Valencia, E. CurbeloRodríguez, aJ., & Ortiz, C. (2019). Feed efficiency of slick hair and wildtype dairy cows under heat stress. *J. Dairy Sci. 102*(1), 311 (Abstract).

RíosSolís[a], C. G., N. R. CidHernández, E. RuizCortés, E. Valencia, J. E. CurbeloRodríguez, and G. Ortiz Colón, (2019). Rectal temperature, respiration rate, and heart rate of slickhair and wildtype lactating Holstein cows under heat stress. *J. Dairy Sci.* Vol. 102, Suppl. 1: 11 (Abstract).

Rojo-Gimeno, C., Postma, M., Dewulf, J., Hogeveen, H., Lauwers, L., & Wauters, E. (2016). Farm-economic analysis of reducing antimicrobial use whilst adopting improved management strategies on farrow-to-finish pig farms. *Preventive Veterinary Medicine*, *129*, 74–87. doi:10.1016/j.prevetmed.2016.05.001 PMID:27317325

Romani, A., Michelin, M., Domingues, L., & Teixeira, J. A. (2018). Valorization of wastes from agrofood and pulp and paper industries within the biorefinery concept: southwestern Europe scenario. In T. Bhaskar, A. Pandey, S. V. Mohan, & D. Lee (Eds.), *J, Khanal, S.K. Waste biorefinery* (pp. 487–504). Elsevier. doi:10.1016/B978-0-444-63992-9.00016-1

Román, L., Saravia, C., Astigarraga, L., Bentancur, O., & La Manna, A. (2019). Shade access in combination with sprinkling and ventilation effects performance of Holstein cows in early and late lactation. *Animal Production Science*, *59*(2), 347. doi:10.1071/AN16571

Roque-borda, C. A., da Silva, P. B., Rodrigues, M. C., Azevedo, R. B., di Filippo, L., Duarte, J. L., Chorilli, M., Vicente, E. F., & Pavan, F. R., (2021). Challenge in the Discovery of New Drugs: Antimicrobial Peptides against WHO-List of Critical and High-Priority Bacteria. *Pharmaceutics 2021*, *13*(6), 773. doi:10.3390/pharmaceutics13060773

Roque-Borda, C. A., Bento da Silva, P., Rodrigues, M. C., di Filippo, L. D., Duarte, J. L., Chorilli, M., Vicente, E. F., Garrido, S. S., & Rogério Pavan, F. (2022). Pharmaceutical nanotechnology: Antimicrobial peptides as potential new drugs against WHO list of critical, high, and medium priority bacteria. *European Journal of Medicinal Chemistry*, *241*, 114640. doi:10.1016/j.ejmech.2022.114640 PMID:35970075

Rosa, J. F., Domingues, L. D. A. S. P., Silva, R. H., De Lima, H. G., & Cereser, N. D. (2017, January-June). Pontos críticos de contaminação na produção leiteira. *Expressa Extensão*, *22*(1), 90–103. doi:10.15210/ee.v22i1.7808

Rossi, R. S., Amarante, A. F., Correia, L. B. N., Guerra, S. T., Nobrega, D. B., Latosinski, G. S., Rossi, B. F., Rall, V. L. M., & Pantoja, J. C. F. (2018). Diagnostic accuracy of Somaticell, California Mastitis Test, and microbiological examination of composite milk to detect Streptococcus agalactiae intramammary infections. *Journal of Dairy Science*, *101*(11), 10220–10229. doi:10.3168/jds.2018-14753 PMID:30146291

Ruegg, P. L. (2017). A 100-Year Review: Mastitis detection, management, and prevention. *Journal of Dairy Science*, *100*(12), 10381–10397. doi:10.3168/jds.2017-13023 PMID:29153171

Ryan, P. D., Prichard, J. F., Kopel, E., & Godke, R. A. (1993). Comparing early embryo mortality in dairy cows during hot and cool seasons of the year. *Theriogenology*, *39*(3), 719737. doi:10.1016/0093-691X(93)90257-6 PMID:16727249

Rydlo, T., Rotem, S., & Mor, A. (2006). Antibacterial Properties of Dermaseptin S4 Derivatives under Extreme Incubation Conditions. *Antimicrobial Agents and Chemotherapy*, *50*(2), 490–497. doi:10.1128/AAC.50.2.490-497.2006 PMID:16436701

Sachi, S., Ferdous, J., & Sikder, M. H. Hussani, (2019). S.M.A.K. Antibiotic residues in milk: Past, present, and future. *J. Adv. Vet. Anim. Res.*, *6*, 315–332. https://www.ncbi.nlm.nih.gov/pmc/articles/PMC6760505/

Saenz-de-Juano, M. D., Silvestrelli, G., Bauersachs, S., & Ulbrich, S. E. (2022). Determining extracellular vesicles properties and miRNA cargo variability in bovine milk from healthy cows and cows undergoing subclinical mastitis. *BMC Genomics*, *23*(1), 1–15. doi:10.118612864-022-08377-z PMID:35255807

Sales, D. C., & Rangel, A. H. do N., Urbano, S. A., Tonhati, H., Galvão Júnior, J. G. B., Guilhermino, M. M., Aguiar, E. M., & Bezerra, M. de F. (2018). Buffalo milk composition, processing factors, whey constituents' recovery and yield in manufacturing Mozzarella cheese. *Food Science and Technology (Campinas)*, *38*(2), 328–334. doi:10.1590/1678-457X.04317

Salovaara, J. J., Pietikainen, J., & Cantell, H. (2021). Perceptions of interconnected sustainability: Students' narratives bridging transition and education. *Journal of Cleaner Production*, *281*, 125336. doi:10.1016/j.jclepro.2020.125336

Sánchez, W. K., McGuire, M. A., & Beede, D. K. (1994). Macromineral Nutrition by Heat Stress Interactions in Dairy Cattle: Review and Original Research. *Journal of Dairy Science*, *77*(7), 2051–2079. https://www.journalofdairyscience.org/article/S0022-0302(94)77150-2/pdf. doi:10.3168/jds.S0022-0302(94)77150-2 PMID:7929965

Sant'Anna, A. C., & Paranhos da Costa, M. J. (2011). The relationship between dairy cow hygiene and somatic cell count in milk. *Journal of Dairy Science*, *94*(8), 3835–3844. doi:10.3168/jds.2010-3951 PMID:21787920

Santajit, S., & Indrawattana, N. (2016). Mechanisms of Antimicrobial Resistance in ESKAPE Pathogens. *BioMed Research International*, *2475067*, 1–8. doi:10.1155/2016/2475067 PMID:27274985

Santana, E. H. W., Beloti, V., Barros, M., Moraes, L. B., Gusmão, V. V., & Pereira, M. S. (2001). de A. F.; Moraes, L. B.; Gusmão, V. V.; Pereira, M. S. Contaminação do leite em diferentes pontos do processo de produção: I. Microrganismos aeróbios mesófilos e psicrotróficos. *Semina. Ciências Agrárias*, *22*(2), 145–154. doi:10.5433/1679-0359.2001v22n2p145

Santo, R. S. E. et al. (2017). Fatores e técnicas de produção e sua influência na produtividade e qualidade do leite bovino.

Santos, A. M., Lopes, E. R. N., & Silva, M. F. Jr. (2017). Percepção ambiental de estudantes do ensino técnico federal em agropecuária e a contribuição da educação ambiental na formação profissional. *Revista Brasileira de Educação Ambiental*, *12*(2), 136–155. doi:10.34024/revbea.2017.v12.2315

Santos, F. F., Queiroz, R. C. S., & Almeida Neto, J. A. (2017). Avaliação da aplicação das técnicas da Produção Mais Limpa em um laticínio no Sul da Bahia. *Gestão & Produção*, *25*(1), 117–131. doi:10.1590/0104-530x2234-16

Santos, H. C. M. Jr, Maranduba, H. L., de Almeida Neto, J. A., & Rodrigues, L. B. (2017). Life cycle assessment of cheese production process in a small-sized dairy industry in Brazil. *Environmental Science and Pollution Research International*, *24*(4), 3470–3482. doi:10.100711356-016-8084-0 PMID:27873115

Santos, J. G., Marion, J. C., & Segatti, S. (2008). *Administração de Custos na Agropecuária* (Vol. 3). Atlas.

Santos, M. J., Teixeira, J. A., & Rodrigues, L. R. (2011). Fractionation and recovery of whey proteins by hydrophobic interaction chromatography. *Journal of Chromatography. B, Analytical Technologies in the Biomedical and Life Sciences*, *879*(7-8), 475–479. doi:10.1016/j.jchromb.2011.01.003 PMID:21292570

Santos, M. V., & Fonseca, L. F. L. (2007). *Estratégias para controle de mastite e melhoria da qualidade do leite*. Manole.

Saqib, Z. A., Zhang, Q., Ou, J., Saqib, K. A., Majeed, S., & Amar Razzaq, A. (2020). Education for sustainable development in Pakistani higher education institutions: An exploratory study of students' and teachers' perceptions. *International Journal of Sustainability in Higher Education, 21*(6), 1249–1267. doi:10.1108/IJSHE-01-2020-0036

Saquib, S. A., Akram, A. S. A., Halim, S. A., & Tassaduq, R. (2017). Sources of β-galactosidase and its applications in food industry. *Biotech, 7*(1), 79. PMID:28500401

Satolo, E. G., Campos, R. S. D., Ussuna, G. D. A., Simon, A. T., Mac-Lean, P. A. B., & Braga Júnior, S. S. (2020). Sustainability Assessment of logistics activities in a dairy: An example of an emerging economy. *Production, 30*, e20190036. doi:10.1590/0103-6513.20190036

Scalco, A. R.; Souza, R. C. (2006). Qualidade na cadeia de produção de leite: diagnóstico e proposição de melhorias. *Organizações Rurais & Agroindustriais, 8*(3).

Scalco, A. R., & De Toledo, J. C. (2002). Gestão da qualidade em laticínios do estado de São Paulo: Situação atual e recomendações [Quality management in dairy products in the state of São Paulo: current situation and recommendations]. *Revista de Administração da Universidade de São Paulo, 37*(2), 17–25.

Scali, F., Camussone, C., Calvinho, L. F., Cipolla, M., & Zecconi, A. (2015). Which are important targets in development of S. aureus mastitis vaccine? *Research in Veterinary Science, 100*, 88–99. doi:10.1016/j.rvsc.2015.03.019 PMID:25975626

Schmaltz, E., Melvin, E. C., Diana, Z., Gunady, E. F., Rittschof, D., Somarelli, J. A., & Dunphy-Daly, M. M. (2020). Plastic pollution solutions: emerging technologies to prevent and collect marine plastic pollution. *Environment international,* 144, 106067. Sgarbieri, V.C., Revisão: Propriedades estruturais e físico-químicas das proteínas do leite. *Brazilian Journal of Food Technology, 8*(1), 43–56.

Schopen, G. C. B., Heck, J. M. L., Bovenhuis, H., Visker, M. H. P. W., van Valenberg, H. J. F., & van Arendonk, J. A. M. (2009). Genetic parameters for major milk proteins in Dutch Holstein-Friesians. *Journal of Dairy Science, 92*(3), 1182–1191. doi:10.3168/jds.2008-1281 PMID:19233812

Schütz, K. E., Rogers, A. R., Cox, N. R., Webster, J. R., & Tucker, C. B. (2011). Dairy cattle prefer shade over sprinklers: Effects on behavior and physiology. *Journal of Dairy Science, 94*(1), 273–283. doi:10.3168/jds.2010-3608 PMID:21183037

Sembada, P., Duteurtre, G., Purwanto, B. P., & Suryahadi. (2016). Improved milk production performance of smallholder farms in West Java (Indonesia). *Tropical Animal Health and Production, 48*(4), 793–799. doi:10.100711250-016-1029-2 PMID:26969314

Servant-Miklos, V. F. C., & Noordzij, G. (2021). Investigating the impact of problem-oriented sustainability education on students' identity: A comparative study of planning and liberal arts students. *Journal of Cleaner Production, 280*, 124846. doi:10.1016/j.jclepro.2020.124846

Sharma, Chetan; Rokana, Namita; Chandra, Mudit; Singh, Brij Pal; Gulhane, Rohini Devidas; Gill, Jatinder Paul Singh; Ray, Pallab; Puniya, Anil Kumar; Panwar, Harsh, (2018). Antimicrobial Resistance: Its Surveillance, Impact, and Alternative Management Strategies in Dairy Animals. *Frontiers in Veterinary Science*, 4(), 237–. doi:10.3389/fvets.2017.00237

Sharma, R. (2019). Whey proteins in functional foods. In H. Deeth & N. Bansal (Eds.), *Whey proteins* (pp. 637–663). Elsevier. doi:10.1016/B978-0-12-812124-5.00018-7

Shen, X., Zhao, C., Lu, J., & Guo, M. (2018). Physicochemical properties of whey-protein-stabilized astaxanthin nanodispersion and its transport via a Caco-2 monolayer. *Journal of Agricultural and Food Chemistry*, *66*(6), 1472–1478. doi:10.1021/acs.jafc.7b05284 PMID:29345928

Shpigelman, A., Cohen, Y., & Livney, Y. (2012). Thermally-induced β-lactoglobulin-EGCG nanovehicles: Loading, stability, sensory and digestive-release study. *Food Hydrocolloids*, *29*(1), 57–67. doi:10.1016/j.foodhyd.2012.01.016

Shryock, T. R. (2004). O futuro dos produtos anti-infecciosos na saúde animal. *Nature Reviews. Microbiology*, *2*(5), 425–430. doi:10.1038/nrmicro887 PMID:15100695

Sidiropoulos, E. (2018). The personal context of student learning for sustainability: Results of a multi-university research study. *Journal of Cleaner Production*, *181*, 537–554. doi:10.1016/j.jclepro.2018.01.083

Sienkiewicz, T., & Riedel, C. L. (1990). *Whey and Whey Utilization*. Verlag Th. Mann.

Sijbers, A. M., Schoemaker, R. J. W., Nauta, A., & Alkema, W. (2020). Revealing new leads for the impact of galacto-oligosaccharides on gut commensals and gut health benefts through text mining. *Beneficial Microbes*, *11*(3), 283–302. doi:10.3920/BM2019.0105 PMID:32299221

Silva, M. F. (2020). *Indicadores de sustentabilidade para a pecuária leiteira*. [Tese Doutorado em Ciências, Universidade de São Paulo].

Silva, M. F., & Gameiro, A. H. (2021). Indicadores de sustentabilidade para a produção de leite: Uma revisão de literatura. *Revista Livre de Sustentabilidade e Empreendedorismo*, *6*, 208–237.

Silva, M. F., & Gameiro, A. H. (2022). Sustainability indicators for Brazilian dairy livestock: The perception of professionals in the sector. *Revista Brasileira de Zootecnia*, *51*, e20210049. doi:10.37496/rbz5120210049

Silva, M. V. M., & Nogueira, J. L. (2010). Mastite: Controle E Profilaxia No Rebanho Bovino. *Revista Científica Eletrônica de Medicina Veterinária*, *15*(VIII), 13.

Silva, M. V., Almeida, G. L. P., Pandorfi, H., Moraes, A. S., Macêdo, G. A. P. A., Batista, P. H. D., Silva, R. A. B., & Guiselini, C. (2021). Influence of meteorological elements on behavioral responses of gir cows and effects on milk quality. *Acta Scientiarum. Animal Sciences*, *43*, e52604. doi:10.4025/actascianimsci.v43i1.52604

Silveira, R. F., Roque-Borda, C. A., & Vicente, E. F. (2021). Antimicrobial peptides as a feed additive alternative to animal production, food safety and public health implications: An overview. *Animal Nutrition*, *7*(3), 896–904. doi:10.1016/j.aninu.2021.01.004 PMID:34632120

Simões, L., Martins, J., Pinheiro, A., Vicente, A., & Ramos, O. (2020). β-lactoglobulin micro- and nanostructures as bioactive compounds vehicle: In vitro studies. *Food Research International*, *131*, 108979. doi:10.1016/j.foodres.2020.108979 PMID:32247463

Singh, T., Øiseth, S., Lundin, L., & Day, L. (2014). Influence of heat and shear induced protein aggregation on the in vitro digestion rate of whey proteins. *Food & Function*, *5*(11), 2686–2698. doi:10.1039/C4FO00454J PMID:25205335

Sisodia, C. S., & Stowe, C. M., (2006). THE MECHANISM OF DRUG SECRETION INTO BOVINE MILK. *Annals of the New York Academy of Sciences*, *111*(2), 650–661. doi:. 1964.tb doi:10.1111/j.1749-6632

Smith, H. W., & Crabb, W. E. (1960a). The effect of diets containing tetracyclines and penicillin on the Staphylococcus aureus flora of the nose and skin of pigs and chickens and their human attendants. *The Journal of Pathology and Bacteriology*, *79*(2), 243–249. doi:10.1002/path.1700790204 PMID:13831990

Smith, H. W., & Crabb, W. E. (1960b). O efeito da quimioterapia sobre o número de colis bacterianas nas fezes de bezerros. *Jornal de Patologia Comparada e Terapêutica*, *70*, 126–134. doi:10.10160368-1742(60)80011

Soares, J. C., Soares, A. C., Popolin-Neto, M., Paulovich, F. V., Oliveira, O. N. Jr, & Mattoso, L. H. C. (2022). Detection of Staphylococcus aureus in milk samples using impedance spectroscopy and data processing with information visualization techniques and multidimensional calibration space. *Sensors and Actuators Reports*, *4*, 100083. doi:10.1016/j.snr.2022.100083

Souza, G. N., Brito, J. R. F., Moreira, E. C., Brito, M. A. V. P., & Silva, M. V. G. B. (2009). Variação da contagem de células somáticas em vacas leiteiras de acordo com patógenos da mastite. *Arquivo Brasileiro de Medicina Veterinária e Zootecnia*, *61*(5), 1015–1020. doi:10.1590/S0102-09352009000500001

Souza, V. M., & Araujo, J. (2015). O currículo verde: Uma discussão sobre a inserção do meio ambiente nas grades curriculares dos cursos de graduação das instituições públicas de ensino superior do estado do Rio de Janeiro. *Revista Brasileira de Educação Ambiental*, *10*(3), 147–163. doi:10.34024/revbea.2015.v10.1909

Sparemberger, A., Büttenbender, P. L., Zamberlan, L., & Hofler, C. E. (2009). *Inovações tecnologicas nas cadeias do agronegócio de alimentos da região fronteira noroeste do Rio Grande do Sul*. COINI - Congreso Argentino de Ingeniería Industria.

Spink, J., & Moyer, D. C. (2011). Defining the Public Health Threat of Food Fraud. *Journal of Food Science*, *76*(9), R157–R163. doi:10.1111/j.1750-3841.2011.02417.x PMID:22416717

Spuria, L., Biasibetti, E., Bisanzio, D., Biasato, I., De Meneghi, D., Nebbia, P., Robino, P., Bianco, P., Lamberti, M., Caruso, C., Di Blasio, A., Peletto, S., Masoero, L., Dondo, A., & Capucchio, M. T. (2017). Microbial agents in macroscopically healthy mammary gland tissues of small ruminants. *PeerJ*, *2017*(11), 1–18. doi:10.7717/peerj.3994 PMID:29152416

Srikok, S., Patchanee, P., Boonyayatra, S., & Chuammitri, P. (2020). Potential role of MicroRNA as a diagnostic tool in the detection of bovine mastitis. *Preventive Veterinary Medicine*, *182*(July), 105101. doi:10.1016/j.prevetmed.2020.105101 PMID:32823253

Stein, R. T., Malinsk, A., da Silva-Reis, C. M., Soares, B. L. C., & Moura, A. S. (2020). *Cadeias produtivas do agronegócio II* [Agribusiness production chains II]. Grupo A.

Stelwagen, K. (2022). Mammary Gland, Milk Biosynthesis and Secretion: Lactose. In P.L.H. McSweeney (Ed.), J. P. McNamara (Ed.), Encyclopedia of Dairy Sciences (3rd Edition), (pp. 184-189). Academic Press. doi:10.1016/B978-0-12-818766-1.00025-8

Stelwagen, K. (2022b). Mammary Gland, Milk Biosynthesis and Secretion: Milk protein. In P.L.H. McSweeney (Ed.), J. P. McNamara (Ed.), Encyclopedia of Dairy Sciences (3rd Edition), (pp. 198-205). Academic Press. doi:10.1016/B978-0-12-818766-1.00026-X

Stevens, A. V., Karges, K., Rezamand, P., Laarman, A. H., & Chibisa, G. E. (2021). Production performance and nitrogen metabolism in dairy cows fed supplemental blends of rumen undegradable protein and rumen-protected amino acids in low- compared with high-protein diets containing corn distillers' grains. *Journal of Dairy Science*, *104*(4), 4134–4145. doi:10.3168/jds.2020-19697 PMID:33612237

Stone, W. C., (2004). Nutritional approaches to minimize subacute ruminal acidosis and laminitis in dairy cattle. *J. Dairy Sci. 87*, E13-E26 doi:10.3168/jds.S0022-0302(04)70057-0

Sullivan, G. M., & Artino, A. R. Jr. (2013). Analyzing and Interpreting Data From Likert-Type Scales. *Journal of Graduate Medical Education*, *5*(4), 541–542. doi:10.4300/JGME-5-4-18 PMID:24454995

Talamini, E., Pedrozo, E. A., & da Silva, A. L. (2005). Gestão da cadeia de suprimentos e a segurança do alimento: Uma pesquisa exploratória na cadeia exportadora de carne suína [Supply chain management and food safety: exploratory research into Brazil's pork export supply chain]. *Gestão & Produção*, *12*(1), 107–120. doi:10.1590/S0104-530X2005000100010

Tekippe, J. A., Tacoma, R., Hristov, A. N., Lee, C., Oh, J., Heyler, K. S., Cassidy, T. W., Varga, G. A., & Bravo, D. (2013). Effect of essential oils on ruminal fermentation and lactation performance of dairy cows. *Journal of Dairy Science*, *96*(12), 7892–7903. doi:10.3168/jds.2013-7128 PMID:24119814

Telles, L. B. (2014). Ferramentas e sistema de custo aplicados a gestão da qualidade no agronegócio [Tools and cost system applied to quality management in agribusiness]. Master's thesis. Universidade Tecnológica Federal do Paraná, Brazil.

The Mastery Institute. (2011). *History of GMP.* http://themasteryinstitute.org/gmpmastery/history.htm

Tian, S., Wang, J., Yu, H., Wang, J., & Zhu, W. (2019). Changes in ileal microbial composition and microbial metabolism by an early-life galacto-oligosaccharides intervention in a neonatal porcine model. *Nutrients, 11*(8), 1753. doi:10.3390/nu11081753 PMID:31366090

Titto, C. G., Negrão, J. A., Titto, E. A., Canaes, T. S., Titto, R. M., & Pereira, A. M. (2013). Effects of an evaporative cooling system on plasma cortisol, IGF-I, and milk production in dairy cows in a tropical environment. *International Journal of Biometeorology, 57*(2), 299–306. doi:10.100700484-012-0554-6 PMID:22580965

Toghdory, A., Ghoorchi, T., Asadi, M., Bokharaeian, M., Najafi, M., & Ghassemi Nejad, J. (2022). Effects of environmental temperature and humidity on milk composition, microbial load, and somatic cells in milk of Holstein dairy cows in northeast Iran. *Animals (Basel), 12*(18), 2484. doi:10.3390/ani12182484 PMID:36139344

Tolosa, T., Verbeke, J., Piepers, S., Tefera, M., Getachew, Y., Supré, K., & Devliegher, S. (2016). *Milk production, quality, and consumption in Jimma (Ethiopia): facts and producers', retailers', and consumers' perspectives* (Vol. 124). Preventive Veterinary Medicine.

Torres, D. P. M., Gonçalves, M., Teixeira, J. A., & Rodrigues, L. R. (2010). Galacto-oligosaccharides: Production, properties, applications, and significance as prebiotics. *Comprehensive Reviews in Food Science and Food Safety, 9*(5), 438–454. doi:10.1111/j.1541-4337.2010.00119.x PMID:33467830

Treiber, F. M., & Beranek-Knauer, H. (2021). Antimicrobial residues in food from animal origin—A review of the literature focusing on products collected in stores and markets worldwide. *Antibiotics (Basel, Switzerland), 10*(5), 534. doi:10.3390/antibiotics10050534 PMID:34066335

Tsang, Y. F., Kumar, V., Samadar, P., Yang, Y., Lee, J., Ok, Y. S., & Jeon, Y. J. (2019). Production of bioplastic through food waste valorization. *Environment International, 127*, 625–644. doi:10.1016/j.envint.2019.03.076 PMID:30991219

Tullo, E., Mattachini, G., Riva, E., Finzi, A., Provolo, G., & Guarino, M. (2019). Effects of Climatic Conditions on the Lying Behavior of a Group of Primiparous Dairy Cows. *Animals (Basel), 9*(11), 869. doi:10.3390/ani9110869 PMID:31717823

U.N. Food and Agriculture Organization and World Health Organization. (2012). *Report of a Joint FAO/WHO Expert Consultation on Evaluation of Health and Nutritional Properties of Probiotics in Food Including Powder Milk with Live Lactic Acid Bacteria*. UN.

U.N. Food and Agriculture Organization. (2016). Probiotics in Animal Nutrition. *FAO Animal Production and Health Paper No. 179*, https://www.fao.org/3/a-i5933e.pdfJoan, S. Jeffrey. (1999). Use of Competitive Exclusion Products in Poultry. *Poultry Fact Sheet No. 30.* Cooperative Extension, University of California http://animalsciencey.ucdavis.edu/avian/pfs30.htm

United States Department of Agriculture (USDA). (2022). *Dairy: World Markets and Trade.* USDA. https://apps.fas.usda.gov/psdonline/circulars/dairy.pdf

US Food and Drug Administration. (2017). *Good Manufacturing Practices for the 21st Century for Food Processing (2004 Study) Section 1: Current Food Good Manufacturing Practices*. FDA. https://www.fda.gov/food/current-good-manufacturing-practices-cgmps-food-and-dietary-supplements/good-manufacturing-practices-21st-century-food-processing-2004-study-section-1-current-food-good

US GAO. (Gov. Account. Off.). (2011). Antibiotic resistance: agencies have made limited progress addressing antibiotic use in animals GAO-11-801, *Comm. Rules*, House Represent, Washington, DC: https://www.gao.gov/assets/330/323090.pdf

USDA - Animal Improvement Programs Laboratory. https://aipl.arsusda.gov

Usha Vyas and Natarajan Ranganathan, (2001). Probiotics, Prebiotics, and Symbiotic: Gut and Beyond. *Gastroenterology Research and Practice*.

Uyeno, Y., Shigemori, S., & Shimosato, T. (2015). Effect of Probiotics/Prebiotics on Cattle Health and Productivity. *Microbes and Environments*, *30*(126), 126–132. doi:10.1264/jsme2.ME14176 PMID:26004794

Vakkamäki, J., Taponen, S., Heikkilä, A. M., & Pyörälä, S. (2017). Bacteriological etiology and treatment of mastitis in Finnish dairy herds. *Acta Veterinaria Scandinavica*, *59*(1), 1–9. doi:10.118613028-017-0301-4 PMID:28545485

Valeeva, N. I., Lam, T. J. G. M., & Hogeveen, H. (2007). Motivation of dairy farmers to improve mastitis management. *Journal of Dairy Science*, *90*(9), 4466–4477. doi:10.3168/jds.2007-0095 PMID:17699068

Van Boeckel, T. P., Brower, C., Gilbert, M., Grenfell, B. T., Levin, S. A., Robinson, T. P., Teillant, A., & Laxminarayan, R. (2015). Global trends in antimicrobial use in food animals. *Proceedings of the National Academy of Sciences of the United States of America*, *112*(18), 5649–5654. doi:10.1073/pnas.1503141112 PMID:25792457

Van Calker, K. J., Berentsen, P. B. M., Giesen, G. W. J., & Huirne, R. B. M. (2005). Identifying and ranking attributes that determine sustainability in Dutch dairy farming. *Agriculture and Human Values*, *22*(1), 53–63. doi:10.100710460-004-7230-3

van Laer, E., Moons, C. P. H., Ampe, B., Sonck, B., Vandaele, L., De Campeneere, S., & Tuyttens, F. A. M. (2015). Effect of summer conditions and shade on behavioural indicators of thermal discomfort in Holstein dairy and Belgian Blue beef cattle on pasture. *Animal*, *9*(9), 1536–1546. doi:10.1017/S1751731115000804 PMID:25994098

Van Soest, P. J. (1994). *Nutritional ecology of the ruminant* (2nd ed.). Cornell University Press. https://www.cornellpress.cornell.edu/book/9780801427725/nutritional-ecology-of-the-ruminant/#bookTabs=1 doi:10.7591/9781501732355

Varney, J., Barrett, J., Scarlata, K., Catsos, P., Gibson, P. R., & Muir, J. G. (2017). FODMAPs: Food composition, defining cutoff values and international application. *Journal of Gastroenterology and Hepatology*, *32*, 53–61. doi:10.1111/jgh.13698 PMID:28244665

Vásquez-García, A., Silva, T. dos S., de Almeida-Queiroz, S. R., Godoy, S. H. S., Fernandes, A. M., Sousa, R. L. M., & Franzolin, R. (2017). Species identification and antimicrobial susceptibility profile of bacteria causing subclinical mastitis in buffalo. *Pesquisa Veterinaria Brasileira*, *37*(5), 447–452. doi:10.15900100-736x2017000500004

Vera, C., Guerrero, C., Aburto, C., Cordova, A., & Illanes, A. (2020). Conventional and nonconventional applications of β-galactosidases. *Biochimica et Biophysica Acta. Proteins and Proteomics*, *1868*(1), 140271. doi:10.1016/j.bbapap.2019.140271 PMID:31494342

Verraes, C., Van Boxstael, S., Van Meervenne, E., Van Coillie, E., Butaye, P., Catry, B., de Schaetzen, M.-A., Van Huffel, X., Imberechts, H., Dierick, K., Daube, G., Saegerman, C., De Block, J., Dewulf, J., & Herman, L. (2013). Antimicrobial Resistance in the Food Chain: A Review. *International Journal of Environmental Research and Public Health*, *10*(7), 2643–2669. doi:10.3390/ijerph10072643 PMID:23812024

Viana, E. (2021). Sistemas de produção na pecuária leiteira. *Esteio Gestão Agropecuária*. . Disponível em: https://esteiogestao.com.br/sistemas-de-producao-na-pecuaria-leiteira/.

Viana, G., & Rinaldi, R. N. (2008). Principais fatores que influenciam no desempenho da cadeia produtiva do leite: um estudo de caso dos produtores de leite do município de Laranjeiras do Sul (PR) [Main factors that influence the performance of the milk production chain: a case study of milk producers in the city of Laranjeiras do Sul (PR)] In *XLVI Congresso da Sociedade Brasileira de Economia, Administração e Sociologia Rural*.

Vieira, F. M. C., Soares, A. A., Herbut, P., Vismara, E. S., Godyń, D., Dos Santos, A. C. Z., Lambertes, T. D. S., & Caetano, W. F. (2021). Spatio-thermal variability and behavior as bio-thermal indicators of heat stress in dairy cows in a compost barn: A case study. *Animals (Basel)*, *11*(5), 1197. doi:10.3390/ani11051197 PMID:33919438

Voelker, J. A., & Allen, M. S. (2003). Pelleted beet pulp substituted for high-moisture corn: 1. Effects on feed intake chewing behavior, and milk production of lactating dairy cows. *Journal of Dairy Science*, *86*(11), 3542–3552. doi:10.3168/jds.S0022-0302(03)73959-9 PMID:14672184

Wang, D., Wang, Z., Yan, Z., Wu, J., Ali, T., Li, J., Lv, Y., & Han, B. (2015). Bovine mastitis Staphylococcus aureus: Antibiotic susceptibility profile, resistance genes and molecular typing of methicillin-resistant and methicillin-sensitive strains in China. *Infection, Genetics and Evolution*, *31*, 9–16. doi:10.1016/j.meegid.2014.12.039 PMID:25582604

Wang, J., Yang, M., & Maresova, P. (2020). Sustainable Development at Higher Education in China: A comparative study of students' perception in public and private universities. *Sustainability*, *12*(6), 2158. doi:10.3390u12062158

Wang, X., Zhang, G., & Choi, C. Y. (2018). Effect of airflow speed and direction on convective heat transfer of standing and reclining cows. *Biosystems Engineering*, *167*, 87–98. doi:10.1016/j.biosystemseng.2017.12.011

Wang, Z., Niu, K., Rushdi, H. E., Zhang, M., Fu, T., Gao, T., Yang, L., Liu, S., & Lin, F. (2022). Heat Stress Induces Shifts in the Rumen Bacteria and Metabolome of Buffalo. *Animals (Basel)*, *12*(10), 1300. doi:10.3390/ani12101300 PMID:35625146

Wang, Z., Yang, D. S., Li, X. Y., Yu, Y. N., Yong, L. Y., Zhang, P. H., He, J. H., Shen, W. J., Wan, F. C., Feng, B. L., Tan, Z. L., & Tang, S. X. (2021). Modulation of rumen fermentation and microbial community through increasing dietary cation-anion difference in Chinese Holstein dairy cows under heat stress conditions. *Journal of Applied Microbiology*, *130*(3), 722–735. doi:10.1111/jam.14812 PMID:32757409

Watson, M. K., Lozano, R., Noyes, C., & Rodgers, M. (2013). Assessing curricula contribution to sustainability more holistically: Experiences from the integration of curricula assessment and students' perceptions at the Georgia Institute of Technology. *Journal of Cleaner Production*, *61*, 106–116. doi:10.1016/j.jclepro.2013.09.010

Weijers, C. A., Franssen, M. C., & Visser, G. M. (2008). Glycosyltransferasecatalyzed synthesis of bioactive oligosaccharides. *Biotechnology Advances*, *26*(5), 436–456. doi:10.1016/j.biotechadv.2008.05.001 PMID:18565714

West, J. W. (2003). Effects of Heat-Stress on Production in Dairy Cattle. *Journal of Dairy Science*, *86*(6), 2131–2144. doi:10.3168/jds.S0022-0302(03)73803-X PMID:12836950

West, J., Mullinix, B. G., & Bernard, J. K. (2003). Effects of hot, humid weather on milk temperature, dry matter intake, and milk yield of lactating dairy cows. *Journal of Dairy Science*, *86*(1), 232–242. doi:10.3168/jds.S0022-0302(03)73602-9 PMID:12613867

Wheelock, J. B., Rhoads, R. P., VanBaale, M. J., Sanders, S. R., & Baumgard, L. H. (2010). Effects of heat stress on energetic metabolism in lactating Holstein cows1. *Journal of Dairy Science*, *93*(2), 644–655. doi:10.3168/jds.2009-2295 PMID:20105536

Wiking, L., Nielsen, J. H., Båvius, A.-K., Edvardsson, A., & Svennersten-Sjaunja, K. (2006). Impact of Milking Frequencies on the Level of Free Fatty Acids in Milk, Fat Globule Size, and Fatty Acid Composition. *Journal of Dairy Science*, *89*(3), 1004–1009. doi:10.3168/jds.S0022-0302(06)72166-X PMID:16507695

Wilkinson, J. M., Lee, M. R. F., Rivero, M. J., & Chamberlain, A. T. (2020). Some challenges and opportunities for grazing dairy cows on temperate pastures. *Grass and Forage Science*, *75*(1), 1–17. doi:10.1111/gfs.12458 PMID:32109974

Willers, C. D. (2014). *Avaliação dos impactos ambientais da pecuária de corte semi-intensiva.*

Wolfenson, D., & Roth, Z. (2019). Impact of heat stress on cow reproduction and fertility. *Animal Frontiers*, *9*(1), 3238. doi:10.1093/af/vfy027 PMID:32002237

World Commission on Environment and Development - WCED. (1987). Report of the World Commission on Environment and Development: Our Common Future. Geneva.

World Health Organization. (2007). *Quality assurance of pharmaceuticals: a compendium of guidelines and related materials. Good manufacturing practices and inspection* (Vol. 2). World Health Organization.

World Health Organization. (2011). WHO Expert Committee on Specifications for Pharmaceutical Preparations. Forty-fifth report. *World Health Organization Technical Report Series*, (961), 1. PMID:21699061

World Health Organization. (2014). *WHO's First Global Report on Antibiotic Resistance Reveals Serious, Worldwide Threat to Public Health.* WHO. https://www.who.int/mediacentre/news/releases/2014/amr-report/en/

World Health Organization. (2021). *Antimicrobial resistance.* WHO. https://www.who.int/news-room/fact-sheets/detail/antimicrobial-resistance

Wu, M. L., & Chan, L. K. (2002). Quality function deployment: A literature review. *European Journal of Operational Research, 143*(3), 463–497. doi:10.1016/S0377-2217(02)00178-9

Xavier, J. R., Ramana, K. V., & Sharma, R. K. (2018). β-galactosidase: Biotechnological applications in food processing. *Journal of Food Biochemistry, 42*(5), 12564. doi:10.1111/jfbc.12564

Ximenes, L. F. (2021). *Lácteos.* https://www.bnb.gov.br/s482-dspace/bitstream/123456789/909/1/2021_CDS_181.pdf

Xudong, Z., Xi, K., Ningning, F., & Gang, L. (2020). Automatic recognition of dairy cow mastitis from thermal images by a deep learning detector. *Computers and Electronics in Agriculture, 178*(May), 105754. Advance online publication. doi:10.1016/j.compag.2020.105754

Yadav, A., & Prakash, A. (2022). Factors influencing sustainable development integration in management education: An Empirical Assessment of management education institutions in India. *International Journal of Management Education, 20*(1), 100604. doi:10.1016/j.ijme.2022.100604

Yadav, J. S. S., Bezawada, J., Ajila, C. M., Yan, S., Tyagi, R. D., & Surampalli, R. Y. (2014). Mixed culture of Kluyveromyces marxianus and Candida krusei for single-cell protein production and organic load removal from whey. *Bioresource Technology, 164*, 119–127. doi:10.1016/j.biortech.2014.04.069 PMID:24844166

Yadav, J. S. S., Yan, S., Pilli, S., Kumar, L., Tyagi, R. D., & Surampalli, R. Y. (2015). Cheese whey: A potential resource to transform into bioprotein, functional/nutritional proteins and bioactive peptides. *Biotechnology Advances, 33*(6), 756–774. doi:10.1016/j.biotechadv.2015.07.002 PMID:26165970

Younas, M., Fuquay, J. W., Smith, A. E., & Moore, A. B. (1993). Estrous and endocrine responses of lactating Holsteins to forced ventilation during summer. *Journal of Dairy Science, 76*(2), 430436. doi:10.3168/jds.S0022-0302(93)77363-4 PMID:8445097

Youssif, N. H., Hafiz, N. M., Halawa, M. A., & Saad, M. F. (2020). Influence of some hygienic measures on the prevalence of subclinical mastitis in a dairy farm. *International Journal of Dairy Science, 15*(1), 38–47. doi:10.3923/ijds.2020.38.47

Zacarchenco, P. B., Van Dender, A. G. F., & Rego, R. A. (2017) Brazil Dairy Trends 2020 - Book, (pp. 373). Instituto de Tecnologia dos Alimentos, Campinas, SP, Brazil. .

Zafalon, L. F., Filho, a N., Oliveira, J. V. De, Resende, F. D. De, Penteado, H., & Odessa, N. (. (2005). Alterações da Composição e da produção de leite oriundo de QUARTOS MAMÁRIOS DE VACAS COM E SEM MASTITE SUBCLÍNICA. *Arquivos do Instituto Biológico, 72*(4), 419–426. doi:10.1590/1808-1657v72p4192005

Zalafon, L. F., Langoni, H., Benvenutto, F., Castelani, L., & Broccolo, C. R. (2008). 56 Aspectos Epidemiológicos Da Mastite Bovina Causada Por. *Veterinária e Zootecnia, 15*(16), 56–65.

Zandona, E., Blažić, M., & Režek Jambrak, A. (2021). Whey utilization: Sustainable uses and environmental Approach. *Food Technology and Biotechnology, 59*(2), 147–161. doi:10.17113/ftb.59.02.21.6968 PMID:34316276

Zanin, A. (2015). Apuração de custos e resultado econômico no manejo da produção leiteira: Uma análise comparativa entre o sistema tradicional e o sistema freestall. *Organizações Rurais & Agroindustdddddriais. Lavras, 17*(4), 431–444.

Zareie, R., Zamani, G. H., & Karami, E. (2016). The effect of dairy farmers' technical knowledge on milk bacterial count. *Academic Web Journal of Agricultural Research, Delta, 1*(1), 9–13.

Zeinhom, M. M., Aziz, R. L. A., Mohammed, A. N., & Bernabucci, U. (2016). Impact of seasonal conditions on quality and pathogens content of milk in Friesian cows. *Asian-Australasian Journal of Animal Sciences, 29*(8), 1207–1213. doi:10.5713/ajas.16.0143 PMID:27165021

Zhang, J., Wang, J., Jin, J., Li, X., Zhang, H., & Zhao, C. (2022). Prevalence, antibiotic resistance, and enterotoxin genes of Staphylococcus aureus isolated from milk and dairy products worldwide: A systematic review and meta-analysis. *Food Research International, 162*, 111969. doi:10.1016/j.foodres.2022.111969 PMID:36461219

Zhao, S., Min, L., Zheng, N., & Wang, J. (2019). Effect of Heat Stress on Bacterial Composition and Metabolism in the Rumen of Lactating Dairy Cows. *Animals (Basel), 9*(11), 925. doi:10.3390/ani9110925 PMID:31694321

Zigo, F., Yasiľ, M., Elečkov, J., Zigová, M., & Farkašová, Z. (2017). Mastitis Pathogens Isolated from Samples of Milk in Dairy Cows and Their Resistance against Antimicrobial Agents. *Journal of Food Science and Engineering, 7*(2). doi:10.17265/2159-5828/2017.02.007

Zoccal, R. (2019). Leite nas grandes regiões brasileiras. [Milk in large Brazilian regions.] In *Anuário Leite 2019: Novos produtos e novas estratégias da cadeia do leite para ganhar competitividade e conquistar os clientes finais*. Texto Comunicação Corporativa.

Zou, J., Chen, X., Hu, Y., Xiao, D., Guo, X., Chang, X., & Zhou, L. (2021). Uncoupling glucose sensing from GAL metabolism for heterologous lactose fermentation in Saccharomyces cerevisiae. *Biotechnology Letters, 43*(8), 1607–1616. doi:10.100710529-021-03136-8 PMID:33937967

Zuin, L. F. S., & Zuin, P. B. (2008). *Produção de Alimentos Tradicionais: Extensão Rural*. Idéias & Letras.

About the Contributors

Eduardo Satolo is an Associate Professor in Faculty of Science and Engineering at São Paulo State University (UNESP), Tupã Campus, lecturing on Business Administration course and Agribussiness and Development Post Graduation Program. Post-Doctorate in Production Engineering from UFF (2018), doctor in Industrial Engineering from the Universidade Metodista de Piracicaba (UNIMEP). Has a master's in Production Engineering from UNIMEP is a graduate in Production Engineering from UNIMEP.

Priscilla Mac-Lean is an Assistant Professor in School of Science and Engineering at Sao Paulo State University (Unesp), Tupã Campus, Brazil, lecturing on Biosystems Engineering course and Agribussiness and Development Post Graduation Program. Doctor in Animal Quality and Productivity by the Postgraduate Program in Animal Science from the University of São Paulo (2012). Master in Animal Production by the Postgraduate Program in Animal Science (2008) and undergraduated in animal science (2005) from Universidade Estadual de Maringá (UEM). Her research interests include the areas of improvement of the animal production chain evaluating good practices related to product quality and quantity, ambience and animal biometeorology and welfare, working mainly in the productive chains of dairy cattle and sericulture.

***

Ferenc Bánkuti holds a degree in Animal Science from the University of São Paulo - USP, Master's and PhD in Production Engineering from the Federal University of São Carlos. In 2015, Ferenc held a post-doctoral internship at Kansas State University - KSU, USA. He is currently an Associate Professor at the State University of Maringá. His main areas of expertise are related to the coordination of agro-industrial systems and the analysis of animal production systems - Farming System.

Julia Barbosa is a graduate student in the Biosystems Engineering course at theSchool of Sciences and Engineering (Unesp).

Josemeyre Bonifacio da Silva Marques received her PhD in Food Science from the Londrina State University (UEL) in 2009. In 2004, during your master, she receveid the training in sensory analysis in University at Illinois-Urbana Champaing. During her master´s and doctoral degree, she worked with soy for human comsuption. Since 2018, she work as a Professor at University Pitágoras Unopar Anhanguera working with Dairy Science research and use of microalgae in agriculture. She writes and presents widely on issues about beta-galactosidase enzyme and its production by yeast and microalgae. She is a scientific editor in Food Engineering at Scientia Plena Journal.

Leonardo Campos is a graduate student in the Biosystems Engineering course at theSchool of Sciences and Engineering (Unesp).

Livia Castelani is a post-doctor in Veterinary Microbiology and Molecular Biology, working in the area of mastitis and alternative treatments, by Instituto de Zootecnia, IZ (2021). PhD in Microbiology, Institute of Biomedical Sciences, University of São Paulo, ICB/USP (2016). Master in Sustainable Animal Production, by Instituto de Zootecnia, IZ (2012). Specialization in Microbiology of Food and Processes, by the University of Campinas, UNICAMP (2022). Bachelor and Degree in Biological Sciences - Medical Modality, Specialist in Clinical Analysis, by Faculdade de Americana, FAM (2008). Line of research: Microbiology, Antimicrobial Resistance and Therapeutic Alternatives, Essential Oils, Bacteriocin, Bovine Mastitis, Staphylococcus spp. milk isolates, Quality and Food Safety of Milk, Molecular Biology.

Marion Costa got 1.401 / 5.000 Resultados de tradução Resultado da tradução Veterinarian graduated from Universidade Federal Fluminense (2011), master (2013) and doctor (2015) in Veterinary Medicine from the Graduate Program in Veterinary Hygiene and Technological Processing of Products of Animal Origin at Universidade Federal Fluminense. Postdoctoral fellow at the Federal University of Rio de Janeiro in 2016 (CAPES grantee) and 2017 (CNPq grantee). Adjunct Professor, professor at the Department of Preventive Veterinary Medicine and Animal Production at the School of Veterinary Medicine and Animal Science, Federal University of Bahia (since 2017), developing Teaching, Research and Extension activities. Coordinating professor of the Laboratory of Inspection and Technology of Milk and Derivatives (LaITLácteos/UFBA) and coordinator of the extension project Provision of Services for analysis of milk and derivatives. Accredited as a Master's and Doctorate advisor in the Graduate Programs in Food Science (PGAli) (since 2017) and Animal Science in the Tropics (PPGCAT) (since 2017) both at the Federal University of Bahia. Has experience in the area of veterinary medicine, and food science and technology, with emphasis on: Inspection and Technology of Milk and derivatives, Physical-Chemical Control of Milk and derivatives, Food Quality Control (GMP, PPHO and HACCP) and Validation of Chromatographic Methods (Carbohydrates, organic acids, ivermectin and biogenic amines).

Ligia dos Santos is a graduate student in the Biosystems Engineering course at theSchool of Sciences and Engineering (Unesp).

Bruna Ferreira graduated in Business Administration from Universidade Estadual Paulista, Brazil. Born in São Roque, São Paulo, Brazil. She is curious, eclectic and, like Marie Curie, believes that in life there is nothing to fear, but to understand.

Layas Freitas is a Researcher in Community Supported Agriculture: Organizational Arrangement to achieve Sustainable Development Goals. In Master Degree at Sao Paulo University - Production Engineering, Graduation in Production Engineering at Federal University of Uberlandia (2016).

Silvana Gonzalez graduated in Veterinary Medicine from the University of Marília (1997) and Master in Animal Science in the area of Preventive Veterinary Medicine, from the State University of Londrina (2002). Since 2003 she has been Professor of Higher Education at UNIFAI-Centro Universitário de Adamantina-Courses in Veterinary Medicine, Environmental Engineering, Biological Sciences, Nursing and Nutrition. She has experience in the area of Preventive Veterinary Medicine, Veterinary Clinical Pathology and Inspection of Products of Animal Origin with an emphasis on Quality Control. Since 2003, it has owned the Emerge Veterinary Medical Diagnostic Center, which operates in laboratory diagnostics for mammals, fish, birds and reptiles.

Juliani Gotardi is a undergraduate in Biological Sciences at Faculdades Adamantinenses Integradas, MBA in Environmental and Business Management and a master's degree from the Graduate Program in Agribusiness and Development at São Paulo State University (UNESP), Tupã Campus, Brazil.

Luiz Ito Morioka graduated from Londrina State University (UEL) in Biological Sciences, specialization in Applied Biochemistry, Master's in Biotechnology in the area of Agroindustrial Biotechnology and PhD in Food Science in the area of Biotechnology/Use of Agricultural and Agroindustrial Residues. He has experience in the field of microalgae biotechnology. He is a Professor at the Biotechnology and Innovation in Health Program at Anhanguera University of Sao Paulo.

Wendell Leite was born in Mato Grosso, on July 30, 1990, in the city of Guiratinga. Bachelor in Animal Science at Federal University Mato Grosso - UFMT, Rondonópolis Campus, 2011-2016. He worked in the field of beef cattle - raising, rearing and fattening systems - in rural properties. He also worked with technical and managerial assistance in rural properties in the field of dairy cattle. Specialist at USP/Esalq – Pecege, MBA in Agribusiness, 2017-2019. Master in Animal Science at UNESP/FCAV, Campus de Jaboticabal, 2018-2021, with research in intensive finishing system for Nellore heifers. Specialist in Teaching for Professional and Technological Education at Federal Institute of Paraíba EaD - IFPB, Campus Cabedelo, 2021-2022. PhD student in Animal Science at UNESP/FCAV, Jaboticabal Campus, with research in the application of biotechnology in the field of animal production.

Thatiane Mitsunaga is an Animal Scientist with Undergraduation degree in Animal Science at University of São Paulo State (UNESP), in Botucatu, São Paulo. During this period, she experienced activities in poultry, swine and dairy production, beef cattle, among others, included in several mandatory activities within the undergraduation program. Her first contact with research was with a project evaluating the use of stable isotopes in animal production. Almost at the end of undergraduation, she applied to a Brazilian program to study abroad in University College Dublin, Ireland, where she improved important tehchniques in scientific reading and writing, as well as in animal production. To conclude undergraduation, she performed her internship in a dairy farm located inside The Animal Science Institute of Nova Odessa, São Paulo, the same place she later started a master's degree in Sustainable Animal Production. In this Institute she developed a master's research project evaluating the use of essential oils from therapeutical plants in dairy cows and its effects on subclinical mastitis and milk quality. At the present time, she is a PhD student in Animal Science and Pastures at Luiz de Queiroz College of Agriculture (ESALQ), from University of São Paulo (USP), with research thesis including the study of incidence and distribution of clinical mastitis causing pathogens in dairy farms and the economic impact of on-farm microbiological culture as a diagnosis tool.

Mayara Mari Murata received her bachelor's degree in Agronomy in 2011 from Londrina State University (UEL) and received her MS in Genetics and Crop Breeding from the Paulista State University (UNESP) in 2013. In 2017, she concluded her PhD in Plant Pathology at University of Florida, followed by a post-doctorate at the same institution in 2018. Since 2019, she works as a Professor at University Pitágoras Unopar Anhanguera, teaching classes for the Agronomy course. She also works as a Researcher in the same institution, conducting experiments in biotechnology innovation and development of bioproducts using microorganisms in agriculture.

João Negrão graduated in Animal Science at the State University of Maringá (UEM, 1991), Master (1993) and Ph.D. (1996) in Biology and Agronomy at Ecole Nationale Superièure Agronomique de Rennes (ENSAR), in the area of Animal Physiology. He is currently Full Professor in Lactation Physiology at the Faculty of Animal Science and Food Engineering (FZEA) of the University of São Paulo (USP), is responsible for the Animal Physiology Laboratory at FZEA, teaches animal physiology for the zootechnics and veterinary medicine, he is also advisor of scientific initiation, master's and doctoral students in Animal Science at the Faculty of Food Engineering (FZEA) and at the Faculty of Agrarian and Veterinary Sciences (FCAV) of the São Paulo State University Júlio de Mesquita Filho (UNESP). He has experience in Animal Science, with an emphasis on Lactation Physiology, Stress Physiology, working mainly on the following topics: endocrinology, milk synthesis, milk ejection, behavior, production and quality of milk.

Matheus Oliveria graduated in Veterinary Medicine from the Federal University of Campina Grande (2022). He has experience in Inspection of Animal Products and food safety. Acting in the area of consulting and technical advice in dairy.

Guillermo Ortiz-Colón, Ph.D., is assistant professor at University of Puerto Rico-Mayaguez. His research is focused on the improvement of the efficiency, economic solvency, and sustainability of the Dairy Industry in Puerto Rico. Specifically, is to promote good pasture management, minimize the dependence on imported grain feeds, and increase the percentage of dairy replacements born and raised in Puerto Rico. His research focus is the development of efficient nutritional/management practices than can permit that raising dairy replacement locally would be a better alternative to the common practice of importing heifers from temperate locations in North America.

Uiara Paim graduated in Veterinary Medicine in the School of Veterinary Medicine and Animal Science of the Universidade Federal da Bahia (2021). Masters student of the Postgraduate Program of Animal Science in the Tropics in the area of Microbiology of Animal Origin Food with emphasis on inspection, technology and microbiological control of milk and dairy products.

Messy Pantoja graduated in Animal Science from the Federal Rural University of Amazônia (2013). She has a master's degree in Animal Science from the Federal University of Pará (2017) and a PhD in Animal Science at the University of

São Paulo-USP / FZEA. She has experience in Animal Science, with an emphasis on Animal Production, working mainly on the following topics: milk production, Biometeorology, Ethology, and animal welfare of buffaloes and sheep.

Alfredo Pereira is a Full Professor in the Animal Science Department of the University of Évora; Graduated in Animal Science Engineering from the University of Évora, Master in Veterinary Medicine and Tropical Animal Science from the Faculty of Veterinary Medicine of Lisbon and PhD in Agricultural Sciences - Animal Bioclimatology, from the University of Évora. Invited Professor at FZEA - University of São Paulo. Senior Researcher at MED (Mediterranean Institute for Agriculture, Environment and Development - University of Évora) and LABE (Laboratory of Ethology and Animal Welfare of FZEA/USP). Acting mainly on the following subjects: adaptive physiology of ruminants, thermal stress, thermotolerance, and animal welfare.

Mohana Priya Rajendram is a zealous educator interested in research, Completed UG & PG with good academic score.

Muthuminal Ramu is a research scholar under structural dynamics with good academic record who is a University rank holder under UG and PG.

Bruna Rekowsky graduated in Veterinary Medicine from the Federal University of Bahia (UFBA) in 2018 with a gain in the research line of interest in the area of Inspection and Technology of Products of Animal Origin and Buffalo Culture. Master's and PhD student at the Graduate Program in Animal Science in the Tropics (PPGCAT) at UFBA with a focus on Technological Innovation and Inspection aimed at the development and diversification of products of animal origin with functional potential.

Luiz Roma Junior has a degree in Agronomic Engineering from the University of São Paulo (2001), Master's degree in Animal Quality and Productivity from FZEA/USP (2004) and PhD in Animal Science and Pastures from ESALQ/USP (2008). Scientific researcher at the Aniimal Science Institute/APTA/SAA and Director of the Dairy Cattle Research Center, IZ/SAA/SP. Experience in Animal Science, with emphasis on Milk Quality and Animal Management, working mainly on the following topics: ambience, dairy cattle, milk quality, medicinal plants.

Isabela Santos holds a degree in biosystems engineering from School of Sciences and Engineering and is studying for a master's degree at Agribussiness and Development Post Graduation Program, both at São Paulo State University.

Isabela Santos received her bachelor's degree in Biosystems Engineering and her Master of Science degree in Agribusiness and Development from São Paulo State University (UNESP).

Mariane Severino was born in Catanduva – SP, on October 29, 1992. Bachelor in Animal Science at Federal University of Mato Grosso (UFMT), Rondonópolis Campus, 2010-2017. Academic exchange with sandwich graduation, in Rhode Island - USA, at University of Rhode Island (URI) and A.C.E. Language Institute, 2013-2014. She worked with management of administrative and zootechnical controls in rural properties, 2017-2018. Master in Animal Science at São Paulo State University (UNESP/FCAV), Jaboticabal Campus, 2019-2021 with research on meat quality of Nelore-Montana cattle using a proteomic approach. Specialist in Teaching for Professional and Technological Education - EaD at Federal Institute of Paraíba (IFPB), Campus Cabedelo, 2021-2022. PhD student in Animal Science at UNESP/FCAV, Jaboticabal Campus with research on the application of antimicrobial peptides for beef cattle.

José Silva veterinarian, Master in Animal Science and Doctor in Animal Bioscience. Professor at the Federal University of Bahia

Mirian Fabiana da Silva graduated in Animal Science from the State University of Goiás, specialization in Agribusiness Management from the State University of Goiás, specialization in Teaching for Professional Education from the Federal Institute of Santa Catarina, master's degree in Animal Science from the Federal University of Viçosa, PhD in Animal Nutrition and Production from the University of São Paulo, Postdoctoral at the State University of Maringá and Agricultural Technician from the Federal Agrotechnical School of Ceres. She is currently a teacher. It has been developing research on: evaluation of the dairy cattle production system, through the use of technical, economic, social and environmental indicators; evaluation of the perception of professionals and students of agricultural sciences about sustainability indicators for Brazilian dairy farming; and also research on training and qualification of professionals linked to dairy farming. She has experience in Animal Science, with an emphasis on animal production management, production and transaction costs, agribusiness, production chain, coordination of agro-industrial systems and evaluation of animal production systems.

Hélio Suguimoto is the Director of Graduate Studies at Cogna Educação and Dean of Research and Graduate Studies at University Pitágoras Unopar Anhanguera. He is a professor at the Biotechnology and Innovation in Health Graduate Program at the Anhanguera University of Sao Paulo and professor at the Sustainable Agribusiness Graduate Program at Anhanguera Uniderp University. He received his bachelor's in Agronomy from Londrina State University in 1989. He has a master's degree (1993) and a doctorate (1998) in Food Sciences from the same institution. He worked actively in the field of dairy technology and he is currently developing research projects with bioactive compounds from microorganisms for use in food, human health and agriculture.

Cynara Mendoca Tinoco is a researcher in Lean Healthcare, Quality Process and Heathy and Safety at Work. PhD in Electrical Engineering - Automation (UNICAMP/2017), Master in Production Engineering (UFSC/2000), Specialist in Occupational Safety Engineering (UFU/2001), Black Belt (FDG/2001), Graduation in Civil Engineering (UFRGS) /1996).

Cristiane Titto is an associate Professor at the Faculty of Animal Science and Food Engineering/USP, with full professorship held in 2016. She teaches classes in Veterinary Medicine, Animal Science and Biosystems Engineering. She has a degree in Animal Science from the Faculty of Animal Science and Food Engineering / USP (2001), Master in Animal Science from FZEA / USP (2007) and a PhD in Animal Science (FZEA / USP - 2010). Her research has an emphasis on Applied Animal Biometeorology and Ethology, working mainly on the following topics: thermal stress, social behavior and welfare of production animals.

Evaldo Titto is a veterinarian graduated from the Faculty of Veterinary Medicine and Animal Science of the University of São Paulo (1978). Master in Animal Nutrition by FMVZ-USP (1989) and PhD in Animal Science at FMVZ-UNESP (1994). Free Teaching in Zootechnical Bioclimatology at FZEA-USP (2007) and Post-Doctorate in Beef Cattle Heat Tolerance at the University of Évora (2008). Full Professor of Faculty of Animal Science and Food Engineering of University of Sao Paulo (since 2011). Was Professor of Faculty of Veterinary Medicine and Animal Science, USP (1983-1992) and of Faculty of Animal Science and Food Engineering, USP (since 1993). Designed, implemented and coordinated the FZEA Veterinary Medicine Undergraduate Course (2006-2016). Researcher and professor in the areas of Biometeorology, Ethology and Animal Welfare, with a focus on the following topics: bioclimatology, behavior, animal welfare, thermoregulation and ambience, for beef and dairy cattle, buffaloes, sheep and horses. Senior Researcher Award "Prof. Dr. Roberto Gomes da Silva" in the area of Behavior and Animal Welfare by the Brazilian Society of Biometeorology, in 2017. He is the Technical Manager at CRMV-SP for the Laboratory of Biometeorology and for the Research Bioterium in Biometeorology, Ethology and Animal Welfare, at the Animal Science's Department at FZEA-USP.

Rafael Titto is an animal scientist with expertise on animal behavior, heat stress and feedlot management.

Eduardo Vicente is an assistant Professor at Department of Biosystem Engineering, School of Sciences and Engineering, Campus of Tupã - São Paulo State University (Unesp). Graduated at Farmacy/Biochemistry at School of Pharmaceutical Sciences, Araraquara, São Paulo State University (2007). Master's degree in Biotechnology in 2009 and P.h.D. in Biotechnology at Institute of Chemistry, Araraquara, Unesp. Has experience in Biochemistry, focusing on peptide membrane interactions, acting on the following subjects: EPR, DEER, peptides, solid phase peptide synthesis and CD, InfraRed, UV/Vis and Fluorescence spectroscopies. Performed a doctorate-sandwich for 9 months at the Department of Chemistry and Biochemistry, Miami University - Oxford, OH - USA, with emphasis on Electronic Paramagnetic Resonance (DEER). Post-doctorate in the Department of Physics of the Faculty of Philosophy, Sciences and Letters of Ribeirão Preto - University of São Paulo (FF-CLRP - USP), in Biophysics and pulsed RPE area. He has experience in the field of Biochemistry and the main studies are the interactions between peptide/protein and membrane mimetics using the techniques of solid phase peptide synthesis (SPFS), Mass Spectrometry and various chromatographic and spectroscopic techniques such as HPLC, UV- VIs, CD, Fluorescence, EPR, NMR and FTIR. Currently works with application of peptides in animal and plant production. He is the leader in the research group Peptides: Synthesis, Optimization and Applied Studies - PeSEAp.

Index

H

I

L

M

N

P

R

S

T

U

V

www.ingramcontent.com/pod-product-compliance
Lightning Source LLC
LaVergne TN
LVHW080638011026
849160LV00014B/1254